Quantum States and Processes

QUANTUM STATES AND PROCESSES

George H. Duffey, Ph.D.
Professor of Physics, South Dakota State University

A Solomon Press Book

PRENTICE HALL PUBLISHERS
Englewood Cliffs, New Jersey 07632

Library of Congress Cataloging-in-Publication Data

Duffey, George H.
  Quantum states and processes : based on symmetry considerations /
George H. Duffey.
    p. cm.
  "A Solomon Press book."
  Includes bibliographical references and index.
  ISBN 0-13-747056-8
  1. Quantum theory. 2. Symmetry (Physics) 3. Chemistry, Physical
and theoretical. I. Title.
  QC174.12.D833 1991   90-27346
  530.1'2—dc20   CIP

Editorial/production supervision: bookworks
Prepress buyer: Kelly Behr
Manufacturing buyer: Susan Brunke
Acquisitions editor: Michael Hays

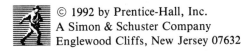 © 1992 by Prentice-Hall, Inc.
A Simon & Schuster Company
Englewood Cliffs, New Jersey 07632

Printed in the United States of America

10 9 8 7 6 5 4 3 2 1

ISBN   0-13-747056-8

Prentice-Hall International (UK) Limited, *London*
Prentice-Hall of Australia Pty. Limited, *Sydney*
Prentice-Hall Canada Inc., *Toronto*
Prentice-Hall Hispanoamericana, S.A., *Mexico*
Prentice-Hall of India Private Limited, *New Delhi*
Prentice-Hall of Japan, Inc., *Tokyo*
Simon & Schuster Asia Pte. Ltd., *Singapore*
Editora Prentice-Hall do Brasil, Ltda., *Rio de Janeiro*

CONTENTS

Preface   *xi*

Chapter 1. Representing Quantum Mechanical States and Observables
1.1.   The Physical Arena   *1*
1.2.   The State Function   *2*
1.3.   Composite States and Mixed States   *6*
1.4.   The Vector the State Function Represents   *7*
1.5.   Matrix Elements Generated by Suitable Operators   *12*
1.6.   A Hilbert-Space Reformulation of the Postulates   *13*
1.7.   The Projection Operator   *16*
1.8.   The Hermitian Condition   *17*
1.9.   Commutativity of Operators for Observables   *21*
1.10. Classical Observables   *23*
1.11. Implications of Indistinguishability   *24*
1.12. The Occupancy of Individual States   *28*
1.13. Highlights   *29*
         Discussion Questions   *30*
         Problems   *32*
         References   *33*

Chapter 2. Allowing for the Translational Symmetries and the
                    Localizability of Particles
2.1.   Introduction   *37*
2.2.   The Galilean Space-Time Continuum   *37*
2.3.   Galilean Transformations of a State Function   *38*
2.4.   Conservation of Particle Mass   *42*
2.5.   Similarity Transformations   *43*

2.6.   Unitary Transformations   *48*
2.7.   The Group Concept   *49*
2.8.   Parameterizing an Operator   *50*
2.9.   Elements of and Generators for the 1-Dimensional Galilean Group   *51*
2.10. The Localizability Requirement   *53*
2.11. Extending the Group to Allow for Particle Localizability   *57*
2.12. Introducing Localized Interactions   *61*
2.13. Representing the Operators for Position and Momentum   *63*
2.14. Operators Governing Temporal Changes   *64*
2.15. Key Concepts   *66*
       Discussion Questions   *69*
       Problems   *71*
       References   *72*

Chapter 3. Introducing the Rotational Symmetries

3.1.   Guiding Considerations   *73*
3.2.   Rotation Modelled as a Composite Translation   *74*
3.3.   An Alternate Construction   *77*
3.4.   Matrices Representing Key Rotations in 3-Dimensional Space   *79*
3.5.   Commutation Relations for Rotations   *82*
3.6.   Commutators for Quantum Particles in 3-Dimensional Space   *83*
3.7.   Hamiltonian Operators for Particles in 3-Dimensional Space   *84*
3.8.   Commutativity among the Hamiltonian Operator and the Angular
       Momentum Operators   *88*
3.9.   Eigenvalues for the Angular Momentum Operators   *90*
3.10. Kinds of Angular Momenta   *94*
3.11. Angular Momentum Matrix Elements   *95*
3.12. Spin Matrices   *96*
3.13. A Spin Hamiltonian   *98*
3.14. Main Features   *99*
       Discussion Questions   *101*
       Problems   *103*
       References   *104*

Chapter 4. Key Eigenfunctions and Kets

4.1.   The Nature of Eigenfunctions and Kets   *107*
4.2.   Translational Eigenfunctions   *108*
4.3.   Separating Angular Motion from the Other Motions   *111*
4.4.   Rotational Eigenfunctions   *113*
4.5.   Resolving the Angular Motion   *114*
4.6.   Simple Vibrational Eigenfunctions   *117*
4.7.   Radial Oscillatory Motion in a Coulomb Field   *119*
4.8.   The Hydrogenlike Atom   *121*
4.9.   Approximate Orbitals for Multielectron Atoms   *124*

4.10. Elements of Photon Transmission, Production, and Absorption   *126*
4.11. Photon Kets   *128*
4.12. Molecular Structure   *131*
4.13. Summary   *131*
    Discussion Questions   *133*
    Problems   *134*
    References   *135*

Chapter 5. Step Operators
5.1. Transformation Operators in a Given Hilbert Space   *137*
5.2. Commutation Relations for a Step Operator   *139*
5.3. Angular-Momentum Step Operators   *142*
5.4. Commutation Relations for the Harmonic Oscillator   *144*
5.5. Harmonic Oscillator Step Operators   *145*
5.6. More General Step Operators   *150*
5.7. The Uniform Normalization Condition   *152*
5.8. Application to the Radial Motion in a Coulomb Field   *153*
5.9. The Morse Potential   *158*
5.10. Energy Levels for the Morse Oscillator   *161*
5.11. Creation and Annihilation Operators for Bosons   *163*
5.12. Creation and Annihilation Operators for Fermions   *165*
5.13. Boson to Fermion and Fermion to Boson Operators   *168*
5.14. Recapitulation   *169*
    Discussion Questions   *171*
    Problems   *172*
    References   *174*

Chapter 6. Variational Procedures
6.1. Using Parameters to Introduce Variations   *177*
6.2. Rate of Change of the State Energy with a Parameter   *178*
6.3. Relating the Kinetic Energy Expectation Value to the Potential
    Energy Expectation Value   *179*
6.4. The Variation Theorem   *181*
6.5. The Ritz Theorem   *185*
6.6. Linear Variation Functions   *187*
6.7. A Secular Equation for Energy   *189*
6.8. Chemical Bonding and the Hückel Procedure   *191*
6.9. Parameters for the Hückel Method   *197*
6.10. Elements of an Extended Hückel Theory   *199*
6.11. Possible Additional Improvements   *200*
6.12. Natural Units   *200*
6.13. An Overview   *202*
    Discussion Questions   *205*
    Problems   *206*
    References   *209*

Chapter 7. Stationary-State Perturbation Theory

7.1. Reckoning with Complicating Influences   *211*
7.2. Development from a Single Simplified State   *212*
7.3. First-Order Description of the Ground State of a Heliumlike
      Atom or Ion   *216*
7.4. The Average Effective Nuclear Charge in the Heliumlike Structure   *220*
7.5. Second and Higher Order Corrections to the Energies   *223*
7.6. Corrections to the Eigenkets   *224*
7.7. Polarizability of the Hydrogen Atom   *227*
7.8. Development from a Cluster of States   *230*
7.9. 2sp Hybridization Caused by an Electric Field   *233*
7.10. The Non-Crossing Rule   *237*
7.11. Perturbation at an Atom in a Hückel-Model Structure   *239*
7.12. Abstract   *242*
      Discussion Questions   *246*
      Problems   *248*
      References   *250*

Chapter 8. Varying-State Perturbation Theory

8.1. Kinds of Processes   *253*
8.2. General Dependence on Time   *254*
8.3. The Sudden Approximation   *255*
8.4. Rates at which Expansion Coefficients Vary   *257*
8.5. Managing Secular (Aperiodic) Perturbing Expressions   *259*
8.6. Perturbation Terms for the Transformed Coefficients   *264*
8.7. Steady Perturbers   *266*
8.8. Transitions to or from a Band of States   *270*
8.9. The Born Approximation for Scattering Processes   *275*
8.10. Beta Decay   *280*
8.11. Basic Ideas   *286*
      Discussion Questions   *288*
      Problems   *290*
      References   *292*

Chapter 9. Absorption, Emission, and Spontaneous Decay

9.1. Interaction with the Electromagnetic Field   *293*
9.2. Hamiltonian Operator for a Charged Particle   *294*
9.3. Matrix Element for Perturbation by Monochromatic Interaction   *295*
9.4. The Density in Energy of Electromagnetic States in a Band   *297*
9.5. Oscillatory Perturbers   *298*
9.6. Transitions to or from a Band of Molecule-Field States   *302*
9.7. Rates of Induced Absorption or Emission   *304*
9.8. Role Played by the Transition Dipole Moment   *306*
9.9. The Bouguer-Beer Law   *308*

9.10. The Principle of Detailed Balance   *311*
9.11. Equilibrium Energy Density in a Radiation Field   *312*
9.12. Spontaneous Emission   *315*
9.13. Parity Considerations   *317*
9.14. Vibrational Transitions   *318*
9.15. Transitions in Axial Angular Momentum   *320*
9.16. Shifts in the Rotational or Azimuthal Quantum Number   *321*
9.17. Synopsis   *322*
      Discussion Questions   *325*
      Problems   *327*
      References   *328*

Chapter 10. Quantum Abruptness and Nonlocality

10.1. Introduction   *331*
10.2. Visible Quantum Jumps   *332*
10.3. Definiteness of Properties   *333*
10.4. Nonlocality in Single Coherent Systems   *336*
10.5. The EPR Paradox   *336*
10.6. Correlations between Incoherent Product Systems   *338*
10.7. Correlations between Spatially Separated Coherent Fermions   *342*
10.8. A Bell Inequality for Pairs of Pass-or-Fail Filters   *345*
10.9. Correlations between Spatially Separated Coherent Photons   *346*
10.10. Altering Coherences   *348*
10.11. The Implied Action at a Distance   *350*
10.12. General Comments   *353*
      Discussion Questions   *353*
      Problems   *354*
      References   *355*

Chapter 11. Quantum Spinors

11.1. Introduction   *359*
11.2. The Nature of Space and Time   *360*
11.3. First and Second Rank Spinors   *362*
11.4. Spinorial Transformations   *367*
11.5. Minkowskian Vectors   *372*
11.6. Spinorial Differentiating Operators   *375*
11.7. A Wave Function that Propagates as Light Does   *378*
11.8. Disturbances that Propagate More Slowly   *380*
11.9. The Dirac Equation   *384*
11.10. Review   *388*
      Discussion Questions   *392*
      Problems   *393*
      References   *394*

Chapter 12. Simple Dirac Systems
12.1.   The One-Particle Dirac Equation   397
12.2.   Spatially Periodic Translational States   399
12.3.   Antiparticles   402
12.4.   The 4-Dimensional Sigma Matrix   402
12.5.   Orbital Angular Momentum   407
12.6.   Total Angular Momentum   408
12.7.   Operators for a Central Field   412
21.8.   Behavior in the Outlying Regions of a Central Field   414
12.9.   Functions *F* and *G* for Hydrogenlike Structures   415
12.10. Energy Levels of Hydrogenlike Structures   418
12.11. Restrictions on the Solutions   420
12.12. Concluding Remarks   422
        Discussions Questions   424
        Problems   425
        References   426

Answers to Problems   429

Index   433

PREFACE

This text develops intermediate quantum mechanics in a language that undergraduate and beginning graduate physics and chemistry students find congenial and meaningful. The first ten chapters deal with Galilean relativistic theory; the last two, with Einstein relativistic quantum theory.

Since the axioms of quantum mechanics are not self-evident, considerable discussion supporting their formulation appears. Why the state of a system is represented by a function, why observables are represented by eigenvalues, how composite states are formulated, the probability interpretation, all are considered. Then the entity that the state function describes, the ket vector, is introduced and the initial axioms are restated.

Characteristic of a particle is its localizability. So we have to consider the behavior of a point during translation in space, in time, in velocity, and in phase. Also, its behavior during rotations. These considerations lead to construction of the commutation relations and to expressions for the key operators. Classical mechanics enters only in identifying the potential terms.

Solutions for free translation, rotation, harmonic oscillation, and motion in a Coulomb field are tabulated and discussed. Step operator theory is developed and employed in explaining the action of creation and annihilation operators.

Variational procedures enable us to establish the virial theorem and a theorem for evaluating candidate functions. For linear variational functions, the secular equation is obtained. This is employed to estimate delocalization effects in molecules.

Perturbational methods are used to introduce complicating interactions. In the stationary state theory, the perturbed form may develop from a single state or from a cluster of states; both possibilities have to be considered. In the varying state theory, both field-induced transitions and spontaneous transitions are treated.

In Chapter 10, evidence that individual processes occur in steps, rather than continuously, is presented. Also, evidence that quantum states are spread out in space as well as in time. The Einstein, Podolsky, Rosen (EPR) effect and violations of the Bell inequalities are described.

When time is bound into the continuum, invariance of space is replaced by invariance of the light cone based on the chosen origin. But spinorial derivatives may then be physically significant. Two coupled differential equations involving two spinorial functions are constructed and applied to free motion. Interactions are introduced and the equations combined to give the Dirac equation. This is applied to hydrogen-like structures.

— George H. Duffey

Quantum States and Processes

1

REPRESENTING QUANTUM MECHANICAL STATES AND OBSERVABLES

1.1
The Physical Arena

Physics is based on our collective knowledge of the physical world, the universe. This knowledge has its source in interactions between identifiable parts of the world and observers or instruments.

During any given investigation, irreversible processes occur and effects are recorded. These may arise as incidents to be located and counted. Or, they may appear as properties to be measured against generally accepted standards.

Physical events are found to occur throughout a vast space and time. From measurements in the laboratory, on earth, and through the observable universe, it is induced that physical space is 3-dimensional and approximately Euclidean. From timing events, it is induced that physical time is 1-dimensional and homogeneous.

Time is not found separate from space, but together the two form a 4-dimensional continuum. Nevertheless, in the first part of this book, we will take this continuum to be Galilean. Thus, we will here consider time to be separable from space, the same everywhere, with the Pythagorean theorem holding in space itself.

1.2
The State Function

Matter is found to be divisible into elements occupying differing portions of space at any given time. Macroscopic and microscopic elements are observed to be separable and independent. Early workers presumed that such independence would persist in the submicroscopic realm also.

However, the mind considers itself as a unit spread out in space and time. A similar extension is observed in nature with molecules, atoms, and their constituents. Even though atoms consist of electrons and nuclei, they exhibit ball-like structures. Thus, an atom in which the occupied shells are filled appears to be spherically symmetric. None is like a miniature solar system. Instead, atoms of the same kind in the same energy and angular momentum state have to be considered indistinguishable in statistical mechanical calculations.

Just as a radioactive nucleus disintegrates at a certain point in time, a fundamental particle (for instance an electron) undergoes interactions, with a field or with other particles, at definite points in space. This fact led initially to the description of a particle as a mobile point with properties. Mathematically, such a point would be a singularity.

But to describe a particle in a nucleus, an atom, a molecule, or whenever it is in motion, requires a function of space and time. For as we have noted, such a system is extended in the space-time continuum. Furthermore, any variation in behavior over neighboring points that is not catastrophic is smooth. As a result, we postulate the following:

(I′) The state of a submicroscopic particle is governed by a function that is represented by a single-valued analytic expression $\Psi(\mathbf{r}, t)$. The state of a system of n submicroscopic particles is similarly governed by a single-valued analytic expression $\Psi(\mathbf{r}_1, \mathbf{r}_2, \ldots, \mathbf{r}_n, t)$.

By analytic, we mean that Ψ is continuous, with continuous spatial and temporal derivatives. The independent variables include the end point of radius vector $\mathbf{r}$ and time t, and the end points of radius vectors $\mathbf{r}_1, \mathbf{r}_2, \ldots, \mathbf{r}_n$ and time t, respectively.

A particle, or system of particles, has a limited range at a given time t. We presume that everywhere beyond the range, the state function Ψ is zero. At the boundary, Ψ and its first derivatives become zero. Particles or units of particles whose ranges do not overlap are mechanically separate.

In elementary calculations, however, one generally assumes that the ranges of the systems are infinite. Note the conventional state functions for translation with periodic boundary conditions, for the harmonic oscillator, for the hydrogen atom, for other atoms, and for molecules. We regard the assumption as a convenient artifice.

As in classical physics, we presume that certain properties are conserved when the appropriate conditions are satisfied. Examples include energy, linear momentum, and angular momentum. Such a property can act at each point over the range of Ψ, at the same strength. On measuring the observable at any of these points, one would then obtain the same result.

But if Ψ governs such a state of a system, the property is derivable from it, by a mathematical operation. Since the property is independent of position, the operation must *not* put a condition on the coordinates. Now, an operator acting on function Ψ yields another function. For the equation to be an identity in the coordinates, the resulting function must equal a function of the observable times Ψ. With the simplest form of the operator, the result is merely the observable times Ψ, when the observable is fixed, that is, when the system is in a definite *eigenstate* with respect to the property. The corresponding Ψ is called an *eigenfunction*. The value of the observable is called an *eigenvalue*. Thus, we postulate the following:

(II') When a state can yield only one value a of an observable, then an operator A exists such that

$$A\Psi = a\Psi. \tag{1.1}$$

For each completely different state we find the same form. If the corresponding eigenvalues are $a_1, a_2, \ldots, a_N$, with $a_{j+1} \geqslant a_j$, we have

$$A\Psi_1 = a_1\Psi_1,$$
$$A\Psi_2 = a_2\Psi_2,$$
$$\cdot$$
$$\cdot$$
$$\cdot$$
$$A\Psi_N = a_N\Psi_N. \tag{1.2}$$

For certain aspects, such as spin, N is a small number. For others, such as energy, N is infinite.

Now, a state that can only yield a single result for one property may yield a spectrum of values for another. Thus, a person can prepare a homogeneous beam of electrons in which each particle has the same momentum. Then a momentum measurement would yield a single value. But if one tried to find the location of an electron in a section containing a single electron, it would appear in an element Δx of the section with a probability equal to the ratio of Δx to the length of the section.

Thus, the eigenstate Ψ_j for operator A may be a composite state Φ for operator B. And an eigenstate Φ_j for B may be a composite state Ψ for operator A. Complementing (1.2) are the eigenvalue equations

$$B\Phi_1 = b_1\Phi_1,$$
$$B\Phi_2 = b_2\Phi_2,$$
$$\cdot$$
$$\cdot$$
$$\cdot$$
$$B\Phi_N = b_N\Phi_N. \tag{1.3}$$

Furthermore, the system may be in a composite state for both operators A and B.

But how is a composite state formed from a set of eigenstates? We note that the equations describing the pure possibilities, Equations (1.2) and (1.3), are linear in the dependent variables. We also consider that operators A and B are linear. Then if $a_{j+1} = a_j$, we find that

$$A(c_j\Psi_j + c_{j+1}\Psi_{j+1}) = a_j(c_j\Psi_j + c_{j+1}\Psi_{j+1}), \qquad (1.4)$$

that each linear superposition of eigenfunctions is an eigenfunction. We are led to make the generalization:

(III') If the possible eigenstates with respect to an operator are described by Ψ_1, $\Psi_2, \ldots, \Psi_N$, each possible composite state for the system is given by a linear superposition

$$\Psi = \Sigma c_j\Psi_j. \qquad (1.5)$$

Expression Ψ describes the most general state function constructible from the independent variables.

Equations (1.2) and (1.3) leave unspecified the magnitude and the phase of a state function at a particular reference point. However, the relative phases and magnitudes are taken into account in (1.5). Thus, we are allowed to introduce the postulate:

(IV') The probability of finding a submicroscopic particle in volume $d^3\mathbf{r}$ of physical space is $\Psi^*\Psi d^3\mathbf{r}$. For a system of particles, the probability for localizing particle 1 in $d^3\mathbf{r}_1$, particle 2 in $d^3\mathbf{r}_2, \ldots$, particle n in $d^3\mathbf{r}_n$ is $\Psi^*\Psi d^3\mathbf{r}_1 d^3\mathbf{r}_2 \ldots d^3\mathbf{r}_n = \Psi^*\Psi d^{3n}\mathbf{r}$.

Here Ψ^* is the complex conjugate of Ψ. The asterisk represents the operation of taking the complex conjugate.

In dealing with a bound system, we let the region in which the particle (or particles) is (are) bound be R. In dealing with a free system of independent particles, one may determine the volume per particle. Then one lets a region representing this volume be R.

Since the total probability of finding a particle is one, we have

$$\int_R \Psi^*\Psi d^3\mathbf{r} = 1 \qquad (1.6)$$

for a one-particle system and

$$\int_R \Psi^*\Psi d^{3n}\mathbf{r} = 1 \qquad (1.7)$$

for an n-particle system. A function Ψ satisfying (1.6) or (1.7) is said to be *normalized* to 1.

The complex conjugate of (1.5) is

$$\Psi^* = \Sigma c_j^*\Psi_j^*, \qquad (1.8)$$

while changing the dummy index in (1.5) yields

$$\Psi = \Sigma c_k \Psi_k.$$ (1.9)

Substituting these sums into (1.6) leads to

$$\int_R \Psi^* \Psi d^3\mathbf{r} = \int_R \Sigma c_j^* \Psi_j^* \Sigma c_k \Psi_k d^3\mathbf{r}$$

$$= \Sigma \Sigma c_j^* c_k \int_R \Psi_j^* \Psi_k d^3\mathbf{r}.$$ (1.10)

If the contribution of $c_k \Psi_k$ to this integral is to be independent of the other terms in Ψ, we must have

$$\int_R \Psi_j^* \Psi_k d^3\mathbf{r} = 0 \qquad \text{when} \qquad j \neq k.$$ (1.11)

If Ψ_j and Ψ_k are all normalized to 1, in addition, Equation (1.10) reduces to

$$\int_R \Psi^* \Psi d^3\mathbf{r} = \Sigma c_j^* c_j.$$ (1.12)

Similar relations exist for (1.7). So we make the following postulate:

(V') For the contributions of the jth and kth states to a composite state to be independent (that is, for these two states to be completely distinct), we must have

$$\int_R \Psi_j^* \Psi_k d^{3n}\mathbf{r} = 0.$$ (1.13)

Restriction (1.13) is referred to as the *orthogonality* condition. Combining (1.7) and (1.13) yields

$$\int_R \Psi_j^* \Psi_k d^{3n}\mathbf{r} = \delta_{jk},$$ (1.14)

where δ_{jk} is the Kronecker delta.

Since the left side of (1.12) is the total probability, the kth term on the right is the probability the system would be found in the kth state. Thus:

(VI') When Ψ_1, Ψ_2, ... are orthogonal and normalized, the statistical weight of the kth eigenstate in

$$\Psi = \Sigma c_j \Psi_j$$ (1.15)

is

$$w_k = c_k^* c_k.$$ (1.16)

In our calculations, we will generally employ othogonal and normalized state functions.

1.3
Composite States and Mixed States

When suitable eigenstates for a given operator and system superpose following Equation (1.5), they produce a composite state that may be an eigenstate for a different operator. When the eigenstates combine without a limitation on their relative phases, they produce a mixed state, which cannot be such an eigenstate. However, statistically weighted means can be obtained in either situation.

Consider the suitable eigenfunctions with respect to a certain determination to be Ψ_1, Ψ_2, . . . , Ψ_N, while the operator for the measurement is A.

When the different eigenfunctions are combined linearly, with definite relative phases and magnitudes, we have

$$\Psi = \Sigma c_j \Psi_j, \qquad (1.17)$$

as noted before. Any possible composite state can be represented thus. Furthermore, with the flexibility available, an arbitrary single-valued function meeting the boundary conditions can be represented in this way. Thus, the eigenfunctions form a *complete* set over R.

Without loss of generality, we may consider the eigenfunctions to be orthogonal to each other and normalized to one. Then act on (1.17) with operator A and introduce the eigenvalue relationship,

$$A\Psi = A\Sigma c_k \Psi_k = \Sigma c_k A \Psi_k = \Sigma c_k a_k \Psi_k. \qquad (1.18)$$

Take the complex conjugate of (1.17), multiply each side by the corresponding side of (1.18), integrate over R, and reduce with (1.16),

$$\begin{aligned}
\int_R \Psi^* A \Psi d^{3n}\mathbf{r} &= \int_R \Sigma c_j^* \Psi_j^* \Sigma c_k a_k \Psi_k d^{3n}\mathbf{r} \\
&= \Sigma \Sigma c_j^* c_k a_k \int_R \Psi_j^* \Psi_k d^{3n}\mathbf{r} \\
&= \Sigma w_k a_k = \langle A \rangle.
\end{aligned} \qquad (1.19)$$

We obtain the properly weighted mean of observable a. This is symbolized by $\langle A \rangle$ and is called the *expectation value* for a. Thus, we have the postulate:

(VII′) The expectation value for an observable whose operator is A is

$$\langle A \rangle = \int_R \Psi^* A \Psi d^{3n}\mathbf{r} \qquad (1.20)$$

when the given system is in the state represented by Ψ.

Any state representable by a single function is said to be *coherent*. Since such a state may be an eigenstate for some operator, it is also said to be *pure* with respect to that operator.

When the relative phasings of the different contributions to a state are not known or definite, one cannot formulate (1.17). The different contributions are then independent and separable; one says the state is *mixed* or *incoherent*.

Nevertheless, if the probabilities (statistical weights) for each constituent are known or obtainable, one can still write

$$\langle A \rangle = \Sigma w_k a_k, \qquad (1.21)$$

where w_k is the weight of the kth eigenstate.

1.4
The Vector the State Function Represents

A function associates a number with each set of values for the independent variables. As a consequence, its expression depends on how such variables are chosen. However, the choice of variables does not influence any physical behavior described by the function. Indeed, a person can represent this invariant aspect by a ket and a conjugate bra in an appropriate Hilbert space.

For the numbers a function associates with the points in a region over which it is taken are analogous to the components of a vector. And since there are an infinite number of points to assess, the function in effect has an infinite number of components.

The functions we have to consider are single valued and smoothly varying. For simplicity, we will first examine ones that depend on a single variable alone. Furthermore, we will look at their representations at a finite number of points. The results will then be appropriately generalized.

Consider two functions $\psi_I(x)$ and $\psi_{II}(x)$ that are single valued and smooth in the interval

$$a \leqslant x \leqslant b, \qquad (1.22)$$

as Figure 1.1 indicates. Let us divide the interval into $n - 1$ equal subintervals and two end intervals half as large:

$$2(x_1 - a) = x_{j+1} - x_j = 2(b - x_n) = \frac{b - a}{n}. \qquad (1.23)$$

Let us construct n orthogonal unit vectors,

$$\mathbf{e}_1, \mathbf{e}_2, \ldots, \mathbf{e}_n, \qquad (1.24)$$

in an n-dimensional Euclidean space and identify the jth component of vector

$$\mathbf{A} = \Sigma A_j \mathbf{e}_j \qquad (1.25)$$

in the space with the value of $\psi_I(x)$ at $x = x_j$:

$$A_j = \psi_I(x_j). \qquad (1.26)$$

Similarly, we set

$$\mathbf{B} = \Sigma B_j \mathbf{e}_j \qquad (1.27)$$

with

$$B_j = \psi_{II}(x_j). \tag{1.28}$$

Vectors **A** and **B** are approximate representations of the functions $\psi_I(x)$ and $\psi_{II}(x)$ in the given interval. Indeed, the vector rule for adding components causes the functions to be added in the conventional manner:

$$A_j + B_j = \psi_I(x_j) + \psi_{II}(x_j). \tag{1.29}$$

Applying the generalized Pythagorean theorem to the complex components yields

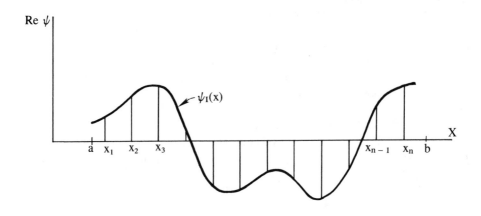

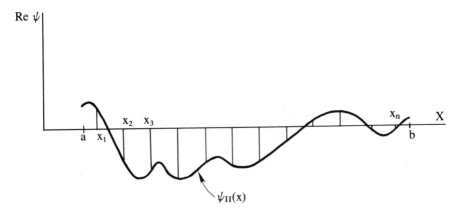

Figure 1.1. The real parts of $\psi_I(x)$ and $\psi_{II}(x)$. The imaginary parts also vary smoothly, but independently.

$$\sum_1^n A_j{}^* A_j = \mathbf{A}^* \cdot \mathbf{A} \qquad (1.30)$$

and

$$\sum_1^n B_j{}^* B_j = \mathbf{B}^* \cdot \mathbf{B} \qquad (1.31)$$

for the squares of the magnitudes of **A** and **B**. Now quantities (1.30) and (1.31) vary with how well the chosen points represent all the points. Errors tend to be reduced by increasing n. However, the sums tend to increase directly with n.

To obtain quantities strictly representative of the functions, we must (a) let the number of points increase without limit and (b) divide by the number of points. We also choose to multiply by the length of the full interval.

For a finite n, we construct

$$\mathbf{A}^* \cdot \mathbf{A} \, \frac{b - a}{n} = \sum_{j=1}^n \psi_I{}^*(x_j) \psi_I(x_j) \Delta x \qquad (1.32)$$

and

$$\mathbf{B}^* \cdot \mathbf{B} \, \frac{b - a}{n} = \sum_{j=1}^n \psi_{II}{}^*(x_j) \psi_{II}(x_j) \Delta x. \qquad (1.33)$$

Then we let n increase without limit. The right side of (1.32) becomes the integral

$$\int_a^b \psi_I{}^*(x) \psi_I(x) \, dx. \qquad (1.34)$$

On the left side of (1.32), expression $[(b - a)/n]^{\frac{1}{2}}$ acts as a normalizing factor on vector **A** and similarly on vector **A***. The limit (1.34) can then be considered as the *scalar product* of a vector representing ψ_I with one representing $\psi_I{}^*$ in the infinite-dimensional Euclidean space. This limiting space is the *Hilbert space* for the system. Note how the interval a to b enters the definition of the space.

The vector representing ψ_I is called the *ket*

$$, \psi_I) = |\,I\rangle; \qquad (1.35)$$

the conjugate vector representing $\psi_I{}^*$ that premultiplies (1.35) is called the *bra*

$$(\psi_I, = \langle\,I\,|. \qquad (1.36)$$

Thus, we write

$$(\psi_I, \psi_I) = \langle I\,|\,I\rangle = \int_a^b \psi_I{}^*(x) \psi_I(x) dx. \qquad (1.37)$$

Similarly, for the limit of (1.33), we write

$$(\psi_{II}, \psi_{II}) = \langle II\,|\,II\rangle = \int_a^b \psi_{II}{}^*(x) \psi_{II}(x) dx. \qquad (1.38)$$

For the *scalar product* of the vector representing $\psi_I{}^*$ with the vector representing ψ_{II}, over interval (1.22), we write

$$(\psi_{\mathrm{I}}, \psi_{\mathrm{II}}) = \langle \mathrm{I} \mid \mathrm{II} \rangle = \int_a^b \psi_{\mathrm{I}}{}^*(x)\psi_{\mathrm{II}}(x)dx. \qquad (1.39)$$

Similarly, for the scalar product of the vector representing $\psi_{\mathrm{II}}{}^*$ with the one representing ψ_{I}, over interval (1.22), we have

$$(\psi_{\mathrm{II}}, \psi_{\mathrm{I}}) = \langle \mathrm{II} \mid \mathrm{I} \rangle = \int_a^b \psi_{\mathrm{II}}{}^*(x)\psi_{\mathrm{I}}(x)dx. \qquad (1.40)$$

Note that

$$\langle \mathrm{I} \mid \mathrm{II} \rangle = \langle \mathrm{II} \mid \mathrm{I} \rangle^*. \qquad (1.41)$$

The arguments we have presented apply to each dimension of a multidimensional system. So they apply in combined form to the system as a whole. When the state of an n-particle system, either pure or composite with respect to an operator, is labeled j and described by a state function Ψ_j, a $3n$-dimensional ket

$$, \Psi_j) = \mid j \rangle \qquad (1.42)$$

and a $3n$-dimensional bra

$$(\Psi_j, = \langle j \mid \qquad (1.43)$$

govern its properties, neglecting spins. If spins are included, each of these is $4n$-dimensional.

Orthonormality condition (1.14), for the jth and kth eigenstates, is written in the form

$$(\Psi_j, \Psi_k) = \langle j \mid k \rangle = \delta_{jk}. \qquad (1.44)$$

Usage of vectors $\mid j \rangle$ and $\langle j \mid$ does not imply usage of a particular function Ψ_j. In discussions where the resolution is irrelevant, the kets and the bras should be employed.

Indeed, just as a vector exists whether any components are known or not, a ket and a bra are objects in themselves. Any function from which they may be constructed is merely a representation of the basic entity.

Example
1.1

How is addition of bras for a given system related to addition of the corresponding kets and to superposition of the state functions?

Let Ψ_a, Ψ_b, Ψ_c, Ψ_d be single-valued smooth functions of the independent variables, constructed to fit the boundary conditions. When normalized, they serve as possible state functions.

An arbitrary superposition of the first two functions has the form

$$\Psi_j = \alpha\Psi_a + \beta\Psi_b,$$

while an arbitrary superposition of the last two is

$$\Psi_k = \gamma\Psi_c + \delta\Psi_d.$$

Substituting the second of these forms into the general scalar product gives us

$$\int_R \Psi_j^*\Psi_k d^{3n}\mathbf{r} = \int_R \Psi_j^*(\gamma\Psi_c + \delta\Psi_d)d^{3n}\mathbf{r} = \gamma\int_R \Psi_j^*\Psi_c d^{3n}\mathbf{r} + \delta\int_R \Psi_j^*\Psi_d d^{3n}\mathbf{r},$$

while substituting the first similarly yields

$$\int_R \Psi_j^*\Psi_k d^{3n}\mathbf{r} = \int_R (\alpha\Psi_a + \beta\Psi_b)^*\Psi_k d^{3n}\mathbf{r} = \alpha^*\int_R \Psi_a^*\Psi_k d^{3n}\mathbf{r} + \beta^*\int_R \Psi_b^*\Psi_k d^{3n}\mathbf{r}$$

In vector language, these equations appear as

$$\langle j | k \rangle = \langle j | (\gamma | c \rangle + \delta | d \rangle)$$
$$= \gamma\langle j | c \rangle + \delta\langle j | d \rangle$$

and

$$\langle j | k \rangle = (\alpha^*\langle a | + \beta^*\langle b |) | k \rangle$$
$$= \alpha^*\langle a | k \rangle + \beta^*\langle b | k \rangle,$$

whence

$$| k \rangle = \gamma | c \rangle + \delta | d \rangle$$

and

$$\langle j | = \alpha^*\langle a | + \beta^*\langle b |.$$

Interchanging the roles of Ψ_j and Ψ_k in the scalar product similarly leads to

$$\langle k | = \gamma^*\langle c | + \delta^*\langle d |$$

and

$$| j \rangle = \alpha | a \rangle + \beta | b \rangle.$$

The law for superposing kets does not differ from that for superposing the corresponding state functions; it is the same linear law. But the formula for superposing bras involves the complex conjugates of the coefficients used in combining the state functions. The operator for transforming from state function to ket is said to be *linear*; the one for transforming from state function to bra, *antilinear*.

Because of the superposition principle, the space of the possible kets for a given system is linear. Therefore, a sum of, or a difference between, two kets is a ket; and any variation of a ket is a ket. Also, a sum of, or a difference between, two bras is a bra; and any variation of a bra is a bra.

1.5
Matrix Elements Generated by Suitable Operators

An explicit mathematical operator acting on an arbitrary state function for a given system may yield an expression that is still single valued, smooth, and meeting the boundary conditions. Then the operation is a representation of a transformation of the ket in its Hilbert space. Among the operators that can serve are those for observables and those that reorient the original system, the symmetry operators.

A person can formulate the scalar product between each ket in a set and each transformed ket. The set of kets that one needs to consider in a particular problem may involve only part of the Hilbert space, or the entire Hilbert space, for the given system. In either situation, the scalar products combine as matrix elements.

Let a set of kets for a system be

$$| 1 \rangle , | 2 \rangle , \ldots , | k \rangle , \ldots \qquad (1.45)$$

and the corresponding bras be

$$\langle 1 | , \langle 2 | , \ldots , \langle k | , \ldots , \qquad (1.46)$$

while the transformation operators are

$$A, \; B, \; \text{and} \; AB. \qquad (1.47)$$

Since $A | k \rangle$, $B | k \rangle$, and $AB | k \rangle$ are kets obtained from $| k \rangle$, one may form the scalar products

$$\langle j | | A | k \rangle = \langle j | A | k \rangle = A_{jk}, \qquad (1.48)$$

$$\langle j | | B | k \rangle = \langle j | B | k \rangle = B_{jk}, \qquad (1.49)$$

$$\langle j | | AB | k \rangle = \langle j | AB | k \rangle = (AB)_{jk}. \qquad (1.50)$$

The vertical double line in each initial expression separates the bra $\langle j |$ from the transformed ket. Since such separation can be inferred from the succession of symbols, the second form involving only single vertical lines is ordinarily employed.

For these expressions to be considered matrix elements, they must combine following the standard matrix rules. Let us see how they do.

The result of any transformation operator acting on any ket in the given set lies in the pertinent space. Consequently, it equals a linear combination of the basis kets:

$$A | k \rangle = \Sigma c_i | i \rangle. \qquad (1.51)$$

Without loss of generality, one can consider the bras and the kets to be orthogonal and normalized. Then multiplying (1.51) by bra $\langle m |$ gives us

$$\langle m | A | k \rangle = \Sigma c_i \langle m | i \rangle = \Sigma c_i \delta_{mi} = c_m. \qquad (1.52)$$

Substituting this result back into (1.51) yields

$$A \mid k\rangle = \sum_m \langle m \mid A \mid k\rangle \mid m\rangle. \qquad (1.53)$$

Similarly, one can obtain

$$B \mid l\rangle = \sum_n \langle n \mid B \mid l\rangle \mid n\rangle. \qquad (1.54)$$

Let A act on (1.54) and expand using (1.53):

$$AB \mid l\rangle = \sum_n \langle n \mid B \mid l\rangle A \mid n\rangle = \sum_m \sum_n \langle n \mid B \mid l\rangle \langle m \mid A \mid n\rangle \mid m\rangle$$

$$= \sum_m \sum_n \langle m \mid A \mid n\rangle \langle n \mid B \mid l\rangle \mid m\rangle. \qquad (1.55)$$

Then multiply by bra $\langle j \mid$ and reduce:

$$\langle j \mid AB \mid l\rangle = \sum_m \sum_n \langle m \mid A \mid n\rangle \langle n \mid B \mid l\rangle \langle j \mid m\rangle$$

$$= \sum_n \langle j \mid A \mid n\rangle \langle n \mid B \mid l\rangle. \qquad (1.56)$$

Introducing (1.50), (1.48), (1.49) now gives us

$$(AB)_{jl} = \sum A_{jn} B_{nl}. \qquad (1.57)$$

In a particular discussion, a person may need to employ only the first p kets and bras. Then the summation in (1.57) would run from 1 to p. Scalar products (1.48), (1.49), and (1.50) could be considered the j-kth elements of $p \times p$ matrices **A**, **B**, **AB**. And, formula (1.57) tells us that **A** and **B** multiply according to the usual matrix rules, for any given p. Because of the linearity in the setup, they also add as matrices. Thus, A_{jk}, B_{jk}, and $(AB)_{jk}$ are matrix elements generated by the operators A, B, and AB.

1.6
A Hilbert-Space Reformulation of the Postulates

Any function Ψ describing a state of a given system is a particular representation of the vector for the state; in effect, it specifies components of the vector in a particular manner. Furthermore, the state vectors for the system combine as linear elements of a Hilbert space. Each possible vector in the space is conveniently described as a ket and as a bra. So let us restate and expand out elementary postulates in terms of the kets and bras.

Indeed, since the state function Ψ represents a ket vector, we replace (I$'$) with the axiom:

(I) The state of a coherent system is governed by a ket $| \, g \, \rangle$ in the appropriate Hilbert space. Conjugate to the ket is the corresponding bra $\langle \, g \, |$.

The *Hilbert space* for a given system allows for each possible set of independent variables and their ranges, no more, no less.

Following (IV'), we interpret the scalar product of the bra vector with the corresponding ket vector as the probability that the state is occupied. A definite state for the system has a probability of 1. Thus, we assume the following:

(II) The bras and kets are normalized so that the probability a system is in the gth state is given by the scalar product $\langle g \, | \, g \rangle$.

An explicit operator A acting on a state function represents a transformation operator A acting on the corresponding ket. Thus (II') is replaced by the axiom:

(III) Whenever a state yields a single value a for an observable, it is an eigenstate whose ket $| \, j \rangle$ obeys the equation

$$A \, | \, j \rangle = a \, | \, j \rangle. \qquad (1.58)$$

We call a the eigenvalue and $| \, j \rangle$ the *eigenket*. In general, the eigenkets for a given operator span the Hilbert space as the unit vectors $\mathbf{i}, \mathbf{j}, \mathbf{k}$ span physical space in classical physics. Thus, (III') is replaced by the statement:

(IV) Each possible ket $| \, g \rangle$ for a given system can be expressed as a linear combination of the eigenkets for a pertinent operator:

$$| \, g \rangle = \Sigma \, c_j \, | \, j \rangle. \qquad (1.59)$$

Statement (IV') is covered by postulate (II). The arguments leading to (V'), (VI'), and (VII') are still relevant. In terms of the vectors in Hilbert space, these three postulates take on the following forms:

(V) For the contributions of the jth and kth eigenstates to be independent of each other, for the two states to be completely separate, we must have

$$\langle j \, | \, k \rangle = \delta_{jk}. \qquad (1.60)$$

The kets are then said to form an orthonormal set.

(VI) When the $| \, j \rangle$'s form an orthonormal set in (1.59), the statistical weight of the kth eigenstate in $| \, g \rangle$ is

$$w_k = c_k{}^* c_k. \qquad (1.61)$$

(VII) The expectation value for an observable whose operator is A is

$$\langle A \rangle = \langle g \, | \, A \, | \, g \rangle \qquad (1.62)$$

when the ket for the state is $| \, g \rangle$.

Here operator A may change the magnitude, the phase, and the direction of $|g\rangle$ within the Hilbert space for the system.

When the relative phasings of the different contributions to a state are not definite, the state is not described by a single ket, such as $|g\rangle$. However, an operator governing the state does still exist.

Consider a system with a set of states described by the orthonormal kets

$$|f_1\rangle, |f_2\rangle, \ldots, |f_j\rangle, \ldots, |f_N\rangle. \tag{1.63}$$

Suppose that from how the system was prepared, or from measurements, one knows that the jth state has, or had, weight w_j. Then to obtain results consistent with (1.21), we postulate the following:

(VIII) The state of an incoherent system formed from kets $|f_1\rangle, \ldots, |f_N\rangle$ is governed by the *density operator*

$$\rho = \sum_1^N |f_j\rangle w_j \langle f_j|. \tag{1.64}$$

(IX) The expectation value for an observable whose operator is A is

$$\langle A \rangle = \sum_k (A\rho)_{kk}. \tag{1.65}$$

Equation (1.64) defines the density operator, or density matrix, for any system, incoherent or coherent. Postulate (IX) applies in general.

Indeed if kets (1.63) are eigenkets of A, then with the eigenvalue equation, we have

$$A\rho = \sum_j A|j\rangle w_j \langle j| = \sum_j a_j |j\rangle w_j \langle j| \tag{1.66}$$

and with (1.50), we obtain

$$\sum_k (A\rho)_{kk} = \sum_k \langle k| (\sum_j a_j |j\rangle w_j \langle j|) |k\rangle$$

$$= \sum_j \sum_k a_j \delta_{kj} w_j \delta_{jk} = \sum_k a_k w_k. \tag{1.67}$$

Otherwise, we use the fact that the eigenkets of A form a complete set in the Hilbert space for the system; so

$$|f_j\rangle = \sum c_i^j |i\rangle. \tag{1.68}$$

Then

$$A\rho = \sum_i \sum_j \sum_h A c_i^j |i\rangle w_j c_h^{j*} \langle h|$$

$$= \sum_i \sum_j \sum_h c_i^j c_h^{j*} a_i w_j |i\rangle \langle h| \tag{1.69}$$

and

$$\sum_k (A\rho)_{kk} = \sum_h \sum_i \sum_j \sum_k c_i{}^j c_h{}^{j*} a_i w_j \langle k \mid i \rangle \langle h \mid k \rangle$$

$$= \sum_j \sum_k c_k{}^{j*} c_k{}^j a_k w_j = \sum_j \langle A \rangle^j w_j. \qquad (1.70)$$

Here $\langle A \rangle^j$ is the expectation value of the observable for state $\mid f_j \rangle$. Note how these results justify postulate (IX).

1.7
The Projection Operator

According to postulate (III), observables are determined by eigenvalue equations. Furthermore, the kets describing possible states exist in a Hilbert space. This space is constructed so that it is just spanned by the eigenkets for each suitable operator.

As a result, an arbitrary coherent state can be represented as a composite of phased fractions of pure states for such an operator. An arbitrary ket can be projected onto each eigenket of a set and the contribution with its phasing obtained.

Let the normalized, orthogonal eigenkets for the chosen operator be

$$\mid 1 \rangle, \ \mid 2 \rangle, \ \mid 3 \rangle, \ \ldots \ . \qquad (1.71)$$

Also, suppose the normalized ket to be analyzed is $\mid g \rangle$. From (1.59), we know that

$$\mid g \rangle = \sum_{j=1}^{\infty} c_j \mid j \rangle = \sum_{k=1}^{\infty} c_k \mid k \rangle. \qquad (1.72)$$

Multiplying the overall equation by $\langle j \mid$ gives us

$$\langle j \mid g \rangle = \sum_{k=1}^{\infty} c_k \langle j \mid k \rangle = \sum_{k=1}^{\infty} c_k \delta_{jk} = c_j. \qquad (1.73)$$

Substituting this result into (1.72) yields

$$\mid g \rangle = \sum_{j=1}^{\infty} \langle j \mid g \rangle \mid j \rangle = \sum_{j=1}^{\infty} \mid j \rangle \langle j \mid g \rangle. \qquad (1.74)$$

Thus the contribution to $\mid g \rangle$ of the jth eigenstate is

$$\mid j \rangle \langle j \mid g \rangle = P_j \mid g \rangle, \qquad (1.75)$$

if we let

$$\mid j \rangle \langle j \mid = P_j. \qquad (1.76)$$

Since P_j acts to remove from the general ket the contribution of the jth eigenket, it is called a *projection operator*.

Adding up the contributions to $|g\rangle$ yields $|g\rangle$ itself. Correspondingly, the sum

$$\sum_{1}^{\infty} P_j = \sum_{1}^{\infty} |j\rangle\langle j| \tag{1.77}$$

acts as an identity operator when operating on the general ket $|g\rangle$.

Example 1.2

What effect does repetition of the projection operator have?

Since projection operator P_j acts to take from an arbitrary ket the contribution of the jth eigenket, we have

$$P_j|g\rangle = c_j|j\rangle$$

and

$$P_j|j\rangle = |j\rangle.$$

Acting on the first equation with P_j then yields

$$P_j(P_j|g\rangle) = P_j c_j|j\rangle = c_j P_j|j\rangle = c_j|j\rangle.$$

By induction, we have

$$P_j^n|g\rangle = c_j|j\rangle.$$

Thus,

$$P_j^2 = P_j$$

and

$$P_j^n = P_j.$$

The projection operator raised to any integral power produces the same effect as P_j itself. Such an operator is said to be *idempotent*.

1.8
The Hermitian Condition

In the Hilbert space for a given system, the operator for an observable is a special transformation operator.

Consider the various possible normalized and phased kets in a given Hilbert space. Any two may be linked by a transformation operator A:

$$A|i\rangle = c|j\rangle. \tag{1.78}$$

Here coefficient c may be complex. Furthermore, A may be altered so that $|j\rangle$ varies continuously from $|i\rangle$.

Now, any two kets $|j\rangle$ and $|k\rangle$, in the Hilbert space, yield the scalar product

$$\langle j|k\rangle = \langle k|j\rangle^*. \tag{1.79}$$

Instead of $|k\rangle$, one may employ the transformed ket $A|k\rangle$. But the result is then equal to the scalar product of an operator B transforming $|j\rangle$ with the original $|k\rangle$. One may construct it either as the bra from $B|j\rangle$ multiplied by $|k\rangle$ or as the complex conjugate of bra $\langle k|$ times $B|j\rangle$:

$$\langle j|A|k\rangle = \langle B|j||k\rangle = \langle k|B|j\rangle^*. \tag{1.80}$$

We call B the *adjoint* $A^\dagger$ of A. Thus by definition, we have

$$\langle j|A|k\rangle = \langle A^\dagger|j||k\rangle = \langle k|A^\dagger|j\rangle^*. \tag{1.81}$$

When

$$\langle j|A|k\rangle = \langle A|j||k\rangle = \langle k|A|j\rangle^* \tag{1.82}$$

for each possible pair of kets in the Hilbert space, A is said to be self-adjoint or *Hermitian*.

From postulate (III), the states yielding sharp values for an observable are eigenstates of the operator for the observable. When the observed value is a_j with certainty and the operator is A, we have

$$A|j\rangle = a_j|j\rangle \tag{1.83}$$

and when the value is a_k with certainty, we have

$$A|k\rangle = a_k|k\rangle. \tag{1.84}$$

Multiplying both sides of (1.83) with the bra vector for the kth state produces

$$\langle k|A|j\rangle = a_j\langle k|j\rangle = a_j\langle j|k\rangle^*, \tag{1.85}$$

while multiplying both sides of (1.84) with the bra vector for the jth state and taking the complex conjugate yields

$$\langle j|A|k\rangle^* = a_k^*\,\langle j|k\rangle^*. \tag{1.86}$$

Whenever A is Hermitian, the left sides of (1.85) and (1.86) are equal. Then when $k = j$, we have

$$a_j = a_j^*. \tag{1.87}$$

Thus, the Hermitian condition on A ensures that all its eigenvalues are real. Since observed quantities are real, we add to our list of postulates the following:

(X) The operator for each observable is Hermitian.

Combining (1.85), (1.86), and (1.87) with (1.82) leads to

$$a_k \langle j \mid k \rangle = a_j \langle j \mid k \rangle. \qquad (1.88)$$

When the jth and kth observables differ,

$$a_k \neq a_j, \qquad (1.89)$$

we must have

$$\langle j \mid k \rangle = 0. \qquad (1.90)$$

Requiring the operator to be Hermitian makes ket $\mid k \rangle$ for one level orthogonal to ket $\mid j \rangle$ for a different level. For convenience, they are generally normalized to 1, also.

Now, n kets are said to be *linearly independent* when no one of them can be expressed as a linear combination of the others. But from such a set, a person can construct n orthonormal kets. See Example 1.3.

Degeneracy is said to occur when two or more linearly independent kets yield the same eigenvalue. The magnitude of the degeneracy equals the number of linearly independent kets one needs to describe, by superposition, any possible ket for the level. These, for simplicity, are often chosen to be orthonormal kets.

We see that, when A is Hermitian, a complete set of base vectors may be chosen with

$$\langle j \mid k \rangle = \delta_{jk} \qquad (1.91)$$

for any j and k in the set. And Equation (1.59) holds with these base kets.

Example 1.3

Show how N orthonormal kets can be constructed from N linearly independent ones.

We are given N kets

$$\mid f_1 \rangle, \mid f_2 \rangle, \ldots, \mid f_j \rangle, \ldots, \mid f_N \rangle$$

such that no one of them equals a linear combination of the others. Construct the normalized ket

$$\mid 1 \rangle = \frac{\mid f_1 \rangle}{\langle f_1 \mid f_1 \rangle^{\frac{1}{2}}} .$$

A ket orthogonal to $|1\rangle$ is

$$|g_2\rangle = |f_2\rangle + a_{21}|1\rangle$$

when

$$\langle 1|(|f_2\rangle + a_{21}|1\rangle) = 0.$$

Solving for the coefficient yields

$$a_{21} = -\langle 1|f_2\rangle.$$

A normalized form of the ket is

$$|2\rangle = \frac{|f_2\rangle - \langle 1|f_2\rangle|1\rangle}{\langle g_2|g_2\rangle^{\frac{1}{2}}}.$$

We proceed similarly up the sequence. For the jth orthogonal ket, we construct

$$|g_j\rangle = |f_j\rangle + a_{j1}|1\rangle + a_{j2}|2\rangle + \ldots + a_{j,j-1}|j-1\rangle$$

$$= |f_j\rangle + \sum_{k=1}^{j-1} a_{jk}|k\rangle$$

with

$$a_{jk} = -\langle k|f_j\rangle.$$

A normalized form is

$$|j\rangle = \frac{|f_j\rangle - \sum_{k=1}^{j-1}\langle k|f_j\rangle|k\rangle}{\langle g_j|g_j\rangle^{\frac{1}{2}}}$$

To see that this is orthogonal to each preceding constructed combination, note that for

$$1 \le i \le j - 1$$

we have

$$\langle i|g_j\rangle = \langle i|f_j\rangle - 0 - \langle i|f_j\rangle\langle i|i\rangle - 0$$

$$= 0.$$

The procedure stops with $j = N$, when N orthonormal kets have been obtained.

We call this the *Gram-Schmidt orthogonalization* procedure. In applications, one may arrange the given kets in any order. Furthermore, one could construct from the $|f_j\rangle$'s innumerable other linearly independent sets and employ each set similarly. Thus, an orthonormal set can be constructed in countless ways from the given set.

1.9
Commutativity of Operators for Observables

In the Hilbert space for a given system, a selected transformation operator may commute with some of, but not all of, the other operators. Those with which it commutes have eigenstates in common.

Where no degeneracy exists, definiteness in one observable implies definiteness in the other. However, a degenerate value for one of the commuting operators may correspond to a spectrum of values for the other. Then definiteness in the first observable goes with the spectrum for the second observable.

First, let us consider two operators with a common complete set of eigenkets. Let the operators be designated A and B, while the common eigenkets are

$$|1\rangle, |2\rangle, \ldots, |j\rangle, \ldots \qquad (1.92)$$

Then we have

$$A|j\rangle = a_j |j\rangle \qquad (1.93)$$

and

$$B|j\rangle = b_j |j\rangle. \qquad (1.94)$$

Let us act on (1.94) with A,

$$AB|j\rangle = b_j A|j\rangle = a_j b_j |j\rangle, \qquad (1.95)$$

on (1.93) with B,

$$BA|j\rangle = a_j B|j\rangle = a_j b_j |j\rangle, \qquad (1.96)$$

and subtract,

$$(AB - BA)|j\rangle = 0. \qquad (1.97)$$

But for an arbitrary ket in the Hilbert space, we have

$$|g\rangle = \sum_j c_j |j\rangle. \qquad (1.98)$$

Let $AB - BA$ act on this equation and introduce (1.97) for $|j\rangle$:

$$(AB - BA)|g\rangle = \sum_j c_j (AB - BA)|j\rangle = 0. \qquad (1.99)$$

Because vector $|g\rangle$ is arbitrary in the given space, the operator itself must act as zero,

$$AB - BA = 0. \qquad (1.100)$$

Thus,

$$AB = BA. \tag{1.101}$$

Whenever two operators have a common complete set of eigenkets, they commute. Secondly, let us assume that A and B commute in the given Hilbert space. Now when operator A acts on the eigenvalue equation for B,

$$B \mid j \rangle = b_j \mid j \rangle, \tag{1.102}$$

the relationship

$$AB \mid j \rangle = Ab_j \mid j \rangle \tag{1.103}$$

results. Operator A commutes with b_j because b_j is a constant and A is linear. Since A also commutes with B, by assumption, we derive

$$BA \mid j \rangle = b_j A \mid j \rangle, \tag{1.104}$$

a form of the eigenvalue equation for B. When the level is not degenerate, a constant times $\mid j \rangle$ satisfies this equation. Then

$$A \mid j \rangle = \lambda_j \mid j \rangle \tag{1.105}$$

and $\mid j \rangle$ is also an eigenket for A.

When n mutually orthogonal, normalized eigenkets exist for eigenvalue b_j of B, we proceed through (1.102)-(1.104) for a typical $\mid j \rangle$, as before. Then we change $\mid j \rangle$ to $\mid k \rangle$, multiply by c_k and sum over the degenerate kets:

$$BA(\Sigma \, c_k \mid k \rangle) = b_j A(\Sigma \, c_k \mid k \rangle). \tag{1.106}$$

A set of coefficients makes the expression in parentheses an eigenket of A if

$$A(\Sigma \, c_k \mid k \rangle) = \lambda(\Sigma \, c_k \mid k \rangle. \tag{1.107}$$

Let us multiply this by bra $\langle j \mid$, rearrange,

$$\Sigma \, \langle j \mid A \mid k \rangle \, c_k - \lambda \, \Sigma \, \langle j \mid k \rangle \, c_k = 0, \tag{1.108}$$

and reduce,

$$\Sigma_k \, (\langle j \mid A \mid k \rangle - \lambda \delta_{jk}) c_k = 0. \tag{1.109}$$

Since

$$j = 1, 2, \ldots, n, \tag{1.110}$$

we have a set of n equations homogeneous in the c_k's. Cramer's rule can be applied in solving them. But one obtains zero for all the c_k's unless

$$
\begin{vmatrix}
\langle 1|A|1\rangle - \lambda & \langle 1|A|2\rangle & \cdots\cdots & \langle 1|A|n\rangle \\
\langle 2|A|1\rangle & \langle 2|A|2\rangle - \lambda & \cdots\cdots & \langle 2|A|n\rangle \\
\cdot & \cdot & \cdot & \cdot \\
\cdot & \cdot & \cdot & \cdot \\
\cdot & \cdot & \cdot & \cdot \\
\langle n|A|1\rangle & \langle n|A|2\rangle & \cdots & \langle n|A|n\rangle - \lambda
\end{vmatrix} = 0.
\qquad (1.111)
$$

This is the *characteristic equation* for the $n \times n$ matrix **A**.

Since (1.111) has n roots, there are n sets of nontrivial c_k's. These yield n eigenkets for A which are eigenkets of B at the degenerate level. The procedure can be repeated for any other degenerate eigenvalue of B. Consequently, operators A and B have a common set of eigenkets.

1.10
Classical Observables

Some properties of a given system appear definite regardless of the state, within bounds, in which the system exists. Thus, a particle exhibits a mass, an electric charge, a spin, a statistics. A composite of particles, bound together, presents a form. Indeed, a molecule has a structure. If the structure is asymmetric, it may exhibit a chirality (D or L). With increasing complexity, other such properties may become evident.

From Section 1.9, the operator for an observable with a *definite* value in the states under consideration commutes with the operators for all other allowed observables. Such an observable is said to be *classical*. On the other hand, an observable is *quantum mechanical* if its operator does not commute with all the others.

Operators for observables can be combined additively and multiplicatively. The combinations can be considered operators for observables. A basic set, in terms of which all others can be constructed, forms an *algebra*. The center of the algebra consists of the operators that commute with all the others, the classical part of the system.

When a classical observable exhibits a single value for all possible quantum mechanical states, then it does not restrict the superpositions and transitions that can occur. Insofar as it is concerned, orthonormal eigenkets spanning the Hilbert space can be superposed without restriction,

$$
|g\rangle = \Sigma\, c_j |j\rangle,
\qquad (1.112)
$$

as postulate (IV) allows, with the provision

$$
\Sigma\, c_j{}^* c_j = 1.
\qquad (1.113)
$$

For a transition going on between $|1\rangle$ and $|2\rangle$, we have

$$|g\rangle = c_1|1\rangle + c_2|2\rangle. \tag{1.114}$$

Postulate (VII) then yields the expectation value for an observable defined by operator A as

$$\begin{aligned}
\langle A \rangle &= (c_1{}^*\langle 1| + c_2{}^*\langle 2|)A(c_1|1\rangle + c_2|2\rangle) \\
&= c_1{}^*c_1\langle 1|A|1\rangle + c_2{}^*c_2\langle 2|A|2\rangle \\
&\quad + c_1{}^*c_2\langle 1|A|2\rangle + c_2{}^*c_1\langle 2|A|1\rangle.
\end{aligned} \tag{1.115}$$

When the states are at different energies, a coordinate representation of $|1\rangle$ would have $e^{-i\omega_1 t}$ as its temporal factor, while the coordinate representation of $|2\rangle$ would have $e^{-i\omega_2 t}$, with $\omega_2 \neq \omega_1$. Then when A is independent of time t, $\langle 1|A|1\rangle$ and $\langle 2|A|2\rangle$ would not contain the time explicitly; but $\langle 1|A|2\rangle$ and $\langle 2|A|1\rangle$ would. For a given c_1 and c_2, these factors would cause $\langle A \rangle$ to change.

So presumably if

$$\langle 1|A|2\rangle = 0, \tag{1.116}$$

the transition would not occur. If Equation (1.116) holds for all observables, then state $|1\rangle$ and state $|2\rangle$ could exhibit different values for a classical observable. The Hilbert space would be split into a sector containing $|1\rangle$ and a sector containing $|2\rangle$.

A person could delineate each of the sectors as a region of the Hilbert space in which all classical observables have fixed values. More than two such sectors may be needed.

Conversely, if transitions between two sectors does not occur, then restriction (1.116) applies for any $|1\rangle$ in the first sector and any $|2\rangle$ in the second one. While the superposition principle would apply to each sector by itself, it would not apply to the union of the two. The sector of Hilbert space involving these forbidden superpositions could not be occupied by the physical system. Equation (1.116) would then be called a *superselection rule*.

1.11
Implications of Indistinguishability

In classical mechanics, particles are labeled by the paths they follow through the space-time continuum. Two particles may be exactly alike, yet they would be distinguishable.

In quantum mechanics, on the other hand, these definite paths do not exist. Whenever their state functions overlap, the two identical particles cannot be distinguished. This indistinguishability leads to a splitting of the Hilbert space for the two into sectors that are separated by a superselection rule. The superposition principle then applies only to each sector by itself; vectors with components in both of these sectors are forbidden. And transitions from one sector to the other cannot occur.

Consider a system of two particles. Let $\mathbf{r}_1$ and s_1 be the position variable and spin variable for the first particle, while $\mathbf{r}_2$ and s_2 are the position variable and spin variable for the second particle. Let state functions in which these coordinates occupy definite places be $\Psi(\mathbf{r}_1, s_1; \mathbf{r}_2, s_2; t)$ and $\Phi(\mathbf{r}_1, s_1; \mathbf{r}_2, s_2; t)$. If the operator interchanging the two particles is Π_{12}, then we have

$$\Pi_{12}\Psi(\mathbf{r}_1, s_1; \mathbf{r}_2, s_2; t) = \Psi(\mathbf{r}_2, s_2; \mathbf{r}_1, s_1; t) \qquad (1.117)$$

and

$$\Pi_{12}\Phi(\mathbf{r}_1, s_1; \mathbf{r}_2, s_2; t) = \Phi(\mathbf{r}_2, s_2; \mathbf{r}_1, s_1; t) \qquad (1.118)$$

Since repeating the interchange restores the original configuration, applying operation Π_{12} twice has the same effect as the identity operation. We express this result as

$$\Pi_{12}\Pi_{12} = 1, \qquad (1.119)$$

whence

$$\Pi_{12}^{-1} = \Pi_{12}. \qquad (1.120)$$

If in the integral representing the scalar product of the bra for Ψ and the ket for Φ, one interchanges the two particles, the result is not affected; for, the coordinates for particle one and for particle two go over the same range. If we replace Φ by $\Pi_{12}\Phi$, we similarly have

$$\langle\, \Psi \mid \Pi_{12}\Phi \,\rangle = \langle\, \Pi_{12}\Psi \mid \Pi_{12}\Pi_{12}\Phi \,\rangle = \langle\, \Pi_{12}\Psi \mid \Phi \,\rangle, \qquad (1.121)$$

or

$$\langle\, \Psi \mid \Pi_{12} \mid \Phi \,\rangle = \langle\, \Phi \mid \Pi_{12} \mid \Psi \,\rangle^*. \qquad (1.122)$$

Thus, Π_{12} is a Hermitian operator.

Next, consider the two particles to be *indistinguishable*. Then the matrix elements for all observables must be invariant under permutation of the particles, since these elements have physical significance. Thus,

$$\langle\Psi \mid A \mid \Phi\rangle = \langle\Pi_{12}\Psi \mid A \mid \Pi_{12}\Phi\rangle. \qquad (1.123)$$

Since A and Π_{12} are Hermitian, (1.123) can be rearranged as

$$\langle A\Psi \mid \Phi\rangle = \langle A\Pi_{12}\Psi \mid \Pi_{12}\Phi\rangle = \langle\Pi_{12}A\Pi_{12}\Psi \mid \Phi\rangle. \qquad (1.124)$$

Equation (1.124) holds for any observable operator A and any state functions Ψ and Φ. Consequently, we must have

$$\Pi_{12} A \Pi_{12} = A. \tag{1.125}$$

Acting on (1.125) with Π_{12} and introducing (1.119) leads to

$$A\Pi_{12} = \Pi_{12} A. \tag{1.126}$$

Since Π_{12} commutes with all the quantum mechanical operators, it must represent a classical observable.

Now, construct the operator

$$P_{\pm} = N(1 \pm \Pi_{12}), \tag{1.127}$$

where N is a normalization constant. Let this operator act on Ψ and let Π_{12} act on the result:

$$\Pi_{12} P_{\pm} \Psi = N(\Pi_{12} \pm 1)\Psi = \pm P_{\pm} \Psi. \tag{1.128}$$

Under the action of Π_{12}, state function

$$\Psi_+ = P_+ \Psi \tag{1.129}$$

is unchanged, while state function

$$\Psi_- = P_- \Psi \tag{1.130}$$

is changed in sign.

When the exchange of two identical particles does not alter the state function, the particles obey Bose-Einstein statistics and they are called *bosons*. When the exchange of two identical particles multiplies the state function by -1, the particles obey Fermi-Dirac statistics and are called *fermions*.

The Hilbert space for two indistinguishable particles is split into a boson sector, a fermion sector, and a mixed statistics sector. For a typical matrix element between the two extreme sectors, we have

$$\langle \Psi_- | A | \Psi_+ \rangle = \langle \Psi_- | A\Pi_{12}\Psi_+ \rangle = \langle \Psi_- | \Pi_{12}A\Psi_+ \rangle$$
$$= \langle \Pi_{12}\Psi_- | A | \Psi_+ \rangle = - \langle \Psi_- | A | \Psi_+ \rangle. \tag{1.131}$$

But when a number equals its negative, it is zero:

$$\langle \Psi_- | A | \Psi_+ \rangle = 0. \tag{1.132}$$

This superselection rule bars any transition

$$\Psi_+ \rightarrow \Psi_- \quad \text{or} \quad \Psi_- \rightarrow \Psi_+. \tag{1.133}$$

Bosons do not change into fermions, nor fermions into bosons. The rule also bars the formation of a mixed statistics function

$$\Psi = c_1\Psi_+ + c_2\Psi_-. \tag{1.134}$$

For a system of n identical particles, a person must consider all possible permutations of the particles among the places in the state function. These permutations may be effected by successive interchanges of particles. Let the operator effecting a permutation be Π_k, with k odd when the permutation involves an odd number of interchanges and even when it involves an even number.

For bosons, the generalization of (1.127) is

$$P_S = N\sum_k \Pi_k. \qquad (1.135)$$

For fermions, the generalization is

$$P_A = N\sum_k (-1)^k \Pi_k. \qquad (1.136)$$

In each formula, the summation proceeds over all possible permutations, while N is the normalization factor.

State function

$$\Psi_S = P_S\Psi \qquad (1.137)$$

is said to be symmetric in the identical particles, while state function

$$\Psi_A = P_A\Psi \qquad (1.138)$$

is antisymmetric in the identical particles.

Example 1.4

One of the identical fermions in a system enters Ψ_I, the other enters Ψ_{II}. Construct a suitable overall state function.

Let the coordinates of the first particle be $\mathbf{r}_1$, s_1, t, those of the second particle $\mathbf{r}_2$, s_2, t. If the first particle were in the first function state, the second particle in the second function state, we would have

$$\Psi_I(\mathbf{r}_1, s_1, t) \equiv \Psi_I(1) \quad \text{and} \quad \Psi_{II}(\mathbf{r}_2, s_2, t) \equiv \Psi_{II}(2).$$

From the significance of multiplying numbers, we realize that independent probabilities combine multiplicatively. To satisfy postulate (IV'), therefore, these functions need to be multiplied to form the overall state function:

$$\Psi = \Psi_I(\mathbf{r}_1, s_1, t)\Psi_{II}(\mathbf{r}_2, s_2, t) \equiv \Psi_I(1)\Psi_{II}(2).$$

This form, however, is not suitable since it distinguishes between the particles. To remove the effect of the labeling, we can act on it with P_+ or P_-. Since the particles are fermions, we choose the latter:

$$\Psi_- = N(1 - \Pi_{12})\Psi$$

$$= N[\Psi_{\mathrm{I}}(1)\Psi_{\mathrm{II}}(2) - \Psi_{\mathrm{I}}(2)\Psi_{\mathrm{II}}(1)]$$

$$= N \begin{vmatrix} \Psi_{\mathrm{I}}(1) & \Psi_{\mathrm{II}}(1) \\ \Psi_{\mathrm{I}}(2) & \Psi_{\mathrm{II}}(2) \end{vmatrix}.$$

1.12
The Occupancy of Individual States

The indistinguishability of particles of the same species leads to restrictions (1.137) and (1.138) on possible overall state functions. The latter form imposes a restriction on the occupancy of individual states.

Consider a system of n identical particles. Suppose that each may occupy the states described by the normalized orthogonal functions

$$\Psi_{\mathrm{I}}(j), \ \Psi_{\mathrm{II}}(j), \ \Psi_{\mathrm{III}}(j), \ \ldots \ . \tag{1.139}$$

From the ways independent choices can combine, we realize that independent probabilities combine multiplicatively. So to satisfy axiom (IV'), these functions multiply in forming a possible overall state function:

$$\Psi = \Psi_{\mathrm{I}}(1)\Psi_{\mathrm{II}}(2) \ldots \Psi_N(n). \tag{1.140}$$

But this form distinguishes each particle from all others. To remove the effect of the labeling, we can act on it with either P_S or P_A.

For bosons, we construct

$$\Psi_S = N\sum_k \Pi_k \Psi. \tag{1.141}$$

With this form any number of the constituent functions listed in (1.139) may be the same. Thus, any number of identical bosons may occupy the same state, or have the same set of quantum numbers.

For fermions, on the other hand, we construct

$$\Psi_A = N\sum_k (-1)^k \Pi_k \Psi. \tag{1.142}$$

Now, if any two of the constituent functions of (1.139) were the same, we could construct an equivalent function with the labeling on the two functions interchanged. But according to (1.142), interchanging the two in Ψ would change the sign of Ψ_A. We would have

$$\Psi_A = -\Psi_A. \tag{1.143}$$

Thus, the corresponding state function would vanish.

For Ψ_A to be different from zero, no two fermions may be in the same state. No two fermions can thus have the same set of quantum numbers. This result is known as the *Pauli exclusion principle*.

A general state function for fermions would be a superposition of functions of type (1.142). Taking the weightings of each contribution into account, the result is that the occupancy of each constituent state, as listed in (1.139), is limited to one.

1.13
Highlights

Submicroscopic particles, such as nuclei, and fundamental particles, such as electrons, follow quantum mechanics, insofar as a person can tell. In this mechanics, every state of a coherent system is represented by an infinite dimensional vector, such as a ket $| g \rangle$ or a bra $\langle g |$. The possible vectors for a given system act as directed linear elements of the Hilbert space that allows for all possible independent variables and their ranges, no more nor less.

A given state generally yields a spectrum of values for a given property. But for a certain set of basis states, $| 1 \rangle$, $| 2 \rangle$, . . . , $| j \rangle$, . . . , definite values are obtained. These are considered to be the eigenvalues in an eigenvalue equation

$$A | j \rangle = a_j | j \rangle. \qquad (1.144)$$

To ensure that a_j is real, we require operator A to be Hermitian. But then, the vectors for different a_j's are orthogonal. Also, the linearly independent vectors for a degenerate a_j can be made orthogonal. With the vectors normalized to one, we then have

$$\langle j | k \rangle = \delta_{jk}. \qquad (1.145)$$

The eigenkets of (1.144) span the Hilbert space for the given system. So a general ket $| g \rangle$ for the system is expressible as a linear combination of them:

$$| g \rangle = \sum_j c_j | j \rangle. \qquad (1.146)$$

With the basis vectors obeying (1.145), the coefficients in this superposition law are given by

$$c_j = \langle j | g \rangle. \qquad (1.147)$$

Also, the average value that an experimenter would obtain for the observable is

$$\langle A \rangle = \langle g | A | g \rangle. \qquad (1.148)$$

If two operators, A and B, have a common complete set of eigenkets, then the operators commute:

$$AB = BA. \tag{1.149}$$

Conversely, if two operators commute, then they have a common complete set of eigenkets.

Equation (1.146) implies an interdependence (a coherence) among the different contributing states. When this is absent or not known, the state is governed by the density operator

$$\rho = \sum_j | j \rangle w_j \langle j | \tag{1.150}$$

in which w_j is the statistical weight of the jth state.

The basic operators in the theory are those that yield observables by (1.144). Can these be constructed rationally? In the next two chapters, we will consider how this may be done.

Discussion Questions

1.1. How does one's knowledge of the physical world depend on processes?

1.2. Why do we consider matter as built up of particles?

1.3. Why can't we treat the particles as points with properties?

1.4. Why is a quantum mechanical particle sometimes called a smearon?

1.5. Why can't the state of a particle be represented by a nonanalytic function?

1.6. When do we expect an observable to be derivable from the state function Ψ?

1.7. If this derivation is effected by an operator, what should the resulting equation be?

1.8. What is the simplest form one may postulate for this equation?

1.9. Why may an eigenstate for one observable be a composite state for another observable?

1.10. What relation governs how possible composite states are obtained from a given complete set of eigenstates?

1.11. How is the magnitude of Ψ at a reference point chosen?

1.12. How is Ψ normalized?

1.13. Explain the condition that ensures that the contributions of the jth and kth states to Ψ are independent (completely distinct)?

1.14. How is the statistical weight of the kth eigenstate in Ψ determined?

1.15. Distinguish between (a) coherent and incoherent superpositions, (b) composite and mixed states.

1.16. Explain how expectation values are calculated.

1.17. To what extent can a finite-dimensional vector represent a function over a given region?

1.18. How can a finite-magnitude infinite-dimensional vector represent the function over the given region?

1.19. Should we consider the state function a representation of its ket or the ket a representation of its state function? Explain.

1.20. What is the fundamental entity in classical-space-time quantum mechanics?

1.21. How are ket $| j \rangle$ and bra $\langle j |$ related to the state function $\Psi_j(r_1, r_2, \ldots, r_n, t)$?

1.22. How is the scalar product of a bra with a ket defined? Is it dependent on the choice of independent variables? Explain.

1.23. Why is taking the ket of a superposition of kets a linear operation while taking the bra of a superposition of bras an antilinear operation?

1.24. Explain why

$$\langle j \,|\, A \,|\, k \rangle = A_{jk}$$

may be considered the j-kth element of a matrix.

1.25. How is a coherent system best described?

1.26. How do the Hilbert spaces for different systems differ?

1.27. How is the operator for an observable represented in Hilbert space?

1.28. How does the superposition principle appear in a Hilbert space?

1.29. How and why are kets normalized?

1.30. What is implied by orthogonality between two kets?

1.31. How is the statistical weight of a ket in a composite state measured?

1.32. How is an incoherent system described?

1.33. Why does the density matrix contain the minimum input data a person needs to calculate expectation values?

1.34. Explain how the projection operator is constructed.

1.35. What does the projection operator do to an arbitrary ket in the pertinent Hilbert space?

1.36. Why is the projection operator idempotent?

1.37. What eigenvalues does the projection operator yield?

1.38. What may an operator acting on a ket produce?

1.39. How is the adjoint of an operator defined?

1.40. How does the extent of the Hilbert space enter the definition of the adjoint?

1.41. When is an operator Hermitian?

1.42. Why do we require operators for observables to be Hermitian?

1.43. What is (a) linear independence, (b) degeneracy?

1.44. In the expansion

$$| g \rangle = \Sigma \, c_j \,| j \rangle,$$

what may serve for the base vectors? Why do we employ orthonormal base vectors?

1.45. How may a set of kets be orthogonalized?

1.46. What is implied by the commutativity of two operators? Explain.

1.47. How is the algebra of two operators affected when they have a common complete set of eigenkets?

1.48. Define (a) a classical observable, (b) a quantum mechanical observable.

1.49. Where does the superposition principle break down?

1.50. How does the superselection rule

$$\langle 1 \,|\, A \,|\, 2 \rangle = 0$$

arise?

1.51. When are identical particles indistinguishable?

1.52. What form of state function distinguishes between particles?

1.53. How does the law for combining probabilities determine how state functions and kets are combined?

1.54. How does one construct a state function that is (a) symmetric, (b) antisymmetric, in its identical particles?

1.55. What is (a) a boson, (b) a fermion?

Problems

1.1. Show that $\psi(\mathbf{r})$ and $e^{i\alpha}\psi(\mathbf{r})$ describe the same submicroscopic state when α is a real number.

1.2. Show when

$$\psi_I = c_1\psi_1 + c_2\psi_2$$

and

$$\psi_{II} = c_1 e^{i\alpha}\psi_1 + c_2 e^{i\beta}\psi_2$$

describe the same state of a coherent system.

1.3. Take the values of the Legendre polynomials of degrees 1 and 3 at the 10 points

$$x = -0.9, -0.7, -0.5, \ldots, 0.9;$$

consider these as components of 10-dimensional vectors $\mathbf{A}$ and $\mathbf{B}$,

$$A_j = P_1(x_j) \quad \text{and} \quad B_j = P_3(x_j);$$

construct $\mathbf{A} \cdot \mathbf{B}$ and multiply the result by

$$\frac{b - a}{n} = \frac{2}{10} = \Delta x.$$

1.4. By integration, show that $P_1(x)$ and $P_3(x)$ are orthogonal to each other over the interval $-1 \le x \le 1$. How does one reconcile this result with that from Problem 1.3?

1.5. Show that the kets for a given system obey the associative law in addition.

1.6. If the energy of the jth state is -2.1 eV and the energy of the completely separate kth state is -4.9 eV, what is the average energy of the composite state

$$|g\rangle = 0.600 |j\rangle + 0.800 |k\rangle \ ?$$

1.7. If each ket in

$$|g\rangle = 0.350 |j\rangle + 0.750 |k\rangle$$

is normalized, what does $\langle j | k \rangle$ equal?

1.8. If

$$\langle j | k \rangle = (\sin \alpha) e^{-i\beta}$$

and $|j\rangle$, $|k\rangle$ are normalized to 1, how is $|j\rangle$ related to $|k\rangle$?

1.9. If $|j\rangle$ and $|k\rangle$ are normalized and orthogonal, construct the ket that is orthogonal to

$$|g\rangle = 0.600 |j\rangle + 0.800 |k\rangle.$$

1.10. From the orthonormal set $|f_1\rangle$, $|f_2\rangle$, $|f_3\rangle$, construct two normalized kets orthogonal to each other and orthogonal to

$$|g\rangle = \cos \alpha |f_1\rangle + \sin \alpha \cos \beta |f_2\rangle + \sin \alpha \sin \beta |f_3\rangle.$$

1.11. In a system containing three identical fermions, one enters Ψ_1, the second enters Ψ_{II}, and the third enters Ψ_{III}. Construct a suitable overall state function.

— — —

1.12. How may Ψ_I and Ψ_{II} differ and still describe the same state of a coherent system?

1.13. Normalize the state functions

$$\psi_{\mathrm{I}} = A \cos \frac{\pi x}{2},$$

$$\psi_{\mathrm{II}} = B(1 - x^2)$$

for a particle translating back and forth between walls at $x = -1$ and $x = 1$.

1.14. Let the values of the Legendre polynomial $P_2(x)$ at 20 points spaced 0.1 unit apart and located symmetrically in the interval $-1 \leq x \leq 1$ be the components of vector **A**. Construct **A*·A** and multiply the result by $\Delta x = 0.1$.

1.15. Compare the result from Problem 1.14 with the integral

$$\int_{-1}^{1} P_2^*(x) P_2(x) dx.$$

1.16. Consider $|j\rangle$ and $|k\rangle$ to be normalized kets in the Hilbert space for a system. When can one write

$$|j\rangle = c_1|k\rangle + c_2|r\rangle,$$

with $|r\rangle$ a possible ket? What does c_1 equal when $|r\rangle$ contains no contribution from $|k\rangle$?

1.17. If each ket in

$$|g\rangle = 0.285|j\rangle + 0.815|k\rangle$$

is normalized, what does $\langle j|k\rangle$ equal, assuming it is real?

1.18. Find the normalized ket orthogonal to $|g\rangle$ in Problem 1.17.

1.19. If

$$\langle j|k\rangle = 0.231$$

and $|j\rangle$, $|k\rangle$ are normalized to 1, how is $|j\rangle$ related to $|k\rangle$?

1.20. Use the appropriate projection operator to remove from the $|g\rangle$ in Problem 1.17 the contribution of ket $|j\rangle$.

1.21. Employ the projection operator technique in determining the contribution of ψ_{I} to ψ_{II} in Problem 1.13.

1.22. With the system in Example 1.4, how is the integral

$$\int_R \Psi_{\mathrm{I}}^*(1) \Psi_{\mathrm{II}}(1) d^3\mathbf{r}_1$$

related to the normalization constant N when the integral is real? Assume that the one particle functions are normalized to 1. What does the integral equal when N is (a) $1/\sqrt{2}$, (b) 1?

References

Books

Bunge, M.: 1985, *Treatise on Basic Philosophy*, vol. 7, part I, *Formal and Physical Sciences*, Reidel, Dordrecht, pp. 165-191. While there is general agreement on the mathematics of quantum mechanics, there is considerable disagreement on its interpretation. Bunge criticizes the various extant views in detail, finally arriving at a tenable position.

Cohen-Tannoudji, C., Diu, B., and Laloe, F. (translated by Hemley, S. R., Ostrowsky, N., and Ostrowsky, D.): 1977, *Quantum Mechanics*, Wiley, New York, pp. 91-367. In Chapter II, the conventional mathematical tools of quantum mechanics are developed, together with useful comments. In Chapter III, the postulates of quantum mechanics are similarly presented.

Dirac, P. A. M.: 1947, *The Principles of Quantum Mechanics*, 3rd edn, Oxford University Press, London, pp. 1-83. In his classic text, Dirac begins with a discussion of the superposition principle and its role. He then develops the theory of bra and ket vectors, and of operators, being guided by a remarkable intuition.

Duffey, G. H.: 1984, *A Development of Quantum Mechanics Based on Symmetry Considerations*, Reidel, Dordrecht, pp. 143-172. The state-function notation is used in this elementary introduction to quantum mechanical operators.

d'Espagnat, B.: 1976, *Conceptual Foundations of Quantum Mechanics*, 2nd edn, Benjamin, Reading, Mass., pp. 1-73. Here the conventional postulates of quantum mechanics are formulated and discussed in detail.

Fano, G.: 1971, *Mathematical Methods of Quantum Mechanics*, McGraw-Hill Book Company, New York, pp. 1-422. Fano develops the mathematics of quantum mechanics in a more rigorous manner than is common.

Jauch, J. M.: 1972, 'On Bras and Kets,' in Salam, A., and Wigner, E. P. (editors), *Aspects of Quantum Theory*, Cambridge University Press, London, pp. 137-167. By refining Dirac's methods, Jauch gives them rigor without undue complication.

Merzbacher, E.: 1970, *Quantum Mechanics*, 2nd edn, Wiley, New York, pp. 294-333. Here one finds a standard presentation of linear vector spaces and quantum mechanical operators. However, the state vector is represented by symbol Ψ, as well as by Dirac's notation.

Messiah, A. (translated by Temmer, G. M.): 1958, *Quantum Mechanics*, Wiley, New York, pp. 3-63, 582-631. The conventional historical approach is developed in the first section cited. In the second section, systems of identical particles are treated.

Schlegel, R.: 1980, *Superposition and Interaction*, University of Chicago Press, Chicago, pp. 24-55, 206-219. Schlegel illuminates the principle of superposition with examples discussed in considerable detail. However, a vector is not equal to a superposition of all its possible projections. (See pp. 29-30.)

Articles

Ballentine, L. E.: 1986, "Probability Theory in Quantum Mechanics," *Am. J. Phys.* **54**, 883-889.

Bunge, M.: 1977, "Quantum Mechanics and Measurement," *Int. J. Quantum Chem.* **12**, Suppl. **1**, 1-13.

Burnel, A., and Caprasse, H.: 1980, "The Symmetry Property in the Hilbert Space Formulation of Quantum Mechanics," *Eur. J. Phys.* **1**, 169-172.

French, S., and Redhead, M.: 1988, "Quantum Physics and the Identity of Indiscernibles," *Brit. J. Phil. Sci.* **39**, 233-246.

Josephson, B. D.: 1988, "Limits to the Universality of Quantum Mechanics," *Found. Phys.* **18**, 1195-1204.

Levy-Leblond, J.-M.: 1977, "The Picture of the Quantum World: From Duality to Unity," *Int. J. Quantum Chem.* **12**, Suppl. **1**, 415-421.

Maxwell, N.: 1988, "Quantum Propensiton Theory: A Testable Resolution of the Wave/Particle Dilemma," *Brit. J. Phil. Sci.* **39**, 1-50.

Muller-Herold, U.: 1985, "A Simple Derivation of Chemically Important Classical Observables and Superselection Rules," *J. Chem. Educ.* **62**, 379-382.

Villars, C. N.: 1984, "Observables, States and Measurements in Quantum Physics," *Eur. J. Phys.* **5**, 177-183.

Weininger, S. J.: 1984, "The Molecular Structure Conundrum: Can Classical Chemistry be Reduced to Quantum Chemistry?" *J. Chem. Educ.* **61**, 939-944.

2

ALLOWING FOR THE TRANSLATIONAL SYMMETRIES AND THE LOCALIZABILITY OF PARTICLES

2.1
Introduction

Physical states exist and evolve in the space-time continuum. As a consequence, all laws governing such states must respect the symmetries and near-symmetries of this continuum. They must not violate the transformation properties imposed by it.

But in quantum mechanics the physical laws appear as eigenvalue equations. So the operator in such an equation must transform in a manner consistent with the fundamental global and local symmetries.

We will find that these symmetries, together with that over phase, lead to construction of general commutation relations for the key operators. Each of these operators corresponds to a physical observable, and so is called a physical operator to distinguish it from other kinds.

While the states of particles will be represented by normalizable kets in Hilbert spaces, the localizability of each particle must be taken into account.

Before developing the pertinent theory, let us survey the Galilean space-time continuum and look at the behavior of a general state function for translation.

2.2
The Galilean Space-Time Continuum

Small systems constrained to undergo the same process again and again, without interruption, may be employed to study time. It is found that when changes in ambient conditions are negligible, one may measure time by the number of cycles traversed. On comparing results from systems with different cycle lengths, and under different conditions, one finds that time is apparently continuous in the standard analytical sense, homogeneous, and 1-dimensional. In Newtonian physics, and in Schrödinger quantum mechanics, one assumes that the time measured at one inertial point can be projected instantaneously on other points in space.

Separated points in space can be identified by material elements located at the points. The distances between may be measured with material tapes or with electromagnetic rays traveling at a known presumably constant speed. A given length of inert material, or a given number of cycles of the wave at a given frequency, is chosen as the unit. Where space is not distorted by forces it is found to be continuous in the standard analytical sense, homogeneous, 3-dimensional, subject to the Pythagorean law, and so Euclidean.

These Galilean aspects of time and space imply that a system's physical properties are not changed by an arbitrary

   (a) translation in time,
   (b) translation in space,
   (c) velocity translation,
   (d) rotation in space,

as long as interactions with other systems are not altered appreciably thereby, as in free space.

According to the postulates in Sections 1.2 and 1.3, a coherent quantum mechanical system is described by a state function Ψ. Furthermore, altering the phase of a state function does not change the probability density at any point, or any expectation value. Consequently, an arbitrary

   (e) translation in phase

does not affect physical properties.

Any transformation that acts on a physical system is said to be *active*. On the other hand, a transformation that acts on the reference axes or bases is said to be *passive*.

Any active transformation that leaves physical properties unchanged, as (a)–(e) would, is called a *symmetry operation*. The resulting invariances must be respected by any acceptable theory.

2.3
Galilean Transformations of a State Function

Particularly important are transformations between inertial frames—those frames that move as nonrotating force-free bodies. For, according to the *relativity principle*, all inertial frames are equivalent; every physical law has the same form in each such frame. Thus, the eigenvalue equations and the expectation value expressions are not altered. However, the phase of a state function, in the coordinate and/or the wavevector representation, is changed. Here we will consider the effect on the general function for particles moving freely in one direction.

We consider a primed inertial frame moving at velocity **v** with respect to an unprimed inertial frame, as Figure 2.1 shows. The origin of the primed system is moving

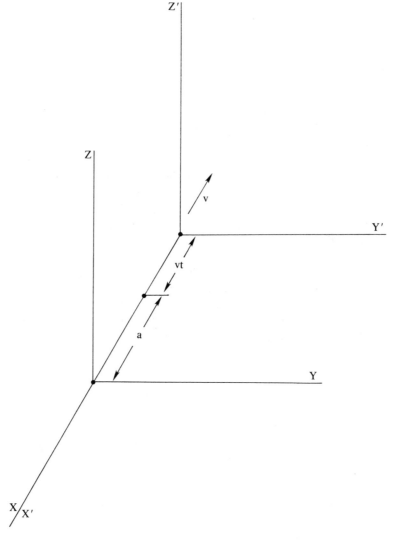

Figure 2.1. Primed Cartesian axes moving at velocity **v** with respect to unprimed Cartesian axes. In the discussion, these are inertial frames.

at constant velocity **v** with respect to the origin of the unprimed one. The x and x' axes are pointed in the direction of $-\mathbf{v}$. The y and z axes are set perpendicular to each other and to the x axis in the right handed sense. The y' and z' axes are oriented parallel to the y and z axes.

From the relativity principle, the motion cannot affect distances in the transverse directions:

$$y' = y, \tag{2.1}$$

$$z' = z. \tag{2.2}$$

If we assume that time is universal and absolute, our relativity is Galilean and

$$t' = t. \tag{2.3}$$

Then distance x' equals x increased by the initial translation a and the translation effected by v acting over the time t:

$$x' = x + vt + a. \tag{2.4}$$

Equations (2.1) through (2.4) are called the *Galilean transformation* equations.

There is no reason for the movement of the primed axes to have any effect on transverse motion, or on the corresponding propagation of a wave function in the y or z directions. However, such movement will affect the longitudinal motion, and propagation in the x direction. So let us consider a general superposition of wavevector motions along the x axis. We also assume that the potential V does not vary with x throughout the region under consideration.

In the unprimed and the primed coordinates, p and p' are the linear momenta, E and E' the energies, k and k' the wavevectors, ω and ω' the angular frequencies, Ψ and Ψ' the state functions in the coordinate representation, ϕ and ϕ' the time-independent state functions in the wavevector representation. Also, $\hbar$ is Planck's constant divided by 2π and μ is the mass of a particle.

The de Broglie equation takes on the forms

$$k = \frac{p}{\hbar} \quad \text{and} \quad k' = \frac{p'}{\hbar}, \tag{2.5}$$

the Einstein equation becomes

$$\omega = \frac{E}{\hbar} \quad \text{and} \quad \omega' = \frac{E'}{\hbar}, \tag{2.6}$$

while the energy equation is

$$E = \frac{p^2}{2\mu} + V \quad \text{and} \quad E' = \frac{p'^2}{2\mu} + V. \tag{2.7}$$

Equation (2.4) yields the temporal derivative

$$\dot{x}' = \dot{x} + v, \tag{2.8}$$

whence

$$p' = \mu \dot{x}' = \mu \dot{x} + \mu v = p + \mu v. \tag{2.9}$$

Substituting into (2.5) yields

$$k' = \frac{p'}{\hbar} = \frac{p}{\hbar} + \frac{\mu v}{\hbar} = k + \frac{\mu v}{\hbar}, \tag{2.10}$$

whence

$$dk' = dk. \tag{2.11}$$

Substituting into (2.6) with (2.7), (2.9), and (2.5) gives us

$$\omega' = \frac{E'}{\hbar} = \frac{(p + \mu v)^2}{2\mu\hbar} + \frac{V}{\hbar} = \frac{p^2}{2\mu\hbar} + \frac{V}{\hbar} + \frac{pv}{\hbar} + \frac{\mu v^2}{2\hbar}$$

$$= \omega + kv + \frac{\mu v^2}{2\hbar}. \tag{2.12}$$

In the primed coordinates, within the chosen region, a person may represent a general state function by the superposition

$$\Psi'(x', t') = \frac{1}{(2\pi)^{\frac{1}{2}}} \int_{-\infty}^{\infty} \phi'(k')e^{i(k'x' - \omega' t')}dk'. \tag{2.13}$$

Now substituting (2.10), (2.4), (2.12), (2.3), and (2.11) in leads to

$$\Psi'(x', t') = \frac{1}{(2\pi)^{\frac{1}{2}}} \int_{-\infty}^{\infty} \phi'(k')e^{i(k + \frac{\mu v}{\hbar})(x + vt + a)}e^{-i(\omega + kv + \frac{\mu v^2}{2\hbar})t}dk$$

$$= e^{i\frac{\mu}{\hbar}(\frac{1}{2}v^2 t + vx)} \frac{1}{(2\pi)^{\frac{1}{2}}} \int_{-\infty}^{\infty} [\phi'(k')e^{i(k + \frac{\mu v}{\hbar})a}]e^{ikx}e^{-i\omega t}dk$$

$$= \exp\left[i\frac{\mu}{\hbar}(\frac{1}{2}v^2 t + vx)\right] \Psi(x, t). \tag{2.14}$$

The expression in brackets in the last integrand written out is identified as $\phi(k)$. With (2.10), we have

$$\phi(k) = e^{ik'a}\phi'(k'). \tag{2.15}$$

The Galilean transformation alters both the coordinate representation and the wavevector representation of the ket. However, at a given point in space and time, the magnitude of Ψ, and the corresponding probability density $\Psi^*\Psi$, are unaltered. The phases of the representations are changed, but these are unobservable.

Thus, the Schrödinger theory is consistent with Galilean relativity. It is not a nonrelativistic theory as often claimed.

2.4
Conservation of Particle Mass

In Schrödinger theory, mass is considered a classical observable. A basis for this treatment is provided by the formulas just derived. Indeed, we will find that a superselection rule forbids the transition from one mass level to another.

Consider a particle moving in a packet described by (2.13) and (2.14). Let us first subject the coordinate system to a translation in space with

$$x' = x + a, \qquad t' = t. \qquad (2.16)$$

Equation (2.14) reduces to

$$\Psi'(x', t') = \Psi(x, t). \qquad (2.17)$$

The wave function is not changed at any point in the continuum.

Let us secondly transform to a coordinate system moving at speed v as in Figure 2.1. Then

$$x'' = x' + vt', \qquad t'' = t', \qquad (2.18)$$

and (2.14) tells us that

$$\Psi''(x'', t'') = \exp\left[i\frac{\mu}{\hbar}\left(\tfrac{1}{2}v^2t' + vx'\right)\right]\Psi'(x', t'). \qquad (2.19)$$

Thirdly, we introduce the inverse of the first translation,

$$x''' = x'' - a, \qquad t''' = t'', \qquad (2.20)$$

for which

$$\Psi'''(x''', t''') = \Psi''(x'', t''). \qquad (2.21)$$

Fourthly, the inverse of the second transformation is introduced:

$$x^{iv} = x''' - vt''', \qquad t^{iv} = t'''. \qquad (2.22)$$

Then

$$\Psi^{iv}(x^{iv}, t^{iv}) = \exp\left[i\frac{\mu}{\hbar}\left(\tfrac{1}{2}v^2t''' - vx'''\right)\right]\Psi'''(x''', t'''). \qquad (2.23)$$

Substituting (2.20), (2.18), (2.16) into (2.22) yields

$$x^{iv} = \{[(x + a) + vt'] - a\} - vt''' = x, \qquad (2.24)$$

$$t^{iv} = t''' = t'' = t' = t. \qquad (2.25)$$

Similarly,

$$x' - x''' = x' - (x' + vt') + a = a - vt. \qquad (2.26)$$

Now, combining (2.17), (2.19), (2.21), (2.23) leads to

$$\Psi^{iv}(x, t) = \exp \left[i\frac{\mu}{\hbar} (\tfrac{1}{2}v^2t - vx''')\right] \exp \left[i\frac{\mu}{\hbar} (\tfrac{1}{2}v^2t + vx')\right] \Psi(x, t)$$

$$= \exp \left\{i\frac{\mu}{\hbar}[v^2t + v(x' - x''')]\right\} \Psi(x, t)$$

$$= \exp \left(i\frac{\mu}{\hbar} va\right) \Psi(x, t). \tag{2.27}$$

Although the series of transformations does not produce any net change in the independent variables, it does shift the phase of the state function at each point. Furthermore, the change in phase angle is proportional to mass μ of the particle, the displacement a, and the speed v. Since a and v are arbitrary, the phase of the normalization constant for a state function is arbitrary.

When a particle can exist in one state described by Ψ_1 and a second state described by Ψ_2, coherent superposition of the states yields

$$\Psi = c_1\Psi_1 + c_2\Psi_2. \tag{2.28}$$

If the mass of the particle in the first state is μ_1 and its mass in the second state is μ_2, then the series of transformations (2.16), (2.18), (2.20), (2.22) would alter a translational (2.28) to

$$\Psi^{tr} = c_1 \exp \left(i\frac{\mu_1}{\hbar} va\right) \Psi_1 + c_2 \exp \left(i\frac{\mu_2}{\hbar} va\right) \Psi_2. \tag{2.29}$$

But since this series returns the system to its initial situation, it amounts to an identity transformation, so it should not affect anything that is observable.

However, the difference in phase between the two terms changes when

$$\mu_1 \neq \mu_2. \tag{2.30}$$

The product $\Psi^{tr*}\Psi^{tr}$ then differs from $\Psi^*\Psi$.

Because the probability density should not be thus changed, we have to conclude that superposition (2.28) is not valid when inequality (2.30) holds. And when a person cannot form a superposition, a transition between the pertinent states cannot occur. Recall Section 1.10. In Schrödinger quantum mechanics, the mass of each particle is conserved; it is a classical observable.

2.5
Similarity Transformations

According to Section 1.4, the state function for a coherent system is a representation of the ket governing the system. Corresponding to the operators that act on the state function are transformation operators acting on the kets. Some of these operators represent observables through eigenvalue equations, as we have noted. Some, when acting on an eigenket, change it to a constant times another eigenket. Others merely transform a ket into a different composite ket.

Consider a coherent system initially in state-ket $|g\rangle$. Introduce an operator A that converts this ket to $|h\rangle$:

$$A|g\rangle = |h\rangle. \qquad (2.31)$$

Also introduce operator U that changes $|g\rangle$ to $|g'\rangle$, $|h\rangle$ to $|h'\rangle$:

$$U|g\rangle = |g'\rangle, \qquad (2.32)$$

$$U|h\rangle = |h'\rangle, \qquad (2.33)$$

Then linking $|g'\rangle$ and $|h'\rangle$ is another operator A':

$$A'|g'\rangle = |h'\rangle. \qquad (2.34)$$

A schematic description appears in Figure 2.2.

As long as the kets remain within the allowed Hilbert space for the system, the inverse operators also exist. Letting exponent -1 on an operator indicate taking the inverse, we rearrange Equations (2.32) and (2.33) to

$$|g\rangle = U^{-1}|g'\rangle, \qquad (2.35)$$

$$|h\rangle = U^{-1}|h'\rangle. \qquad (2.36)$$

Substituting (2.35) into (2.31), acting on the result with U, and introducing (2.33) yields

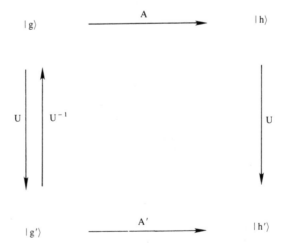

Figure 2.2. Relationships among kets $|g\rangle$, $|h\rangle$, $|g'\rangle$, $|h'\rangle$ in the Hilbert space for a coherent system.

$$UAU^{-1}|g'\rangle = U|h\rangle = |h'\rangle. \tag{2.37}$$

Since (2.34) and (2.37) hold for an arbitrary $|g'\rangle$ within the Hilbert space, we see that

$$UAU^{-1} = A'. \tag{2.38}$$

Note how this equality can be inferred from Figure 2.2. The change in the operator A is called a *similarity transformation* of A by U.

A person may let operator U represent a geometric operation that does not distort the physical system, while operator A acts to alter the physical system. Then A' represents the same physical operation as A does. And Equation (2.38) describes how the geometric operation transforms the physical operator.

Alternatively, an evolving system may be under consideration. This system may be in state $|g\rangle$ at zero time and in state $|g'\rangle$ at time t. Then U would be the evolution operator for the system, for the path from $|g\rangle$ to $|g'\rangle$.

Example 2.1

How does the geometric operation, the coordinate displacement

$$x' = x + a, \qquad t' = t,$$

affect the translational state

$$\Psi' = A\,e^{i(k'x' - \omega't')}?$$

We have here a Galilean transformation with $v = 0$. From (2.10) and (2.12), we see that

$$k' = k \qquad \text{and} \qquad \omega' = \omega.$$

Substituting into the representation of the ket yields

$$\Psi' = A\,e^{i[k(x + a) - \omega t]} = e^{ika}A\,e^{i(kx - \omega t)} = e^{ika}\Psi.$$

On comparing this result with (2.32), we see that

$$U = e^{ika} \qquad \text{and} \qquad U^{-1} = e^{-ika}$$

in the coordinate representation of the kets.

Note that the same equations arise whether (a) the primed coordinate system is displaced by distance a in the negative x direction or (b) the physical system is displaced by distance a in the positive x direction. Thus, the operation may be (a) passive or (b) active.

Example 2.2

Construct an operator for rotating the orbital

$$\psi = R(r)\Theta(\vartheta)e^{iM\varphi}$$

about the z axis by angle α.

In the prescribed rotation, (r, ϑ, φ) goes to $(r', \vartheta', \varphi')$ with

$$\varphi' = \varphi + \alpha, \qquad \vartheta' = \vartheta, \qquad r' = r.$$

But the transformed orbital is

$$\psi' = R(r')\Theta(\vartheta')e^{iM\varphi'}.$$

Substituting into this form yields

$$\psi' = R(r)\Theta(\vartheta)e^{iM(\varphi + \alpha)} = e^{iM\alpha}R(r)\Theta(\vartheta)e^{iM\varphi}$$

$$= e^{iM\alpha}\psi.$$

Comparing this result with (2.32), we see that

$$U = e^{iM\alpha} \qquad \text{and} \qquad U^{-1} = e^{-im\alpha}$$

in the coordinate representation of the kets.

Example 2.3

Show that one can rewrite the transformation equation

$$A|g\rangle = |h\rangle$$

as

$$\langle g|A^\dagger = \langle h|$$

where $A^\dagger$ is the adjoint of A.

Choose an arbitrary bra $\langle f|$ in the Hilbert space of the system. Form the scalar product of it with the first equation to get

$$\langle f|A|g\rangle = \langle f|h\rangle.$$

Apply adjoint formula (1.81) to the left side, obtaining

$$\langle g|A^\dagger|f\rangle^* = \langle f|h\rangle = \langle h|f\rangle^*.$$

Take the complex conjugate of the overall equation,

$$\langle g \mid A^\dagger \mid f \rangle = \langle h \mid f \rangle,$$

whence

$$\langle g \mid A^\dagger = \langle h \mid,$$

since $\mid f \rangle$ is arbitrary.

Example 2.4

What is the adjoint of the product of two operators that act in the Hilbert space for a given system?

Let the operators be A and B, the product AB. Also choose two arbitrary kets $\mid f \rangle$ and $\mid g \rangle$ in the Hilbert space. Construct the matrix element of AB with $\langle f \mid$ and $\mid g \rangle$ and apply definition (1.81), considering AB as a single operator:

$$\langle f \mid AB \mid g \rangle = \langle (AB)^\dagger f \mid \mid g \rangle.$$

Next consider that $B \mid g \rangle$ is the ket on which A acts and switch A to the adjoint position:

$$\langle f \mid AB \mid g \rangle = \langle A^\dagger \mid g \mid \mid B \mid g \rangle.$$

Then consider $\mid g \rangle$ as the ket on which B acts and switch B to the adjoint position:

$$\langle A^\dagger \mid f \mid \mid B \mid g \rangle = \langle B^\dagger A^\dagger f \mid \mid g \rangle.$$

For the two results to be the same regardless of the choice of $\mid f \rangle$ and $\mid g \rangle$, we must have

$$(AB)^\dagger = B^\dagger A^\dagger.$$

Example 2.5

Find the adjoint of a constant complex number c.

Here the operator is c. We again choose two arbitrary kets $\mid f \rangle$ and $\mid g \rangle$ in the Hilbert space for the given system, with representations Ψ_f and Ψ_g.

Since c is a constant, it can be taken inside the integral in the pertinent generalization of (1.39); thus

$$c \langle f \mid g \rangle = \langle f \mid c \mid g \rangle.$$

Similarly,

$$c \langle g \mid f \rangle^* = \langle g \mid c^* \mid f \rangle^*.$$

But since the left sides of these equations are equal, the right sides are also:

$$\langle f | c | g \rangle = \langle g | c^* | f \rangle^*.$$

On comparing with (1.81), we see that

$$c^\dagger = c^*,$$

the adjoint of a constant equals its complex conjugate.

2.6
Unitary Transformations

For the evolving system, and for the geometrically changing, undistorted, system, operator U obeys a special condition.

Consider that over time t, a given coherent system evolves from the state described by $| g \rangle$ to the state described by $| g' \rangle$. Since the process cannot change the total probability, we have

$$\langle g' | g' \rangle = \langle g | g \rangle = \langle g | g \rangle^*. \tag{2.39}$$

However, introducing U through (2.32) and $U^\dagger$ through (1.81), then using (2.32) again, yields

$$\langle g' | g' \rangle = \langle g' | U | g \rangle = \langle g | U^\dagger | g' \rangle^* = \langle g | U^\dagger U | g \rangle^*. \tag{2.40}$$

These equations have to hold regardless of where $| g \rangle$ is in the Hilbert space. So to make (2.40) consistent with (2.39), we must have

$$U^\dagger U = 1. \tag{2.41}$$

Any operator U satisfying (2.41) is said to be *unitary*. Thus, we have the postulate:

(XI) The evolution of a coherent quantum mechanical system is effected by a unitary operator.

A nondistortive geometric operation cannot affect physical properties that are independent of the coordinates. Consequently, it cannot affect the product of a bra with a ket:

$$\langle g' | h' \rangle = \langle g | h \rangle = \langle h | g \rangle^*. \tag{2.42}$$

Now, introduce U through (2.33), the definition of the adjoint $U^\dagger$, another U through (2.32):

$$\langle g' \mid h' \rangle = \langle g' \mid U \mid h \rangle = \langle h \mid U^\dagger \mid g' \rangle^* = \langle h \mid U^\dagger U \mid g \rangle^*. \qquad (2.43)$$

These equations have to hold regardless of where $\mid g \rangle$ and $\mid h \rangle$ are in the Hilbert space. To make (2.43) consistent with (2.42), we must have

$$U^\dagger U = 1. \qquad (2.44)$$

Thus, we have the postulate:

(XII) The change in a state ket caused by a symmetry operation is effected by a unitary operator.

From (2.41) and (2.44), we see that an operator is unitary when its adjoint equals its inverse.

2.7
The Group Concept

Transformations and the operators that represent them may form complete sets called groups.

By definition, a *group* is a set of elements that contains

(a) an identity,
(b) an inverse for each element,
(c) the binary combination of any two elements,

and for which, in a larger combination, successive pairs can be replaced by their equivalents. Thus,

(d) the associative law holds.

We may let I be the identity element, while A^{-1} is the inverse of element A. The binary combination of element A and element B is written as AB or BA, depending on the order in which we consider them.

Specification (a) implies that I is present in any group. Specification (b) implies that if A is present, A^{-1} is also there and the combination $A^{-1}A$ equals I. Specification (c) implies that if A and B are in the group, then both AB and BA are elements of the group. Specification (d) implies that if A, B, C are elements, the set contains

$$ABC = (AB)C = A(BC). \qquad (2.45)$$

The transformations that cover the Hilbert space for a given system, and closed subsets of these, form groups. Operators representing the transformations in a closed set belong to the same group.

2.8
Parameterizing an Operator

Analogously to a function, an operator may vary in its action, with one or more linearly independent parameters. Furthermore, this variation may be continuous in the standard analytical sense. Each such parameter may be a change or shift in an independent variable, or in a function of the chosen independent variables.

For a translation in space, passive or active, the parameters could be the components of the introduced displacement **a**. For a velocity translation, the parameters could be the components of the introduced change **v**. For a translation in time, the parameter could be the introduced change b. For a rotation in space about a given axis, the parameter could be the angle α. Thus in Example 2.1, where the operator is

$$U = e^{ika}, \tag{2.46}$$

the parameter is a; in Example 2.2, where the operator is

$$U = e^{iM\alpha}, \tag{2.47}$$

the parameter is α.

A general operation may be compounded of operations involving single parameters. So let us consider the operator $U(g)$, where g is the parameter. Let us choose the origin of g so that when $g = 0$, U is the identity operator 1. We also consider that U varies continuously and smoothly with g.

The derivative of U with respect to g then exists. For an infinitesimal change $\delta g = 0$, we have

$$U(\delta g) = 1 + \left(\frac{\partial U}{\partial g}\right)_{g=0} \delta g = 1 + iG\delta g. \tag{2.48}$$

In the second equality, the product iG has been introduced by the definition

$$\left(\frac{\partial U}{\partial g}\right)_{g=0} \equiv iG. \tag{2.49}$$

One may call iG the *infinitesimal operator* with respect to the parameter g.

Over the allowed range of the parameter, the infinitesimal operations can be repeated. Then from the definition of the exponential, we have

$$U(g) = \lim_{k \to \infty} (1 + iG\frac{g}{k})^k = e^{iGg}. \tag{2.50}$$

This repetition without limit is referred to as *exponentiation* of the infinitesimal operator. Since thereby various group elements are generated, iG is also called an *infinitesimal generator* for the group.

The inverse of the operator in (2.48) is

$$U^{-1}(\delta g) = 1 - iG\,dg \tag{2.51}$$

whence

$$U^{-1}(g) = e^{-iGg}. \tag{2.52}$$

The group of operators of form $U(g)$, where U varies continuously with parameter g, is called a *Lie group*. Instead of depending on one parameter alone, such an operator may depend continuously on n linearly independent parameters $g_1, \ldots, g_n$. Thus, it may possess the form $U(g_1, \ldots, g_n)$. A complete set of such operators would be called an n-parameter Lie group.

The Lie groups that we will consider here are the 1-D and the 3-D Galilean groups, and extensions of these groups that include phase translations.

2.9
Elements of and Generators for the 1-Dimensional Galilean Group

Spatial displacements, velocity displacements, and time displacements make up the Galilean group. These elements are conveniently represented by square matrices. For simplicity, we will here limit ourselves to one space dimension. The matrices will then be 3×3 ones. From the form for finite operations, the infinitesimal generators will be obtained.

Consider the coordinates of a particular point to be (x, t) before a transformation and (x', t') afterwards. As in (2.4), we introduce at time $t = 0$, a translation in space of a, a translation in velocity of v. Then

$$x' = x + vt + a. \tag{2.53}$$

We also introduce a displacement in time b:

$$t' = t + b. \tag{2.54}$$

Since we want to represent the operation with a matrix and this matrix needs to be square to combine with other such matrices as a group element, we add the identity equation

$$1 = 1. \tag{2.55}$$

A matrix form of (2.53), (2.54), (2.55) is

$$\begin{pmatrix} x' \\ t' \\ 1 \end{pmatrix} = \begin{pmatrix} 1 & v & a \\ 0 & 1 & b \\ 0 & 0 & 1 \end{pmatrix} \begin{pmatrix} x \\ t \\ 1 \end{pmatrix}. \tag{2.56}$$

To describe a similar transformation of (x', t') to (x'', t''), we add a prime to each of the quantities in (2.56), except to the 1's. Then we introduce (2.53) and (2.54), multiply the matrices, and factor out the column matrix for the unprimed coordinates:

$$
\begin{pmatrix} x'' \\ t'' \\ 1 \end{pmatrix} = \begin{pmatrix} 1 & v' & a' \\ 0 & 1 & b' \\ 0 & 0 & 1 \end{pmatrix} \begin{pmatrix} x' \\ t' \\ 1 \end{pmatrix} = \begin{pmatrix} 1 & v' & a' \\ 0 & 1 & b' \\ 0 & 0 & 1 \end{pmatrix} \begin{pmatrix} x + vt + a \\ t + b \\ 1 \end{pmatrix}
$$

$$
= \begin{pmatrix} x + vt + a + v't + v'b + a' \\ t + b + b' \\ 1 \end{pmatrix} = \begin{pmatrix} 1 & v + v' & a + a' + v'b \\ 0 & 1 & b + b' \\ 0 & 0 & 1 \end{pmatrix} \begin{pmatrix} x \\ t \\ 1 \end{pmatrix}
$$

$$\tag{2.57}$$

The overall equation describes the direct transformation from (x, t) to (x'', t''). Note how the parameters combine.

Now, the final square matrix also results from premultiplying the square matrix in (2.56) by the initial one in (2.57). Thus, the square matrices combine as the operations; they serve to *represent* the corresponding operations of the group.

For operations infinitesimally close to the identity, we replace a, b, and v in (2.56) with δa, δb, δv, respectively. And we consider the transformation matrix as a generalization of the U in (2.48).

When only δa differs from zero, we obtain the infinitesimal operator

$$
\begin{pmatrix} 0 & 0 & 1 \\ 0 & 0 & 0 \\ 0 & 0 & 0 \end{pmatrix} = iP_0, \text{ for distance.} \tag{2.58}
$$

Similarly, with only δb different from zero, we get

$$
\begin{pmatrix} 0 & 0 & 0 \\ 0 & 0 & 1 \\ 0 & 0 & 0 \end{pmatrix} = iH_0, \text{ for time.} \tag{2.59}
$$

With only δv different from zero, we have

$$
\begin{pmatrix} 0 & 1 & 0 \\ 0 & 0 & 0 \\ 0 & 0 & 0 \end{pmatrix} = iK_0, \text{ for velocity.} \tag{2.60}
$$

For the identity operation (1), we employ the infinitesimal operator

$$\begin{pmatrix} 0 & 0 & 0 \\ 0 & 0 & 0 \\ 0 & 0 & 0 \end{pmatrix} = iI_0. \tag{2.61}$$

Here the symbols P_0, H_0, K_0, and I_0 have been introduced by definition.

Note that the infinitesimal operator for the identity is a null matrix. Also, $(iP_0)^2$, $(iH_0)^2$, and $(iK_0)^2$ equal this null matrix. Any operation A satisfying

$$A^n = 0 \tag{2.62}$$

is said to be *nilpotent*; the matrices in (2.58), (2.59), (2.60) are nilpotent.

Since the square matrix in (2.56) and the initial square matrix in (2.57) act to transform the column matrices

$$\begin{pmatrix} x \\ t \\ 1 \end{pmatrix} \quad \text{and} \quad \begin{pmatrix} x' \\ t' \\ 1 \end{pmatrix}, \tag{2.63}$$

they are not formulations of U's for transforming any kets. Nevertheless, these square matrices combine as such U's would; so these matrices represent their algebraic behavior. Independent aspects of this behavior are represented by the infinitesimal operators. Indeed, they serve as infinitesimal generators of the group.

The scheme here is incomplete since it does not allow for an explicit phase translation. Also, the mass of the particle under consideration is not involved. In the following sections, we will see how the group needs to be extended to take these ingredients into account.

2.10
The Localizability Requirement

Characteristic of a particle is its possible localization; a particle can act at a point in space and time. Consequently, a physical operator for position should exist. And this operator must transform in a spatial translation following similarity law (2.38). Furthermore, the localizability property is unaffected by an instantaneous translation of velocity. But when the position is moving, a time translation introduces a shift. These conditions lead to the basic commutation relations.

Let the physical operator for a position along the x axis be X before the given transformation and X' after. For an instantaneous translation by distance a, we must then have

$$X' = X + a(1), \tag{2.64}$$

where (1) is the identity operator. When the translation is by an infinitesimal amount δa, we write

$$X' = X + \delta a(1). \tag{2.65}$$

Let iP be the infinitesimal operator for translation in space. Furthermore, express X' as the similarity transform of X, following (2.38), introduce the appropriate forms of (2.48) and (2.51), and multiply the result out to obtain

$$\begin{aligned} X' &= U(\delta a)XU^{-1}(\delta a) = (1 + iP\delta a)X(1 - iP\delta a) \\ &= X - i(XP - PX)\delta a + PXP(\delta a)^2 \\ &= X - i[X, P]\delta a + 0(\delta a)^2. \end{aligned} \tag{2.66}$$

In the last step, the notation

$$XP - PX = [X, P] \tag{2.67}$$

has been introduced for the *commutator* of X and P. For the final right side of (2.66) to reduce to that of (2.65), we must have the operator relationship

$$[X, P] = i(1). \tag{2.68}$$

Over an instantaneous velocity translation, the position in space and time is not affected, whether the shift is finite or infinitesimal. So the physical operator is unchanged in form:

$$X' = X. \tag{2.69}$$

We let iK be the infinitesimal operator for the velocity translation. Expressing X' as the infinitesimal similarity transform, as before, introducing the expansion of U and U^{-1}, and multiplying out, yields

$$\begin{aligned} X' &= U(\delta v) XU^{-1}(\delta v) = (1 + iK\delta v) X(1 - iK\delta v) \\ &= X - i(XK - KX)\delta v + KXK(\delta v)^2 \\ &= X - i[X, K]\delta v + 0(\delta v)^2. \end{aligned} \tag{2.70}$$

For result (2.70) to reduce to (2.69), we must have the operator relationship

$$[X, K] = 0. \tag{2.71}$$

Condition (2.71) is met by setting X equal to a constant times K. We write

$$X = -\frac{1}{\mu}K \tag{2.72}$$

with μ to be identified.

Substituting (2.72) into (2.68) gives us

$$\left[\frac{1}{\mu} K, P \right] = -i(1) \tag{2.73}$$

or

$$[K, P] = -i\mu(1). \tag{2.74}$$

A point moving at velocity v subjected to a time translation of δb travels distance $v \delta b$ and the operator relationship becomes

$$X' = X + v\delta b(1). \tag{2.75}$$

Now we let iH be the pertinent infinitesimal operator. The corresponding transform of physical operator X, with geometric operator U expressed as in (2.48), geometric operator U^{-1} as in (2.51), reduces as follows:

$$\begin{aligned} X' &= U(\delta b) X U^{-1}(\delta b) = (1 + iH\delta b) X (1 - iH\delta b) \\ &= X - i(XH - HX)\delta b + HXH(\delta b)^2 \\ &= X - i[X, H]\delta b + 0(\delta b)^2. \end{aligned} \tag{2.76}$$

For the final form in (2.76) to agree with that in (2.75), we must have the operator relationship

$$[X, H] = iv(1). \tag{2.77}$$

Introducing (2.72) and multiplying through by $-\mu$ yields

$$[K, H] = -i\mu v(1). \tag{2.78}$$

But from (2.59), (2.60), and (2.58), we have

$$[K_0, H_0] = -iP_0. \tag{2.79}$$

Equations (2.78) and (2.79) are made similar by identifying μv with P; then

$$[K, H] = -iP. \tag{2.80}$$

Also from (2.58) and (2.59), we obtain the condition

$$[H_0, P_0] = 0, \tag{2.81}$$

which we generalize to

$$[H, P] = 0. \tag{2.82}$$

Restriction (2.71) led us to introduce the scalar constant μ. In (2.78), this quantity appears multiplying velocity operator $v(1)$. The result we have labeled P. Example 2.7 shows that this P behaves as the operator for *momentum* of the particle under consideration. So we are led to consider μ as the *mass* of the particle in some system of units.

Now, condition (2.68) is the canonical commutation rule in a system of units where

$$\hbar = 1. \tag{2.83}$$

We can go from this system, which we call a *natural* one, to a general system by replacing parameter μ with mass μ divided by $\hbar$ wherever it occurs. Correspondingly, P is replaced by $P/\hbar$.

In summary, note how rule (2.68) follows from (a) the localizability property of a particle and (b) the Galilean transformation properties of space-time.

Example 2.6

How does a velocity translation by v affect the physical operator for velocity V? When

$$x' = x + vt$$

and

$$t' = t,$$

a physically significant point traveling at speed V in the unprimed system travels at speed $V + v$ in the primed system. So we write for the operators

$$V' = V + v(1).$$

Example 2.7

How is operator P affected by a velocity translation?

Consider an infinitesimal translation in velocity by δv. Let iK be the corresponding infinitesimal operator, employ it in (2.48) and (2.51), introduce the results into the similarity transform (2.38), and reduce:

$$
\begin{aligned}
P' = U(\delta v)\, PU^{-1}(\delta v) &= (1 + iK\delta v)P(1 - iK\delta v) \\
&= P + i(KP - PK)\,\delta v + KPK\,(\delta v)^2 \\
&= P + i[K, P]\,\delta v + 0(\delta v)^2 \\
&= P + \mu\,\delta v(1).
\end{aligned}
$$

In the penultimate step, the definition of the commutator is introduced; in the last step, (2.74) is used.

With the choices to be made in the next section, the final result above is true to all orders and

$$P' = P + \mu v(1).$$

This corresponds to the classical result that the velocity translation v causes the momentum of the particle to be increased by μv.

2.11
Extending the Group
to Allow for Particle Localizability

The infinitesimal operators (2.58), (2.59), (2.60), and (2.61) do not contain the parameter μ, the mass of the particle under consideration. As a result, they have to be expanded to 4×4 matrices. These will be formulated so that the conditions in Section 2.10 are met.

We assume without difficulty that the commutator of H_0 and P_0 is unchanged in the generalization; thus

$$[H, P] = 0, \tag{2.84}$$

as in (2.82).

Since a particle can be localized to a point, a physical operator expressing this fact must exist. In 1-dimensional space this operator is the X whose transformation properties we have investigated. We found that it must satisfy (2.68), (2.71), and (2.77). Equation (2.71) was satisfied by employing

$$K = -\mu X. \tag{2.85}$$

Substituting into (2.68) and introducing I as the generalization of the identity operator I_0, we write

$$[K, P] = -i\mu I. \tag{2.86}$$

Finally, we note that, for (2.78), the transformed localization position moves at velocity v. So we set

$$v(1) = V \tag{2.87}$$

and identify V as the operator for the velocity v of a particle in the original inertial frame. Also, we set

$$\mu V = P, \tag{2.88}$$

with P the operator for the particle's linear momentum. Equation (2.88) holds in free space. From (2.77) with (2.85) and (2.88), we obtain

$$[K, H] = -iP. \tag{2.89}$$

Since IP_0 and iH_0 already lead to formula (2.84), we extend these matrices with a 4th row and 4th column of zeroes. Then to get formula (2.86), we introduce μ in the 41 position of iK, with the other added elements zero. And we introduce 1 in the 43 position of iI.

Thus, we construct the infinitesimal generators

$$\begin{pmatrix} 0 & 0 & 1 & 0 \\ 0 & 0 & 0 & 0 \\ 0 & 0 & 0 & 0 \\ 0 & 0 & 0 & 0 \end{pmatrix} = iP, \text{ for distance,} \tag{2.90}$$

$$\begin{pmatrix} 0 & 0 & 0 & 0 \\ 0 & 0 & 1 & 0 \\ 0 & 0 & 0 & 0 \\ 0 & 0 & 0 & 0 \end{pmatrix} = iH, \text{ for time,} \tag{2.91}$$

$$\begin{pmatrix} 0 & 1 & 0 & 0 \\ 0 & 0 & 0 & 0 \\ 0 & 0 & 0 & 0 \\ \mu & 0 & 0 & 0 \end{pmatrix} = iK, \text{ for velocity,} \tag{2.92}$$

$$\begin{pmatrix} 0 & 0 & 0 & 0 \\ 0 & 0 & 0 & 0 \\ 0 & 0 & 0 & 0 \\ 0 & 0 & 1 & 0 \end{pmatrix} = iI, \text{ for phase.} \tag{2.93}$$

The corresponding geometric operators are obtained with formulas (2.48) and (2.50). Carrying out the exponentiation on each infinitesimal generator and multiplying the results yields

$$U = e^{iPa} e^{iHb} e^{iKv} e^{iI\vartheta}. \tag{2.94}$$

Here a is the increase in distance from the origin along the x axis, b is the increase in time, v is the increase in velocity along the x axis, and ϑ is the increase in phase.

The 4×4 matrices are nilpotent; indeed from (2.90), (2.91), (2.92), (2.93), we find that

$$-P^2 = 0, \tag{2.95}$$

$$-H^2 = 0, \tag{2.96}$$

$$-iK^3 = 0, \tag{2.97}$$

$$-I^2 = 0. \tag{2.98}$$

Furthermore,

$$-K^2 = \begin{pmatrix} 0 & 0 & 0 & 0 \\ 0 & 0 & 0 & 0 \\ 0 & 0 & 0 & 0 \\ 0 & \mu & 0 & 0 \end{pmatrix}. \tag{2.99}$$

Introducing the expansions of the exponentials and these nilpotencies into (2.94) yields

$$U = (1 + iPa)(1 + iHb)(1 + iKv - \tfrac{1}{2}K^2v^2)(1 + iI\vartheta). \tag{2.100}$$

Into (2.100) substitute (2.90), (2.91), (2.92), (2.93) and multiply out to obtain

$$U = \begin{pmatrix} 1 & v & a & 0 \\ 0 & 1 & b & 0 \\ 0 & 0 & 1 & 0 \\ \mu v & \tfrac{1}{2}\mu v^2 & \vartheta & 1 \end{pmatrix}. \tag{2.101}$$

This matrix acts to transform the vector

$$\begin{pmatrix} x \\ t \\ 1 \\ \xi \end{pmatrix} \tag{2.102}$$

where x is the position coordinate, t the time coordinate, while ξ is identified as the phase coordinate.

When (2.101) acts on (2.102), the vector

$$\begin{pmatrix} x' \\ t' \\ 1 \\ \xi' \end{pmatrix} = \begin{pmatrix} x & + & vt & + & a \\ & & t & + & b \\ & & & & 1 \\ \mu vx & + & \tfrac{1}{2}\mu v^2\, t & + & \vartheta + \xi \end{pmatrix} \tag{2.103}$$

results. Thus, Equations (2.53), (2.54), and (2.55) are still satisfied.

If we do not introduce any translation in phase, we have $\vartheta = 0$ and

$$\xi' = \mu vx + \tfrac{1}{2}\mu v^2 t + \xi. \tag{2.104}$$

The expressions here are in natural units, with $\hbar = 1$. For a general system of units, μ is replaced by $\mu/\hbar$. Then (2.104) becomes

$$\xi' = \frac{\mu}{\hbar}\,(vx + \tfrac{1}{2}v^2 t) + \xi. \tag{2.105}$$

The phase of a classical wave is not altered by a Galilean transformation. For such a wave, we would have

$$F'(x', t) = F(x, t). \tag{2.106}$$

But for the state function of a particle moving freely along the x axis, the phase must change as in (2.105). We have

$$e^{i\xi'} F'(x', t') = e^{i\mu(vx + \frac{1}{2}v^2 t)/\hbar}\, e^{i\xi} F(x, t) \tag{2.107}$$

whence

$$\Psi'(x', t') = e^{i\mu(vx + \frac{1}{2}v^2 t)/\hbar}\, \Psi(x, t). \tag{2.108}$$

Note that this result agrees with (2.14).

Example 2.8

Determine how the parameters for the extended Galilean group combine under successive operations of the group.

Components $(x, t, 1, \xi)$ are transformed to $(x', t', 1, \xi')$ by the U of (2.101). Similarly, components $(x', t', 1, \xi')$ are transformed to $(x'', t'', 1, \xi'')$ by

$$U' = \begin{pmatrix} 1 & v' & a' & 0 \\ 0 & 1 & b' & 0 \\ 0 & 0 & 1 & 0 \\ \mu v' & \tfrac{1}{2}\mu v'^2 & \vartheta' & 1 \end{pmatrix}.$$

And the overall transformation, from $(x, t, 1, \xi)$ to $(x'', t'', 1, \xi'')$, is effected by the matrix product of U and U', which is

$$U'U = \begin{pmatrix} 1 & v+v' & a+a'+bv' & 0 \\ 0 & 1 & b+b' & 0 \\ 0 & 0 & 1 & 0 \\ \mu(v+v') & \tfrac{1}{2}\mu(v+v')^2 & \vartheta+\vartheta'+\mu v'a+\tfrac{1}{2}\mu v'^2 b & 1 \end{pmatrix}$$

From the elements in this product, we see that the parameters combine according to the law:

$$(a',\ b',\ v',\ \vartheta')(a,\ b,\ v,\ \vartheta) =$$

$$(a+a'+bv',\ b+b',\ v+v',\ \vartheta+\vartheta'+\mu v'a+\tfrac{1}{2}\mu v'^2 b).$$

2.12
Introducing Localized Interactions

The Galilean symmetry employed in Sections 2.10 and 2.11 is that which a freely moving particle would experience. But in general, a given particle is subjected to interactions with other particles. These must be taken into account.

In classical mechanics each particle of the pertinent system moves along a definite curve in physical space. The field that it generates depends on its position and its velocity over the pertinent times. The fields from the different sources add up at the point where the particle under study is at the given time. And the effects depend on this particle's position and velocity.

How should this scheme be generalized? The existence of each particle at a point with a definite velocity may be replaced by the probability that the particle is there with a certain velocity. Or, as we have done for the free particle, we may employ the localizability of the particle with a definite velocity together with operators for the pertinent properties.

When a given particle is localized at position (x, t) with velocity v, the interaction with a given field presumably depends only on these variables. Then on applying an infinitesimal transformation from the extended group, the changes in the interactions are infinitesimal and the effects of these changes are infinitesimals of higher order. Let us neglect these secondary effects over the infinitesimal transformations. Then the key commutation relations for the particle hold as they stand.

Now, formula (2.77) applies to a localization position traveling at velocity v. Through (2.87), we introduce the operator for particle velocity, V, and rearrange to get

$$i\,[H,\,X] \;=\; V. \tag{2.109}$$

In free space, operator P was identified with μV. Substituting this into (2.68) yields

$$[X,\,\mu V] \;=\; i(1). \tag{2.110}$$

Multiply (2.110) by operator V from the left,

$$VX\mu V \;-\; \mu V^2 X \;=\; iV, \tag{2.111}$$

and by operator V from the right,

$$X\mu V^2 \;-\; \mu VXV \;=\; iV. \tag{2.112}$$

Adding these two equations gives us

$$X\mu V^2 \;-\; \mu V^2 X \;=\; 2iV, \tag{2.113}$$

whence

$$i\,[\tfrac{1}{2}\mu V^2,\,X] \;=\; V \tag{2.114}$$

Subtracting (2.114) from (2.109) and canceling i yields

$$[H \;-\; \tfrac{1}{2}\mu V^2,\,X] \;=\; 0. \tag{2.115}$$

But from (2.68), we have

$$[P,\,X] \;\neq\; 0. \tag{2.116}$$

Thus, operator $H - \tfrac{1}{2}\mu V^2$ cannot contain P. It may, however, depend on X and t. We therefore write

$$H \;=\; \tfrac{1}{2}\mu V^2 \;+\; U(X,\,t). \tag{2.117}$$

From Example 2.7, we have

$$P' \;=\; P \;+\; \mu v(1) \tag{2.118}$$

when the localization position is subjected to a velocity translation by v. The corresponding change in the velocity operator is

$$V' \;=\; V \;+\; v(1). \tag{2.119}$$

Eliminating $v(1)$ from (2.118), (2.119) and rearranging leads to

$$P' \;-\; \mu V' \;=\; P \;-\; \mu V. \tag{2.120}$$

In the interacting system, Equation (2.88) need not hold. But we have $P - \mu V$ invariant in the velocity translation. This expression may, however, depend on X and t; so

$$P - \mu V = A(X, t) \tag{2.121}$$

or

$$P - A(X, t) = \mu V. \tag{2.122}$$

Substituting (2.122) into (2.117) gives us the operator relationship

$$H = \frac{1}{2\mu} [P - A(X, t)]^2 + U(X, t). \tag{2.123}$$

This has the same form as the 1-dimensional Hamiltonian function of classical physics

$$H = \frac{1}{2\mu} [p - A(x, t)]^2 + U(x, t), \tag{2.124}$$

in which p is the canonical momentum, A the vector potential, U the scalar potential, μ the particle mass, and H the energy. So in quantum mechanics, we employ the form of the scalar potential for the given system in $U(X, t)$ and the pertinent form of the vector potential in $A(X. t)$. A ket $| j \rangle$ for which

$$H | j \rangle = E_j | j \rangle \tag{2.125}$$

then yields the energy E_j of the eigenstate.

The fundamental argument in this section is that one cannot discriminate over any infinitesimal increment of time between a free and an interacting system. Consequently, the particle under consideration is subject to the same commutation relations. And, the form for its Hamiltonian operator is induced from the symmetry of the space-time continuum and the localizability of the particle.

A 3-dimensional generalization of this reasoning will be developed in the next chapter. The *time-independent Schrödinger equation* will still be of form (2.125).

2.13
Representing the Operators for Position and Momentum

Any mathematical representation of the operators must be consistent with the commutator relationships. For a free particle, and for a particle which can only interact locally as described in Section 2.12, we consider (2.68) to be valid.

Thus, the operator condition

$$XP - PX = i(1) \tag{2.126}$$

is to be satisfied, with $\hbar = 1$. For a general set of units, μ is replaced by $\mu/\hbar$. Then (2.126) becomes

$$XP - PX = i\hbar(1). \tag{2.127}$$

A person may employ either position or momentum coordinates with the time coordinate. In the space-time description the state function, $\Psi(x, t)$, becomes the object on which the operators act. We then choose coordinate x to represent the operator X. To satisfy (2.127), we correspondingly let $\hbar/i$ times differentiation with respect to the coordinate represent operator P. Thus, we introduce

$$X = x, \tag{2.128}$$

$$P = \frac{\hbar}{i} \frac{\partial}{\partial x}. \tag{2.129}$$

Substituting these into (2.123) leads to the customary Hamiltonian operator.

2.14
Operators Governing Temporal Changes

In Galilean relativity, all elements of a system are subject to the same uniform, inexorable, march of time. Each particle, whether free or under interaction, experiences a symmetry over time similar to the symmetry of a free particle over space.

The free particle is subject to translational symmetries that lead to a particular phase change during a given Galilean transformation. The result will enable us to link the variation over time of a state function to the variation over space that we have established.

With operator (2.129), a person can construct an eigenvalue equation for a free particle moving in the x direction. A typical eigenfunction for the operator is

$$\Psi = A e^{i(p_x/\hbar)x} T(t) = A e^{ikx} T(t). \tag{2.130}$$

But when

$$T(t) = e^{-i(E/\hbar)t} = e^{-i\omega t}, \tag{2.131}$$

the eigenfunction has the form of each constituent of the wave packet in Section 2.3.

This wave packet yielded the phase change during a Galilean transformation described by Equation (2.14). However, the *same* phase change was obtained from our symmetry arguments, in Equation (2.108). Thus, these arguments indicate that (2.131) is valid as long as (2.130) is valid.

Now, (2.131) is a typical eigenfunction for the operator

$$E = -\frac{\hbar}{i} \frac{\partial}{\partial t}, \tag{2.132}$$

in the space-time description. Since we accept (2.129), the source of (2.130), we are led to accept (2.132).

So when the system is in the jth eigenstate, the state function satisfies the equation

$$- \frac{\hbar}{i} \frac{\partial}{\partial t} \Psi_j = E_j \Psi_j. \tag{2.133}$$

Eliminating the eigenvalue between (2.133) and (2.125) yields

$$H\Psi_j = - \frac{\hbar}{i} \frac{\partial}{\partial t} \Psi_j. \tag{2.134}$$

We see no reason why this equation should not hold for systems as they occur in practice.

A general state is obtained by the superposition

$$\Psi = \Sigma \ c_j \ \Psi_j. \tag{2.135}$$

When the eigenfunctions are made orthogonal where they need not be and are normalized to 1, we have

$$\Sigma \ c_j{}^* c_j = 1. \tag{2.136}$$

Since (2.134) is valid for each of the eigenfunctions, it is valid for any definite superposition. We therefore write

$$H\Psi = - \frac{\hbar}{i} \frac{\partial}{\partial t} \Psi. \tag{2.137}$$

Now, recall how a radioactive substance decays; also, how materials absorb or emit photons. Submicroscopic particles appear to go between definite superpositions, between definite states, by jumps. Our argument indicates that (2.137) holds before and after each such jump. If we consider every jump to be instantaneous, then we may apply (2.137) at effectively all times.

In practice, one may consider an ensemble of similar systems and use (2.137) to determine the expected rate of change. Then (2.137) is called the *time-dependent Schrödinger equation*. A person may formally rearrange this equation to

$$\frac{d\Psi}{\Psi} = - \frac{iH}{\hbar} dt. \tag{2.138}$$

When the Hamiltonian operator is independent of time, as in a spontaneous process, (2.138) integrates to

$$\ln \Psi = - \frac{iH}{\hbar} t + \ln \Psi_0 \tag{2.139}$$

or

$$\Psi(\mathbf{r}, t) = \exp \left(- \frac{it}{\hbar} H \right) \Psi(\mathbf{r}, 0). \tag{2.140}$$

In formula (2.140), the exponential is considered to be an *evolution operator* acting on the state function for time $t = 0$. Note that this operator is unitary.

2.15
Key Concepts

esponding to each geometric change to which a coherent physical system m. subjected is a transformation of its state vector, its ket. Such a transformation is implemented by an operator designated U. This operator varies continuously from the identity operator whenever the geometric change may vary continuously from naught.

A 1-dimensio translation in space, in time, in velocity, or in phase depends continuously on a single parameter. The corresponding operator is denoted $U(g)$, with g an appropriate parameter that equals zero when no change is effected. Differentiating U in the neighborhood of the identity then yields the infinitesimal operator

$$\left(\frac{\partial U}{\partial g}\right)_{g=0} = iG. \tag{2.141}$$

We let g be the displacement introduced by the chosen translation. Then for a displacement of δg, the operator U is

$$U(\delta g) = 1 + \delta U = 1 + iG\delta g. \tag{2.142}$$

Exponentiation of this expression yields

$$U(g) = e^{iGg}. \tag{2.143}$$

Since a particle can act at a point in space and time, a physical operator for a position at rest exists. For points x and x' along an axis, there are operators X and X'. When x' is distance δa further from the origin than x, we have

$$x' = x + \delta a \tag{2.144}$$

and

$$X' = X + \delta a(1). \tag{2.145}$$

But the infinitesimal geometric operator for free spatial translation, not allowing for a change in phase, is

$$\begin{pmatrix} 0 & 0 & 1 \\ 0 & 0 & 0 \\ 0 & 0 & 0 \end{pmatrix} = iP. \tag{2.146}$$

Furthermore, operator X' equals the similarity transform of operator X by the corresponding U:

$$X' = (1 + iP\delta a) X (1 - iP\delta a)$$

$$= X - i[X, P]\delta a + PXP(\delta a)^2. \tag{2.147}$$

For this to equal (2.145) to first order in δa, we must have

$$[X, P] = i(1). \tag{2.148}$$

Similarly treating an instantaneous velocity shift acting on operator X leads to

$$[X, K] = 0 \tag{2.149}$$

whence

$$X = - \frac{1}{\mu} K. \tag{2.150}$$

A velocity translation by v followed by a time translation of δb yields

$$[X, H] = iv(1). \tag{2.151}$$

Substituting (2.150) into (2.148) gives us

$$[K, P] = -i\mu(1). \tag{2.152}$$

An infinitesimal velocity translation acting on operator P also leads to (2.152) if

$$P' = P + \mu\delta v(1). \tag{2.153}$$

The operators are constructed so that (2.153) is still true when δv is finite. Then μ is identified as the particle mass and P the physical operator for linear momentum.

To allow properly for the mass of the particle and the phase of the state function, we have to expand the 3×3 matrices obtained from the Galilean transformation equations to the 4×4 matrices:

$$\begin{pmatrix} 0 & 0 & 1 & 0 \\ 0 & 0 & 0 & 0 \\ 0 & 0 & 0 & 0 \\ 0 & 0 & 0 & 0 \end{pmatrix} = iP, \text{ for the translation in space,} \tag{2.154}$$

$$\begin{pmatrix} 0 & 0 & 0 & 0 \\ 0 & 0 & 1 & 0 \\ 0 & 0 & 0 & 0 \\ 0 & 0 & 0 & 0 \end{pmatrix} = iH, \text{ for translation in time,} \tag{2.155}$$

$$\begin{pmatrix} 0 & 1 & 0 & 0 \\ 0 & 0 & 0 & 0 \\ 0 & 0 & 0 & 0 \\ \mu & 0 & 0 & 0 \end{pmatrix} = iK, \text{ for translation in velocity,} \qquad (2.156)$$

$$\begin{pmatrix} 0 & 0 & 0 & 0 \\ 0 & 0 & 0 & 0 \\ 0 & 0 & 0 & 0 \\ 0 & 0 & 1 & 0 \end{pmatrix} = iI, \text{ for translation in phase.} \qquad (2.157)$$

These serve as infinitesimal generators of a Lie group. They yield the canonical commutation relations in a natural set of units where $\hbar = 1$.

For a general set of units, μ is replaced by $\mu/\hbar$ and P by $P/\hbar$. Equation (2.148) becomes

$$XP - PX = i\hbar(1). \qquad (2.158)$$

If we represent X with multiplication by coordinate x, so

$$X = x, \qquad (2.159)$$

then

$$P = \frac{\hbar}{i} \frac{\partial}{\partial x} . \qquad (2.160)$$

Particles are generally subject to interactions (often loosely described as forces) due to imposed fields. We presume that such fields act locally and smoothly, For infinitesimal shifts we presume that effects of their changes are infinitesimals of higher order. Over the infinitesimal transformations considered these effects are neglected. Thus, the key commutation relations are applied as they stand.

From (2.151) and (2.148), we find that

$$[H - \tfrac{1}{2}\mu V^2, X] = 0. \qquad (2.161)$$

But from (2.148), the relation

$$[P, X] \neq 0 \qquad (2.162)$$

holds. As a consequence, operator $H - \tfrac{1}{2}\mu V^2$ should not contain P.

Equation (2.153) together with

$$V' = V + v(1) \qquad (2.163)$$

lead to $P - \mu V$ being invariant in a velocity translation.

Considering what these expressions still depend on, we then find that

$$H = \frac{1}{2\mu} [P - A(X, t)]^2 + U(X, t). \tag{2.164}$$

This is called the Hamiltonian operator since it has the same general form as the classical Hamiltonian function. It is the quantum mechanical operator for energy.

As the various operators that we have been considering do, the Hamiltonian H acts in the Hilbert space for the given system. A state of the system described by normalized ket $|g\rangle$ yields, on the average, the energy

$$E = \langle g | H | g \rangle. \tag{2.165}$$

When this ket is an eigenket $|j\rangle$, we obtain a definite energy E_j such that

$$H | j \rangle = E_j | j \rangle. \tag{2.166}$$

Discussion Questions

2.1. How are the attributes of time induced? How are the attributes of space induced?

2.2. In Schrödinger quantum mechanics, what assumptions are made about the space-time continuum? Are these consistent with the relativity principle?

2.3. Explain the nature of (a) translation in time, (b) translation in space, (c) velocity translation, (d) rotation in space, (e) translation in phase.

2.4. What is a Galilean transformation?

2.5. Explain how one constructs a general 1-dimensional translational state function.

2.6. Explain how relative motion affects the de Broglie (a) wavevector, (b) angular frequency.

2.7. How does relative motion affect a ket's (a) coordinate representation, (b) wavevector representation?

2.8. Why is the phase of a Schrödinger state function arbitrary?

2.9. Explain how the superselection rule forbidding change of mass arises.

2.10. What induces a similarity transformation in a physical operator?

2.11. Distinguish between (a) an active translation, (b) a passive translation, (c) an active rotation, (d) a passive rotation.

2.12. How does one parameterize (a) a translation operator, (b) a rotation operator?

2.13. How does the eigenvalue equation appear in bra form?

2.14. Explain the formula

$$(AB)^\dagger = B^\dagger A^\dagger.$$

2.15. What is a unitary operator? Explain why (a) an evolution operator, (b) a symmetry operator, is unitary.

2.16. How may operators depend continuously and smoothly, in the standard analytic sense, on parameters?

2.17. How is the infinitesimal operator with respect to a given parameter defined?

2.18. Construct a form for an operator differing infinitesimally from the identity.

2.19. From the form in Question 2.17 generate an operator differing by a finite amount from the identity. How is this amount measured?

2.20. What is a group? What is a Lie group?

2.21. Why in a matrix representation of a group do the matrices need to be square?

2.22. Explain how the elements in the 1-D Galilean group do not commute.

2.23. How are the square matrices for transforming x, t, and 1 related to the U's for transforming kets?

2.24. What aspects of the behavior are represented by the infinitesimal generators (operators) iP_0, iH_0, iK_0, iI_0?

2.25. How are these infinitesimal operators for distance, for time, and for velocity obtained?

2.26. What characterizes a particle?

2.27. How does the localization requirement, with $\hbar = 1$, lead to

(a) $[X, P] = i(1)$,
(b) $[X, K] = 0$,
(c) $[X, H] = iv(1)$?

2.28. How do we justify

$$[K, P] = -i\mu(1)?$$

2.29. When and how do we identify P with μv and μ with the particle mass?

2.30. How are the infinitesimal generators enlarged to agree with

$$[K, P] = -i\mu(1)?$$

2.31. How do we obtain the representation matrix

$$\begin{pmatrix} 1 & v & a & 0 \\ 0 & 1 & b & 0 \\ 0 & 0 & 1 & 0 \\ \mu v & \frac{1}{2}\mu v^2 & \vartheta & 1 \end{pmatrix} ?$$

2.32. Explain how the parameters for the extended Galilean group combine.

2.33. Explain how the phase of the state function for free translation changes in a Galilean transformation.

2.34. What is a localized interaction?

2.35. If interactions were not localized, would causality be violated?

2.36. Can a person discriminate over an infinitesimal interval of time between a free and an interacting system?

2.37. Under what conditions are (a) $H - \frac{1}{2}\mu v^2$, (b) $P - \mu V$, invariant?

2.38. How are invariance principles employed in constructing the Hamiltonian operator?

2.39. Explain what the explicit form for P must be when the operator for X is taken to be coordinate x.

2.40. How do symmetry arguments support setting the temporal factor in Ψ equal to $\exp[-i(E/\hbar)t]$?

2.41. How do we obtain the operator for energy E from this temporal factor?

2.42. How can one formulate the Schrödinger equation for a general state from the Schrödinger equation for a fixed-energy state?

2.43. Does the time-dependent Schrödinger equation have the same fundamental status as the fixed-state Schrödinger equation?

Problems

2.1. If an operator is both Hermitian and unitary, what powers of it are equivalent to the operator itself?

2.2. How does the velocity translation

$$x' = x + vt, \qquad t' = t$$

affect the translational state

$$\Psi' = A\, e^{i(k'x' - \omega't')}?$$

2.3. Show when the geometric operator

$$U(g) = e^{Ag}$$

is unitary.

2.4. Show by multiplication that

$$U^{-1}(\delta g) = U(-\delta g).$$

2.5. Exponentiate the operators iP_0, iH_0, and iK_0. Multiply the results to obtain a general U. Then show how this reduces to the transformation matrix in (2.56).

2.6. What do (2.58) and (2.60) yield for

$$[K_0, P_0]?$$

2.7. Carry out the matrix multiplication described by (2.100).

2.8. Determine the operator for position in the momentum-time description, where the operator for momentum is multiplication by p.

— — —

2.9. Calculate the probability density $\Psi^{tr*}\Psi^{tr}$ for (2.29). How does this reduce when $\mu_1 = \mu_2$?

2.10. How does the time translation

$$x' = x, \qquad t' = t + b,$$

affect the translational state for which

$$\Psi' = A\, e^{i(k'x' - \omega't')}?$$

2.11. What is the adjoint of a complex number times a Hermitian operator?

2.12. Show that in the phase angle operator

$$U = e^{i\gamma},$$

parameter γ must be real.

2.13. Show that (2.59) and (2.60) lead to

$$[K_0, H_0] = -iP_0.$$

2.14. If we required $P' = P$ in Galilean transformations, what would $[K, P]$ equal?

2.15. Carry out the matrix multiplications that lead to (2.95) and (2.97).

2.16. Using the operator for p_x, construct the eigenvalue equation for linear momentum in free space. Solve this equation for the spatial dependence of the state function.

References

Books

Duffey, G. H.: 1984, *A Development of Quantum Mechanics Based on Symmetry Considerations*, Reidel, Dordrecht, pp. 1-14, 113-116. In these sections, a simple description of the symmetry basis for quantum mechanics is developed.

Kaempffer, F. A.: 1965, *Concepts in Quantum Mechanics*, Academic Press, New York, pp. 341-346. A more extensive treatment of Galilean transformation theory appears here.

Articles

Bernstein, H. J.: 1967. "New Characterization of the Ray Representations of the Galilei Group." *J. Math. Phys.* **8**, 406-408.

Frieden, B. R.: 1989. "Fisher Information as the Basis for the Schrödinger Wave Equations." *Am. J. Phys.* **57**, 1004-1008.

Hassoun, G. Q., and Kobe, D. H.: 1989. "Synthesis of the Planck and Bohr Formulations of the Correspondence Principle." *Am. J. Phys.* **57**, 658-662.

Landsberg, P. T.: 1988. "Why Quantum Mechanics?" *Found. Phys.* **18**, 969-982.

Levy-Leblond, J.-M.: 1974. "The Pedagogical Role and Epistemological Significance of Group Theory in Quantum Mechanics." *Riv. Nuovo Cimento* **4**, 99-143.

Peres, A.: 1980, "Zeno Paradox in Quantum Theory." *Am. J. Phys.* **48**, 931-932.

Woo, C. H.: 1989. "Chaos, Ineffectiveness, and the Contrast between Classical and Quantal Physics." *Found. Phys.* **19**, 57-76.

3

INTRODUCING THE
ROTATIONAL SYMMETRIES

3.1
Guiding Considerations

Adding two spatial dimensions to the one employed in Chapter 2 does not negate the arguments developed there. However, it does introduce additional phenomena that must be treated.

In 3-dimensional space, a structureless point exhibits rotational symmetry. Now, we have seen that a particle, for instance an electron, is localizable to such a point; in various processes it can act at a point in space and time. As a consequence, a particle possesses intrinsic propensities subject to commutation relations. For reasons that will become evident, we call these apparent properties components of spin and spin itself.

In 3-dimensional space, an isolated rigid array of points possesses a rotationally symmetric surroundings. Correspondingly, a stable multiparticle system is subject to a similar symmetry. This imposes commutation relations among propensities of the system. In particular measurement orientations, certain of these propensities appear as properties. These include the components of rotational angular momentum and of total angular momentum, together with the net values.

According to classical mechanics the rotation of a linear system is modelled by the revolution of a single particle about the axis of rotation. A similar model can be employed in quantum mechanics. Thus, a hydrogen atom is modelled by a single particle of reduced mass moving in a central field (a spherically symmetric one).

For a particular system, a person may employ either rectangular coordinates or spherical coordinates. Rectangular coordinates are useful in that rotation about any one of the coordinate axes involves the coordinates similarly. Spherical coordinates are useful because they fit the symmetry of a spherically symmetric field whenever the origin is placed at the center of symmetry.

3.2
Rotation Modelled as a Composite Translation

The rotation that is represented by a single particle revolving around an axis can be considered in some respects as a translation. Here we suppose the axis of rotation to be the z axis of a Cartesian system. The origin is identified with the center of symmetry, and the particle is presumed to be localizable as it was in Section 2.10.

A given point (x, y, z) is located as (r, ϑ, φ), where r is the distance from the origin to the point, ϑ the angle between the z axis and the radius vector drawn from the origin to the point, φ the angle from the x axis to the projection of the radius vector on the xy plane. See Figure 3.1.

Let us consider transformations at a given r and ϑ. If the remaining coordinates of a point are (φ, t) before and (φ', t') after the operation, then

$$\varphi' = \varphi + \omega t + \alpha, \tag{3.1}$$

$$t' = t + b, \tag{3.2}$$

$$1 = 1, \tag{3.3}$$

Here α is the translation in angle, b the translation in time, and ω the translation in angular velocity.

Proceeding as in Section 2.9, we construct the infinitesimal operators. But now we let iL_z be the one for angular distance.

We let ϕ be the physical operator for the angle before, and ϕ' after, the transformation. Paralleling (2.65), we have

$$\phi' = \phi + \delta\alpha(1). \tag{3.4}$$

And similar to (2.66), we have

$$\phi' = (1 + iL_z\delta\alpha)\phi(1 - iL_z\delta\alpha)$$
$$= \phi - i[\phi, L_z]\delta\alpha + 0(\delta\alpha)^2. \tag{3.5}$$

On comparing (3.4) and (3.5), we see that

$$[\phi, L_z] = i(1). \qquad (3.6)$$

In the coordinate representation, ϕ is given by multiplication by angle φ:

$$\phi = \varphi. \qquad (3.7)$$

Then to satisfy (3.6), we must have

$$L_z = \frac{1}{i} \frac{\partial}{\partial \varphi} \qquad (3.8)$$

in natural units, where $\hbar = 1$. For a general set of units (3.8) becomes

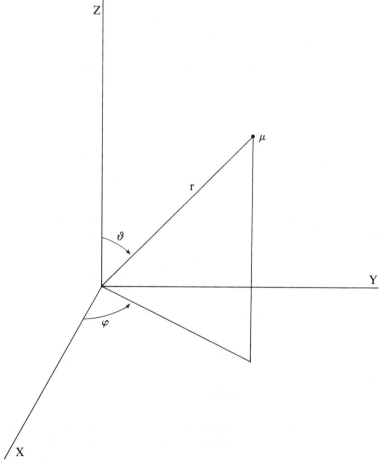

Figure 3.1. Spherical coordinates for a particle of mass μ.

$$L_z = \frac{\hbar}{i} \frac{\partial}{\partial \varphi} .$$

(3.9)

Example 3.1

What is operator L_z, for revolution of the model particle, in rectangular coordinates?

Rectangular coordinates are related to spherical coordinates by the equations

$$x = r \sin \vartheta \cos \varphi,$$

$$y = r \sin \vartheta \sin \varphi,$$

$$z = r \cos \vartheta.$$

Differentiate these with respect to φ,

$$\frac{\partial x}{\partial \varphi} = -r \sin \vartheta \sin \varphi = -y,$$

$$\frac{\partial y}{\partial \varphi} = r \sin \vartheta \cos \varphi = x,$$

$$\frac{\partial z}{\partial \varphi} = 0,$$

and substitute into

$$\frac{\partial}{\partial \varphi} = \frac{\partial x}{\partial \varphi} \frac{\partial}{\partial x} + \frac{\partial y}{\partial \varphi} \frac{\partial}{\partial y} + \frac{\partial z}{\partial \varphi} \frac{\partial}{\partial z} ,$$

to obtain

$$L_z = \frac{\hbar}{i} \frac{\partial}{\partial \varphi} = \frac{\hbar}{i} \left(-y \frac{\partial}{\partial x} + x \frac{\partial}{\partial y} \right)$$

$$= \frac{\hbar}{i} \left(x \frac{\partial}{\partial y} - y \frac{\partial}{\partial x} \right) .$$

Example 3.2

Construct representations of L_x and L_y for revolution of the model particle.

It is given that the model particle moves in a field spherically symmetric about the origin. But replacing x with y, y with z, and z with x corresponds to rotation by 1/3 turn about a diagonal, movement to an equivalent position.

So we may introduce such a substitution into the result from Example 3.1. We obtain

$$L_x = \frac{\hbar}{i} \left(y \frac{\partial}{\partial z} - z \frac{\partial}{\partial y} \right).$$

Then apply the same substitution again. We obtain

$$L_y = \frac{\hbar}{i} \left(z \frac{\partial}{\partial x} - x \frac{\partial}{\partial z} \right).$$

3.3
An Alternate Construction

Over an infinitesimal interval of time, the model particle of Section 3.2 undergoes a leveraged linear translation. Classically, its angular momentum about the symmetry axis equals the cross product of its position vector, measured from a point on the axis, with its instantaneous linear momentum. In quantum mechanics, one may employ the corresponding operators combined similarly.

As in the previous section, we consider a single particle revolving around an axis passing through the center of symmetry. Its mass is set equal to the classical reduced mass for the rotation.

Let the physical operator for a possible position of the particle with respect to the reference point be $\mathbf{X}$, while the physical operator for its linear momentum is $\mathbf{P}$. The angular momentum operator is then constructed as

$$\mathbf{L} = \mathbf{X} \times \mathbf{P}. \tag{3.10}$$

If ϵ_{jkl} is the permutation symbol defined by the conditions

$$\epsilon_{jkl} = 1 \quad \text{if } jkl \text{ is } 123,\ 231,\ 312, \tag{3.11}$$

$$\epsilon_{jkl} = -1 \ \text{if } jkl \text{ is } 213,\ 321,\ 132, \tag{3.12}$$

$$\epsilon_{jkl} = 0 \quad \text{if } jkl \text{ has any two numbers the same}, \tag{3.13}$$

then the jth Cartesian component of (3.10) is

$$L_j = \sum_k \sum_l \epsilon_{jkl} X_k P_l. \tag{3.14}$$

Let us employ position and time coordinates in our representation of the ket. Generalizing the result in Section 2.13 then leads to the explicit forms

$$X_k = x_k, \tag{3.15}$$

$$P_l = \frac{\hbar}{i} \frac{\partial}{\partial x_l} \, . \qquad (3.16)$$

Substituting (3.15), (3.16) into (3.14) yields

$$L_1 = \frac{\hbar}{i} \left(x_2 \frac{\partial}{\partial x_3} - x_3 \frac{\partial}{\partial x_2} \right), \qquad (3.17)$$

$$L_2 = \frac{\hbar}{i} \left(x_3 \frac{\partial}{\partial x_1} - x_1 \frac{\partial}{\partial x_3} \right), \qquad (3.18)$$

$$L_3 = \frac{\hbar}{i} \left(x_1 \frac{\partial}{\partial x_2} - x_2 \frac{\partial}{\partial x_1} \right). \qquad (3.19)$$

By direct computation, we see that these satisfy the commutation relations

$$[L_j, L_k] = - \frac{\hbar}{i} \sum_l \epsilon_{jkl} L_l. \qquad (3.20)$$

Note that (3.17), (3.18), (3.19) agree with the results obtained in Examples 3.1 and 3.2.

Constructing the commutators of (3.17), (3.18), (3.19) with (3.15) and reducing yields

$$[L_j, X_k] = - \frac{\hbar}{i} \sum_l \epsilon_{jkl} X_l. \qquad (3.21)$$

Employing (3.16) in place of (3.15) in these commutators leads to

$$[L_j, P_k] = - \frac{\hbar}{i} \sum_l \epsilon_{jkl} P_l. \qquad (3.22)$$

With K_j proportional to X_j, we obtain from (3.21) the form

$$[L_j, K_k] = - \frac{\hbar}{i} \sum_l \epsilon_{jkl} K_l. \qquad (3.23)$$

In natural units, with $\hbar = 1$, these may be rewritten as

$$[L_j, A_k] = \sum_l i \, \epsilon_{jkl} A_l \quad \text{for} \quad \mathbf{A} = \mathbf{L}, \mathbf{P}, \mathbf{K}. \qquad (3.24)$$

From (2.134), the corresponding representation of H is

$$H = - \frac{\hbar}{i} \frac{\partial}{\partial t}. \qquad (3.25)$$

This combines with (3.17), (3.18), (3.19) so that

$$[H, L_j] = 0. \qquad (3.26)$$

Example 3.3

Interpret the binary combination of operators

$$\left(x_1 \frac{\partial}{\partial x_2} \right) \left(x_2 \frac{\partial}{\partial x_3} \right)$$

which occurs in the evaluation of $(L_3, L_1]$.

From the form of this combination, we see that the independent variables include the rectangular coordinates x_1, x_2, x_3. So the ket for the particles is being represented in the coordinate-time system; we are concerned with

$$\left(x_1 \frac{\partial}{\partial x_2} \right) \left(x_2 \frac{\partial}{\partial x_3} \right) \Psi(x_1, x_2, x_3, t).$$

With the rule for differentiating a product, this expression yields

$$x_1 \frac{\partial x_2}{\partial x_2} \frac{\partial \Psi}{\partial x_3} + x_1 x_2 \frac{\partial^2 \Psi}{\partial x_2 \partial x_3} = \left(x_1 \frac{\partial}{\partial x_3} + x_1 x_2 \frac{\partial^2}{\partial x_2 \partial x_3} \right) \Psi.$$

Thus, the given combination expands to

$$x_1 \frac{\partial}{\partial x_3} + x_1 x_2 \frac{\partial^2}{\partial x_2 \partial x_3}.$$

3.4
Matrices Representing Key Rotations in 3-Dimensional Space

We have to consider not only the revolution discussed in Sections 3.2 and 3.3 but also consequences of the localizability of the particle to a point. A general approach to spherical symmetry is needed and will now be supplied.

As preparation, consider the rotation of axes in Figure 3.2. Applying the standard trigonometric definitions to the construction leads to the transformation equations

$$x' = x \cos \varphi + y \sin \varphi, \qquad (3.27)$$

$$y' = -x \sin \varphi + y \cos \varphi, \qquad (3.28)$$

$$z' = z. \qquad (3.29)$$

Let us add subscript z to φ to designate the axis about which the rotation is effected. A matrix form of (3.27), (3.28), (3.29) becomes

$$\begin{pmatrix} x' \\ y' \\ z' \end{pmatrix} = \begin{pmatrix} \cos \varphi_z & \sin \varphi_z & 0 \\ -\sin \varphi_z & \cos \varphi_z & 0 \\ 0 & 0 & 1 \end{pmatrix} \begin{pmatrix} x \\ y \\ z \end{pmatrix}. \qquad (3.30)$$

An equation for rotation by angle φ_x around the x axis is obtained on replacing x with y, y with z, and z with x (both unprimed and primed). We then rearrange the matrices to get

$$\begin{pmatrix} x' \\ y' \\ z' \end{pmatrix} = \begin{pmatrix} 1 & 0 & 0 \\ 0 & \cos \varphi_x & \sin \varphi_x \\ 0 & -\sin \varphi_x & \cos \varphi_x \end{pmatrix} \begin{pmatrix} x \\ y \\ z \end{pmatrix} . \qquad (3.31)$$

The corresponding equation for rotation by angle φ_y around the y axis is obtained on repeating the substitutions and rearranging the result:

$$\begin{pmatrix} x' \\ y' \\ z' \end{pmatrix} = \begin{pmatrix} \cos \varphi_y & 0 & -\sin \varphi_y \\ 0 & 1 & 0 \\ \sin \varphi_y & 0 & \cos \varphi_y \end{pmatrix} \begin{pmatrix} x \\ y \\ z \end{pmatrix} . \qquad (3.32)$$

When the rotation is by an infinitesimal $\delta\varphi$, we employ $\cos \delta\varphi = 1$ and $\sin \delta\varphi = \delta\varphi$. Then (3.31), (3.32), (3.30) reduce to

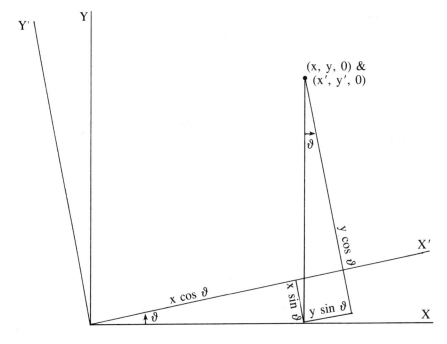

Figure 3.2. Coordinates before and after rotation of the coordinate axes by angle ϑ around the z axis.

$$\begin{pmatrix} x' \\ y' \\ z' \end{pmatrix} = \begin{pmatrix} 1 & 0 & 0 \\ 0 & 1 & \delta\varphi_x \\ 0 & -\delta\varphi_x & 1 \end{pmatrix} \begin{pmatrix} x \\ y \\ z \end{pmatrix} , \qquad (3.33)$$

$$\begin{pmatrix} x' \\ y' \\ z' \end{pmatrix} = \begin{pmatrix} 1 & 0 & -\delta\varphi_y \\ 0 & 1 & 0 \\ \delta\varphi_y & 0 & 1 \end{pmatrix} \begin{pmatrix} x \\ y \\ z \end{pmatrix} , \qquad (3.34)$$

$$\begin{pmatrix} x' \\ y' \\ z' \end{pmatrix} = \begin{pmatrix} 1 & \delta\varphi_z & 0 \\ -\delta\varphi_z & 1 & 0 \\ 0 & 0 & 1 \end{pmatrix} \begin{pmatrix} x \\ y \\ z \end{pmatrix} . \qquad (3.35)$$

In each of these equations, the 3×3 matrix effects the rotation. Successive rotations are described by combining the corresponding equations. The square matrix effecting the overall operation is obtained on multiplying the matrices for the constituent rotations in order.

These equations have the form

$$\mathbf{r}' = [\mathbf{U}(\delta\varphi)]\mathbf{r}. \qquad (3.36)$$

Introducing the definition of an infinitesimal operator $i\mathbf{J}$ through

$$\mathbf{U}(\delta\varphi) = \mathbf{1} + i\mathbf{J}\,\delta\varphi \qquad (3.37)$$

leads to the 3×3 matrices

$$\begin{pmatrix} 0 & 0 & 0 \\ 0 & 0 & 1 \\ 0 & -1 & 0 \end{pmatrix} = iJ_x, \qquad (3.38)$$

$$\begin{pmatrix} 0 & 0 & -1 \\ 0 & 0 & 0 \\ 1 & 0 & 0 \end{pmatrix} = iJ_y, \qquad (3.39)$$

$$\begin{pmatrix} 0 & 1 & 0 \\ -1 & 0 & 0 \\ 0 & 0 & 0 \end{pmatrix} = iJ_z. \qquad (3.40)$$

Here iJ_x, iJ_y, iJ_z have been introduced as labels for the matrices.

The group elements are generated from these using formulas (2.48) and (2.50). The exponentiation can be carried out on each infinitesimal generator and the results multiplied.

In applications, we may consider the primed coordinates axes to be fixed with respect to a possible localization of parts of the given physical system. Then the group elements would be associated with possible rotations of the system with respect to the unprimed reference frame.

3.5
Commutation Relations for Rotations

The proper rotations about a point of symmetry in 3-dimensional space make up a group labelled **SO**(3). All elements of this group can be represented by 3 × 3 matrices. From those for infinitesimal rotations about the Cartesian coordinate axes meeting at the singular point, a person can construct all possibilities.

How the rotations combine are governed by commutation relations among the infinitesimal operators. With (3.38), (3.39), (3.40), and the conventional matrix rules, we find that

$$-[iJ_x, iJ_y] = -(iJ_x)(iJ_y) + (iJ_y)(iJ_x) = iJ_z \tag{3.41}$$

or

$$[J_x, J_y] = iJ_z. \tag{3.42}$$

Similarly,

$$[J_y, J_z] = iJ_x, \tag{3.43}$$

$$[J_z, J_x] = iJ_y. \tag{3.44}$$

If we consider x as the 1st coordinate, y as the 2nd coordinate, and z as the 3rd coordinate, we may replace (3.42), (3.43), (3.44) by

$$[J_j, J_k] = \sum_l i\epsilon_{jkl}J_l, \tag{3.45}$$

with $\hbar = 1$. For a general set of units, we have

$$[J_j, J_k] = -\frac{\hbar}{i} \sum_l \epsilon_{jkl}J_l. \tag{3.46}$$

Note that the components of **J** satisfy the same commutation relations as the components of **L** in (3.20) and (3.24). Consequently, their algebras are similar.

Since **L** is the operator for the angular momentum due to motion through space, the orbital motion, **J** may also be interpreted as an operator for angular momentum. But since the rotations corresponding to the components of **J** involve the whole system, the pertinent propensity is the total angular momentum (orbital plus spin).

3.6
Commutators for Quantum Particles in 3-Dimensional Space

In Chapter 2 we introduced Galilean transformations over a single spatial coordinate and time. Adding two orthogonal spatial coordinates adds the corresponding distance translations and velocity translations. Furthermore, it allows three independent rotations to appear. These affect the propagation properties and intrinsic properties.

For 3-dimensional space and 1-dimensional time, a set of independent operators includes

iH	generating time translations,
iP_j ($j = 1, 2, 3$)	generating distance translations along three orthogonal axes,
iK_j ($j = 1, 2, 3$)	generating velocity translations along the three axes,
iJ_j ($j = 1, 2, 3$)	generating rotations around the three axes,
iI	generating phase translations.

How these combine must reflect the transformability of space-time and the localizability of the particle under consideration. In absence of interaction, the system is governed by the phase extension of the full Galilean group. In presence of interaction, we assume that the prevailing field acts locally, as in Section 2.12. We also consider the origin to be the center of symmetry when one is present.

Formulas (2.82), (2.80), (2.74), and (2.72) apply to each spatial dimension in turn. Furthermore, representations (3.15) and (3.16) may be employed in determining further commutation relations. And since the J_j's behave algebraicly as the L_j's, we may replace L_j by J_j in formulas (3.26) and (3.24). Thus, we obtain the relations

$$[H, P_j] = 0 \tag{3.47}$$

$$[H, J_j] = 0, \tag{3.48}$$

$$[K_j, H] = -iP_j, \tag{3.49}$$

$$[K_j, P_k] = -i\mu\delta_{jk} I, \tag{3.50}$$

$$[K_j, K_k] = 0 \tag{3.51}$$

$$[P_j, P_k] = 0, \tag{3.52}$$

$$[J_j, A_k] = \sum_l i\epsilon_{jkl} A_l, \text{ for } \mathbf{A} = \mathbf{J}, \mathbf{K}, \mathbf{P}. \tag{3.53}$$

which govern the transformation properties in the extended full Galilean group. Note that δ_{jk} is the Kronecker delta and ϵ_{jkl} the permutation symbol.

3.7
Hamiltonian Operators for Particles
in 3-Dimensional Space

In Section 2.12, a quantum mechanical basis for the 1-dimensional Hamiltonian operator was developed. The argument for any given particle in 3-dimensional space proceeds similarly. This can then be generalized to cover a system of N particles.

As before, we start with the fundamental postulate:

(XIII) Over any infinitesimal increment of time, a person cannot discriminate between a free and an interacting system.

As a consequence, each particle is subject to the same commutation relations as if it were free. And, the form for the Hamiltonian operator can be induced from the symmetry of the space-time continuum and the localizability of the particles.

Consider one of the particles in the field of the others in the system. Since the given particle can act at a point in space and time the localizability requirements must be satisfied. Thus, one obtains the generalization of (2.68), (2.88),

$$[X_j, \; \mu V_k] = i\delta_{jk}(1), \tag{3.54}$$

of (2.72),

$$\mathbf{X} = -\frac{1}{\mu} \mathbf{K}, \tag{3.55}$$

and of (2.77),

$$i[H, \mathbf{X}] = \mathbf{V}. \tag{3.56}$$

Here $\mathbf{X}$ is the operator for the radius vector locating the point of localization and $\mathbf{V}$ is the operator for this point's velocity.

Now, multiply (3.54) from the left by V_k and sum over k,

$$\sum_k (V_k X_j \mu V_k - \mu V_k V_k X_j) = iV_j; \tag{3.57}$$

then multiply (3.54) from the right by V_k and sum over k,

$$\sum_k (X_j \mu V_k V_k - \mu V_k X_j V_k) = iV_j. \tag{3.58}$$

Add these and divide by $2i$ to get

$$\frac{1}{i} \left[X_j, \; \frac{\mu}{2} \mathbf{V \cdot V} \right] = V_j, \tag{3.59}$$

whence

$$i[\tfrac{1}{2}\mu \mathbf{V}^2, \mathbf{X}] = \mathbf{V}. \tag{3.60}$$

Subtracting (3.60) from (3.56) now yields

$$i\left[H - \frac{1}{2\mu} (\mu\mathbf{V} \cdot \mu\mathbf{V}), \mathbf{X} \right] = 0. \tag{3.61}$$

But from (3.50) and (3.55), we know that

$$[P_j, X_j] \neq 0. \tag{3.62}$$

Therefore, the first expression in brackets in (3.61) cannot depend on $\mathbf{P}$ and we have

$$H - \frac{1}{2\mu} (\mu\mathbf{V} \cdot \mu\mathbf{V}) = U(\mathbf{X}, t). \tag{3.63}$$

As before, $\mathbf{P} - \mu\mathbf{V}$ is invariant in a pure velocity translation. So

$$\mathbf{P} - \mu\mathbf{V} = \mathbf{A}(\mathbf{X}, t) \tag{3.64}$$

and

$$\mu\mathbf{V} = \mathbf{P} - \mathbf{A}(\mathbf{X}, t). \tag{3.65}$$

Substituting into (3.63) leads to the statement:

(XIV) The operator for the energy of a particle in a definite state is the Hamiltonian:

$$H = \frac{1}{2\mu} [\mathbf{P} - \mathbf{A}(\mathbf{X}, t)] \cdot [\mathbf{P} - \mathbf{A}(\mathbf{X}, t)] + U(\mathbf{X}, t). \tag{3.66}$$

Note that (3.66) is valid in any consistent set of units; constant $\hbar$ need not equal 1 unit.

In dealing with coordinates and momenta, I acts as the identity operator. So combining (3.55), (3.50), and going to a general set of units leads to

$$[X_j, P_k] = i\hbar\delta_{jk}(1). \tag{3.67}$$

For the coordinate representation of the state ket, we would employ multiplication by coordinate x_j for operator X_j. Then to satisfy (3.67), we would have to represent P_j by operator $(\hbar/i)\partial/\partial x_j$. Thus, we have the axiom:

(XV) In the coordinate representation the basic operators are

$$X_j = x_j \tag{3.68}$$

and

$$P_j = \frac{\hbar}{i} \frac{\partial}{\partial x_j}. \tag{3.69}$$

These forms are employed in formulating the Hamiltonian for state function $\Psi(\mathbf{x}, t)$. Here x_j is the jth rectangular coordinate and t is the time for the localizability position of the particle under consideration, while μ is its mass.

Next, consider the behavior of all particles in an N-particle system. Let the mass of the ith particle be μ_i. And let the operator for the localizability position of the

ith particle be X_i, while the operator for the position's velocity is V_i. The argument from (3.54) to (3.60) proceeds as before for each particle in turn; we obtain

$$i[\tfrac{1}{2}\mu_i V_i^2, X_i] = V_i. \qquad (3.70)$$

One can construct a $3N$-dimensional space in which the localizability position of the ith particle is given by the Cartesian coordinates x_{3i-2}, x_{3i-1}, x_{3i}. The corresponding velocity components are v_{3i-2}, v_{3i-1}, v_{3i}. And for the operators, we would have

$$\sum X_i = X, \qquad \sum V_i = V. \qquad (3.71)$$

So summing (3.70) over all the particles in the system yields

$$i\left[\sum \frac{1}{2\mu_i} (\mu_i V_i) \cdot (\mu_i V_i), X \right] = V. \qquad (3.72)$$

We presume that (3.56) is valid in this space. Subtracting (3.72) from (3.56) then gives us

$$i\left[H - \sum \frac{1}{2\mu_i} (\mu_i V_i \cdot \mu_i V_i), X \right] = 0. \qquad (3.73)$$

Because (3.62) holds for each particle, we argue that the first expression in brackets cannot depend on P and

$$H - \sum \frac{1}{2\mu_i} (\mu_i V_i \cdot \mu_i V_i) = U(X_1, X_2, \ldots, X_n, t). \qquad (3.74)$$

Since a velocity translation of the ith particle leaves $P_i - \mu_i V_i$ invariant, we write

$$P_i - \mu_i V_i = A_i(X_i, t), \qquad (3.75)$$

whence

$$\mu_i V_i = P_i - A_i(X_i, t). \qquad (3.76)$$

As before, U is the operator for the scalar potential. And, operator A_i corresponds to the vector potential to which the ith particle is subjected. Consequently, A_i may depend on the propensities of the other particles also. Operator P_i represents the canonical momentum of the ith particle. So H has the form of the classical Hamiltonian function.

These results based on symmetry considerations are summarized in the statement:

(XVI) The operator for the energy of an N-particle system in a definite state is the Hamiltonian:

$$H = \sum \frac{1}{2\mu_i}[P_i - A_i(X_i, t)] \cdot [P_i - A_i(X_i, t)]$$

$$+ U(X_1, X_2, \ldots, X_N, t). \qquad (3.77)$$

When this is to act on a state function $\Psi(\mathbf{x}_1, \mathbf{x}_2, \ldots, \mathbf{x}_N, t)$, we also employ the axiom:

(XVII) In the coordinate representation the basic operators for an N particle system are

$$(X_i)_j = (x_i)_j \tag{3.78}$$

and

$$(P_i)_j = \frac{\hbar}{i} \frac{\partial}{\partial(x_i)_j} . \tag{3.79}$$

Following axiom (X), these must be introduced into (3.77) so that the Hamiltonian is Hermitian. Furthermore, the state function must be formulated so that the prevailing permutation symmetries are satisfied.

Example 3.4

Construct the Schrödinger equation for a particle of mass μ in a static scalar field. Employ a coordinate representation.

For the potentials, we are given that

$$\mathbf{A} = 0 \qquad \text{and} \qquad U = U(x, y, z).$$

With (3.69), we also have

$$\mathbf{P} \cdot \mathbf{P} = \Sigma \frac{\hbar}{i} \frac{\partial}{\partial x_j} \frac{\hbar}{i} \frac{\partial}{\partial x_j} = -\hbar^2 \Sigma \frac{\partial^2}{\partial x_j^2} = -\hbar^2 \nabla^2.$$

Substituting into (3.66) yields

$$H = -\frac{\hbar^2}{2\mu} \nabla^2 + U.$$

One can easily show that this operator is Hermitian. Now, a coordinate representation of the ket is $\psi(x, y, z)$. So (2.125) becomes

$$-\frac{\hbar^2}{2\mu} \nabla^2 \psi + U\psi = E\psi.$$

This rearranges to

$$\nabla^2 \psi + \frac{2\mu}{\hbar^2} (E - U)\psi = 0.$$

or

$$\nabla^2 \psi + k^2(x, y, z)\psi = 0$$

where

$$k^2 = \frac{2\mu}{\hbar^2} (E - U).$$

3.8
Commutativity among the Hamiltonian Operator and the Angular Momentum Operators

When the orientation of a system does not affect any external interactions to which the system is subjected, a rotation about any axis passing through its center of mass does not alter its Hamiltonian. Then the Hamiltonian commutes with the corresponding angular momentum operator. And to a given eigenvalue of the Hamiltonian correspond sets of eigenvalues for the angular momentum operators.

Let the Hamiltonian for the given system be H while an angular momentum operator for some motion about the center of mass is J_k. The infinitesimal operator is then iJ_k, while the corresponding rotation operator is

$$U_k = 1 + iJ_k \delta\varphi. \tag{3.80}$$

The Hamiltonian operator transforms following law (2.38); here

$$
\begin{aligned}
H' = U_k(\delta\varphi) H U_k^{-1}(\delta\varphi) &= (1 + iJ_k\delta\varphi) H (1 - iJ_k\delta\varphi) \\
&= H - i(HJ_k - J_kH)\delta\varphi + J_kHJ_k(\delta\varphi)^2 \\
&= H - i[H, J_k]\delta\varphi + 0(\delta\varphi)^2.
\end{aligned}
\tag{3.81}
$$

Since the rotation does not alter the Hamiltonian, the first order term must vanish and

$$[H, J_k] = 0, \tag{3.82}$$

as (3.48) states.

The commutator of the Hamiltonian with an angular momentum operator applied twice is

$$
\begin{aligned}
[H, J_k^2] &= HJ_kJ_k - J_kHJ_k + J_kHJ_k - J_kJ_kH \\
&= [H, J_k]J_k + J_k[H, J_k] = 0.
\end{aligned}
\tag{3.83}
$$

So if we construct

$$J^2 = J_x^2 + J_y^2 + J_z^2, \tag{3.84}$$

then

$$[H, J^2] = [H, J_x^2] + [H, J_y^2] + [H, J_z^2] = 0. \qquad (3.85)$$

Note that

$$[J_x^2, J_x] = (J_x J_x) J_x - J_x (J_x J_x) = 0. \qquad (3.86)$$

Also, with (3.42) we find that

$$
\begin{aligned}
[J_y^2, J_x] &= J_y J_y J_x - J_y J_x J_y + J_y J_x J_y - J_x J_y J_y \\
&= J_y [J_y, J_x] + [J_y, J_x] J_y = -iJ_y J_z - iJ_z J_y; \qquad (3.87)
\end{aligned}
$$

and with (3.44), we have

$$
\begin{aligned}
[J_z^2, J_x] &= J_z J_z J_x - J_z J_x J_z + J_z J_x J_z - J_x J_z J_z \\
&= J_z [J_z, J_x] + [J_z, J_x] J_z = iJ_z J_y + iJ_y J_z. \qquad (3.88)
\end{aligned}
$$

Adding (3.86), (3.87), (3.88) then yields

$$[J^2, J_x] = 0. \qquad (3.89)$$

Similarly,

$$[J^2, J_y] = 0 \qquad (3.90)$$

and

$$[J^2, J_z] = 0. \qquad (3.91)$$

Since J_z commutes with J^2 and H, and J^2 commutes with H, these operators have a common set of eigenkets. Similar statements hold for J_x, J^2, H and for J_y, J^2, H. But since J_x, J_y, and J_z do not commute, the three sets are different.

Example 3.5

Show that if A is a Hermitian operator, then any diagonal matrix element of its square, $(A^2)_{jj}$, is positive or zero.

Employ formula (1.57) to expand the diagonal element. Then introduce the Hermitian property of A and examine the resulting sum. Thus, we obtain

$$(A^2)_{jj} = \sum_k A_{jk} A_{kj} = \sum_k A_{kj}^* A_{kj} = \sum_k |A_{kj}|^2 \geqslant 0.$$

This result also applies to the sum of the squares of two Hermitian operators:

$$(A^2 + B^2)_{jj} \geqslant 0.$$

3.9
Eigenvalues for the Angular Momentum Operators

The commutation rules together with the Hermiticity requirement determine the possible eigenvalues for J^2 and for J_z.

Let a typical common eigenket of H, J^2, and J_z be labeled $|jm\rangle$. Also let the eigenvalues for J^2 and J_z be labeled η_j and m; thus

$$J^2|jm\rangle = \eta_j|jm\rangle , \tag{3.92}$$

$$J_z|jm\rangle = m|jm\rangle , \tag{3.93}$$

and

$$J_z{}^2|jm\rangle = mJ_z|jm\rangle = m^2|jm\rangle . \tag{3.94}$$

But from (3.84), we have

$$J^2 - J_z{}^2 = J_x{}^2 + J_y{}^2. \tag{3.95}$$

Subtracting (3.94) from (3.92) and introducing (3.95) leads to

$$(J_x{}^2 + J_y{}^2)|jm\rangle = (\eta_j - m^2)|jm\rangle , \tag{3.96}$$

whence

$$\langle jm|J_x{}^2 + J_y{}^2|jm\rangle = \eta_j - m^2, \tag{3.97}$$

as long as the kets are normalized to 1. Now, J_x and J_y are Hermitian. So from Example 3.5,

$$\eta_j - m^2 \geqslant 0. \tag{3.98}$$

We see that the possible values for m are bounded for each value of η_j.

Introduce the combination

$$J_\pm = J_x \pm iJ_y, \tag{3.99}$$

which with (3.89) and (3.90) yields

$$[J^2, J_\pm] = [J^2, J_x] \pm i[J^2, J_y] = 0. \tag{3.100}$$

Furthermore with (3.44), (3.43), it yields

$$[J_z, J_\pm] = [J_z, J_x] \pm i[J_z, J_y] = iJ_y \pm i(-iJ_x)$$
$$= \pm J_x + iJ_y = \pm(J_x \pm iJ_y) = \pm J_\pm. \tag{3.101}$$

From the definition of the commutator of J^2 and $J_\pm$, rearranged, we have

$$J^2 J_\pm = J_\pm J^2 + [J^2, J_\pm]. \tag{3.102}$$

Let both sides act on eigenket $|jm\rangle$ and bring in (3.92) and (3.100):

$$J^2 J_\pm | jm\rangle = J_\pm J^2 | jm\rangle + [J^2, J_\pm] | jm\rangle$$

$$= J_\pm \eta_j | jm\rangle + 0 = \eta_j J_\pm | jm\rangle . \qquad (3.103)$$

Thus, $J_\pm | jm\rangle$ is an eigenket of J^2 when this ket is not zero.

From the definition of the commutator of J_z and $J_\pm$ rearranged, we have

$$J_z J_\pm = J_\pm J_z + [J_z, J_\pm]. \qquad (3.104)$$

Let both sides act on the eigenket $|jm\rangle$ and introduce (3.93) and (3.101):

$$J_z J_\pm | jm\rangle = J_\pm J_z | jm\rangle + [J_z, J_\pm] | jm\rangle$$

$$= J_\pm m | jm\rangle \pm J_\pm | jm\rangle = (m \pm 1) J_\pm | jm\rangle . \qquad (3.105)$$

We see that $J_\pm | jm\rangle$ is an eigenket (unnormalized) of J_z with the eigenvalue $m \pm 1$, when this ket is not zero. Thus, $J_\pm$ acts to transform eigenket $| jm\rangle$ into a constant times eigenket $| j\, m \pm 1\rangle$ when this new eigenket exists. So J_+ is called a *step-up operator*, J_- a *step-down operator*, for angular momentum.

From (3.98), m has an upper limit which we label m_2, and a lower limit which we label m_1. At the upper limit,

$$J_+ | jm_2\rangle = 0, \qquad (3.106)$$

while at the lower limit,

$$J_- | jm_1\rangle = 0. \qquad (3.107)$$

Since

$$J_\mp J_\pm = (J_x \mp iJ_y)(J_x \pm iJ_y) = J_x^2 + J_y^2 \mp iJ_y J_x \pm iJ_x J_y$$

$$= J^2 - J_z^2 \mp J_z = J^2 - J_z(J_z \pm 1), \qquad (3.108)$$

J_- acting on (3.106) yields

$$J_- J_+ | jm_2\rangle = [J^2 - J_z(J_z + 1)] | jm_2\rangle$$

$$= [\eta_j - m_2(m_2 + 1)] | jm_2\rangle = 0, \qquad (3.109)$$

while J_+ acting on (3.107) yields

$$J_+ J_- | jm_1\rangle = [J^2 - J_z(J_z - 1)] | jm_1\rangle$$

$$= [\eta_j - m_1(m_1 - 1)] | jm_1\rangle = 0. \qquad (3.110)$$

Since the kets in (3.109) and (3.110) are not zero, we have

$$\eta_j - m_2(m_2 + 1) = 0, \qquad (3.111)$$

$$\eta_j - m_1(m_1 - 1) = 0. \qquad (3.112)$$

whence

$$m_2(m_2 + 1) = m_1(m_1 - 1). \tag{3.113}$$

Since $m_1 \leqslant m_2$, the solution of (3.113) is

$$m_1 = -m_2. \tag{3.114}$$

Designating m_2 as j leads to

$$\eta_j = j(j + 1) \tag{3.115}$$

and

$$m = -j, -j + 1, \ldots, j - 1, j. \tag{3.116}$$

Note that there are $2j + 1$ values of m for each j.

In a general set of units, Equations (3.92) and (3.93) are replaced by

$$J^2 \,|\, jm \rangle = \eta_j \hbar^2 \,|\, jm \rangle \tag{3.117}$$

and

$$J_z \,|\, jm \rangle = m\hbar \,|\, jm \rangle . \tag{3.118}$$

Nevertheless, parameter $\hbar$ enters the commutation relations so that one still ends with formulas (3.115) and (3.116).

Example 3.6

Construct the operators L_x, L_y, L_z, for revolution of the model particle, in spherical coordinates.

From Example 3.2, we have

$$L_x = -i\hbar \left(y \frac{\partial}{\partial z} - z \frac{\partial}{\partial y} \right),$$

$$L_y = -i\hbar \left(z \frac{\partial}{\partial x} - x \frac{\partial}{\partial z} \right).$$

But from Figure 3.1,

$$x = r \sin \vartheta \cos \varphi, \qquad\qquad \vartheta = \cos^{-1} \frac{z}{r},$$

$$y = r \sin \vartheta \sin \varphi,$$

$$z = r \cos \vartheta, \qquad\qquad\quad \varphi = \tan^{-1} \frac{y}{x}.$$

Also for the motion,

$$r = \text{const.}$$

So we obtain

$$\frac{\partial}{\partial x} = \frac{\partial \vartheta}{\partial x} \frac{\partial}{\partial \vartheta} + \frac{\partial \varphi}{\partial x} \frac{\partial}{\partial \varphi} = -\frac{\sin \varphi}{r \sin \vartheta} \frac{\partial}{\partial \varphi} ,$$

$$\frac{\partial}{\partial y} = \frac{\partial \vartheta}{\partial y} \frac{\partial}{\partial \vartheta} + \frac{\partial \varphi}{\partial y} \frac{\partial}{\partial \varphi} = \frac{\cos \varphi}{r \sin \vartheta} \frac{\partial}{\partial \varphi} ,$$

$$\frac{\partial}{\partial z} = \frac{\partial \vartheta}{\partial z} \frac{\partial}{\partial \vartheta} + \frac{\partial \varphi}{\partial z} \frac{\partial}{\partial \varphi} = -\frac{1}{r \sin \vartheta} \frac{\partial}{\partial \vartheta} .$$

Substituting into the Cartesian forms for L_x and L_y yields

$$L_x = i\hbar \left(\sin \varphi \frac{\partial}{\partial \vartheta} + \cot \vartheta \cos \varphi \frac{\partial}{\partial \varphi} \right) ,$$

$$L_y = i\hbar \left(-\cos \varphi \frac{\partial}{\partial \vartheta} + \cot \vartheta \sin \varphi \frac{\partial}{\partial \varphi} \right) .$$

And from (3.9), we have

$$L_z = -i\hbar \frac{\partial}{\partial \varphi} .$$

Example 3.7

Construct the operator for L^2 in spherical coordinates.
By definition, we have

$$L^2 = L_x^2 + L_y^2 + L_z^2.$$

Substitute in the results from Example 3.6 and reduce to get

$$L^2 = -\hbar^2 \left(\frac{\partial^2}{\partial \vartheta^2} + \cot \vartheta \frac{\partial}{\partial \vartheta} + \csc^2 \vartheta \frac{\partial^2}{\partial \varphi^2} \right)$$

$$= -\hbar^2 \left[\frac{1}{\sin \vartheta} \frac{\partial}{\partial \vartheta} \left(\sin \vartheta \frac{\partial}{\partial \vartheta} \right) + \frac{1}{\sin^2 \vartheta} \frac{\partial^2}{\partial \varphi^2} \right]$$

3.10
Kinds of Angular Momenta

From the way its components were constructed in Section 3.2 and Examples 3.1, 3.2, operator **L** represents the angular momentum associated with orbital motion. For each eigenvalue set, a well behaved single valued eigenfunction of the coordinates exists. Operation **J**, on the other hand, involves the entire system. And, each particle may have an intrinsic contribution (said to be due to its spin) as well as an extrinsic contribution (said to be due to its orbital motion).

In Equations (3.20) and (3.46) we see that the components of **L** exhibit the same commutation properties as the components of **J**. So the arguments of Sections 3.8 and 3.9 apply to both **L** and **J**; successive projections of either on a chosen axis differ by $\hbar$. Because of symmetry the maximum and the minimum on this axis merely differ in sign. For **L**, the extremes are $\pm L\hbar$; for **J**, these are $\pm J\hbar$. Since the state function for orbital motion must be single valued, L is integral. Since there is no similar limitation on **J**, J may be half integral.

Since **J** and **L** are governed by similar commutation relations, the components of the difference operator

$$\mathbf{S} = \mathbf{J} - \mathbf{L} \qquad (3.119)$$

commute with all the infinitesimal generators of the group. As a consequence, the operator

$$\mathbf{S}^2 = (\mathbf{J} - \mathbf{L})^2 = (\mathbf{J} - \mathbf{X} \times \mathbf{P})^2 \qquad (3.120)$$

is an invariant of the Lie algebra. A person may also choose one of the components of **S** as an invariant.

Representing an angular momentum, spin **S** has components that obey the commutation rule

$$[S_j, S_k] = -\frac{\hbar}{i} \sum_l \epsilon_{jkl} S_l. \qquad (3.121)$$

The spin property derives from the simultaneous existence of translational and rotational degrees of freedom and the localizability of the constituent particles. It is not dependent on the validity of Einstein relativity.

The projections of **S** on a chosen axis must differ by $\hbar$ since those of **L** and **J** do. By symmetry, the maximum and the minimum along this axis are equal in magnitude; we label them $\pm S\hbar$. As with J, number S may be half integral. By experiment, one finds that S for a fundamental particle, such as an electron, a proton, a neutron, equals ½.

3.11
Angular Momentum Matrix Elements

From the results in Section 3.9, the elements of matrices corresponding to the eigenkets common to J^2 and a component of J can be constructed. These will be particularly useful in further studies of spin.

Let us choose a complete set of orthonormal eigenkets common to J^2 and J_z. From (3.92), (3.115), and (3.93), we then have

$$\langle j' \, m' \mid J^2 \mid j \, m \rangle = j(j + 1) \delta_{j'j} \delta_{m'm} \qquad (3.122)$$

and

$$\langle j' \, m' \mid J_z \mid j \, m \rangle = m \, \delta_{j'j} \delta_{m'm}. \qquad (3.123)$$

Thus, the eigenkets diagonalize the matrices for J^2 and J_z. But they do not diagonalize the matrices for J_x and J_y.

However, the step operator links neighboring eigenkets in the set. Let us first look at the normalization integral for $J_\pm \mid j \, m \rangle$. We use the fact that in

$$J_\pm = J_x \pm iJ_y, \qquad (3.124)$$

both J_x and J_y are Hermitian. So we have

$$\mid \Gamma_\pm \mid^2 = \langle J_\pm \mid j \, m \mid \mid J_\pm \mid j \, m \rangle = \langle j \, m \mid J_\mp J_\pm \mid j \, m \rangle \qquad (3.125)$$

where $\Gamma_\pm$ is the normalization constant for the shifted ket.

Next, introduce (3.108) and (3.115) to obtain

$$\begin{aligned} J_\mp J_\pm \mid j \, m \rangle &= [J^2 - J_z(J_z \pm 1)] \mid j \, m \rangle \\ &= [j(j + 1) - m(m \pm 1)] \mid j \, m \rangle \\ &= (j \mp m)(j \pm m + 1) \mid j \, m \rangle . \end{aligned} \qquad (3.126)$$

Combining (3.125) and (3.126) leads to

$$\mid \Gamma_\pm \mid^2 = (j \mp m)(j \pm m + 1). \qquad (3.127)$$

Now, we choose the phase of the shifted ket by taking $\Gamma_\pm$ real:

$$\Gamma_\pm = [(j \mp m)(j \pm m + 1)]^{1/2}. \qquad (3.128)$$

Then

$$J_\pm \mid j \, m \rangle = \Gamma_\pm \mid j \, m \pm 1 \rangle \qquad (3.129)$$

and

$$\langle j \, m \pm 1 \mid \mid J_\pm \mid j \, m \rangle = [(j \mp m)(j \pm m + 1)]^{1/2} \qquad (3.130)$$

Since the different eigenkets are orthogonal, the general expression is

$$\langle j' \; m' \,||\, J_\pm \,|\, j \; m \rangle = \Gamma_\pm \delta_{j'j} \, \delta_{m'm \pm 1}. \tag{3.131}$$

3.12
Spin Matrices

When j equals ½, the J of Section 3.11 reduces to S and the angular momentum matrices become those for a fundamental particle with no orbital angular momentum. These we will consider here.

From Equation (3.123), we obtain

$$\langle \tfrac{1}{2} \; \tfrac{1}{2} \,|\, S_z \,|\, \tfrac{1}{2} \; \tfrac{1}{2} \rangle = \tfrac{1}{2}, \tag{3.132}$$

$$\langle \tfrac{1}{2} \; -\tfrac{1}{2} \,|\, S_z \,|\, \tfrac{1}{2} \; -\tfrac{1}{2} \rangle = -\tfrac{1}{2}, \tag{3.133}$$

$$\langle \tfrac{1}{2} \; \pm\tfrac{1}{2} \,|\, S_z \,|\, \tfrac{1}{2} \; \mp\tfrac{1}{2} \rangle = 0. \tag{3.134}$$

The corresponding complete matrix is

$$\mathbf{S}_z = \begin{pmatrix} \tfrac{1}{2} & 0 \\ 0 & -\tfrac{1}{2} \end{pmatrix} . \tag{3.135}$$

From (3.131), we also have

$$\langle \tfrac{1}{2} \; \tfrac{1}{2} \,||\, S_+ \,|\, \tfrac{1}{2} \; -\tfrac{1}{2} \rangle = 1, \tag{3.136}$$

$$\langle \tfrac{1}{2} \; -\tfrac{1}{2} \,||\, S_- \,|\, \tfrac{1}{2} \; \tfrac{1}{2} \rangle = 1, \tag{3.137}$$

with all other elements zero. The complete matrices are

$$\mathbf{S}_x + i\mathbf{S}_y = \mathbf{S}_+ = \begin{pmatrix} 0 & 1 \\ 0 & 0 \end{pmatrix} . \tag{3.138}$$

and

$$\mathbf{S}_x - i\mathbf{S}_y = \mathbf{S}_- = \begin{pmatrix} 0 & 0 \\ 1 & 0 \end{pmatrix} . \tag{3.139}$$

Add (3.139) to (3.138) and divide by 2 to get

$$\mathbf{S}_x = \begin{pmatrix} 0 & \tfrac{1}{2} \\ \tfrac{1}{2} & 0 \end{pmatrix} . \tag{3.140}$$

Subtract (3.139) from (3.138) and divide by $2i$ to get

$$\mathbf{S}_y = \begin{pmatrix} 0 & -\tfrac{1}{2}i \\ \tfrac{1}{2}i & 0 \end{pmatrix}. \tag{3.141}$$

These spin matrices are normalized on multiplying by 2. The results are the *Pauli spin matrices*

$$\sigma_x = \begin{pmatrix} 0 & 1 \\ 1 & 0 \end{pmatrix}, \tag{3.142}$$

$$\sigma_y = \begin{pmatrix} 0 & -i \\ i & 0 \end{pmatrix}, \tag{3.143}$$

$$\sigma_z = \begin{pmatrix} 1 & 0 \\ 0 & -1 \end{pmatrix}, \tag{3.144}$$

for which

$$\sigma_k\sigma_l + \sigma_l\sigma_k = 2\delta_{kl} \tag{3.145}$$

and

$$[\sigma_k,\ \sigma_l] = \sum_m 2i\,\epsilon_{klm}\,\sigma_m. \tag{3.146}$$

One can also deduce from (3.142), (3.143), (3.144) the relation

$$\sigma_k\sigma_l = \delta_{kl} + \sum_m i\,\epsilon_{klm}\sigma_m. \tag{3.147}$$

So if operators $\mathbf{A}$ and $\mathbf{B}$ commute with σ, we have

$$\sigma{\cdot}\mathbf{A}\ \sigma{\cdot}\mathbf{B} = \sum\sum \sigma_k\sigma_l A_k B_l = \sum\sum (\delta_{kl} + \sum i\,\epsilon_{klm}\sigma_m)\,A_k B_l$$
$$= \mathbf{A}{\cdot}\mathbf{B} + i\,\sigma\,\dot{}\ \mathbf{A}{\times}\mathbf{B}. \tag{3.148}$$

Example 3.8

Reduce the operator

$$\sigma{\cdot}\mathbf{P}\ \sigma{\cdot}\mathbf{P}.$$

The momentum operator **P** commutes with σ; so (3.148) applies. We have

$$\sigma \cdot \mathbf{P} \, \sigma \cdot \mathbf{P} = \mathbf{P} \cdot \mathbf{P} + i\,\sigma \cdot \mathbf{P} \times \mathbf{P} = P^2.$$

3.13
A Spin Hamiltonian

Considerations of symmetry and simplicity led us to Hamiltonians (3.66) and (3.77). But as they stand, these forms do not properly allow for spin. So we have to introduce an operator for it in a way that does not destroy the symmetry properties of H.

This operator must reflect the properties of spin. Yet it must combine with the momentum in such a way that the Hamiltonian of a free particle is unaffected.

The result in Example 3.8 suggests how this may be done. One may replace

$$H = \frac{1}{2\mu} \, \mathbf{P} \cdot \mathbf{P} \qquad (3.149)$$

with

$$H = \frac{1}{2\mu} \, \sigma \cdot \mathbf{P} \, \sigma \cdot \mathbf{P} \qquad (3.150)$$

since the two forms are equivalent. The corresponding change of (3.66) involves adopting the postulate:

(XVIII) The operator for the energy of a spin ½ particle carrying charge q is the Hamiltonian

$$H = \frac{1}{2\mu} \, \sigma \cdot (\mathbf{P} - q\mathbf{A}) \, \sigma \cdot (\mathbf{P} - q\mathbf{A}) + q\phi. \qquad (3.151)$$

Here **A** is the conventional vector potential and ϕ the conventional scalar potential of electromagnetism. This form does allow for the interaction of the spin of the particle with a magnetic field.

Indeed, multiplying (3.151) out and rearranging leads to

$$H - q\phi =$$
$$P^2/2\mu - q\mathbf{A} \cdot \mathbf{P}/\mu + iq\hbar\nabla \cdot \mathbf{A}/2\mu - q\hbar\,\sigma \cdot (\nabla \times \mathbf{A})/2\mu + q^2\mathbf{A}^2/2\mu. \qquad (3.152)$$

In the reduction, we have employed

$$i\mathbf{P} = \hbar\nabla \qquad (3.153)$$

and

$$\nabla \times \mathbf{A} = (\nabla \times \mathbf{A}) - \mathbf{A} \times \nabla. \qquad (3.154)$$

The last term in (3.152) can be neglected in weak fields. The next to last term is

$$-q\hbar\ \sigma \cdot \mathbf{B}/2\mu = -\mu_s \cdot \mathbf{B}. \qquad (3.155)$$

The third from last term can be rewritten as

$$-q\mathbf{L} \cdot \mathbf{B}/2\mu = -\mu_l \cdot \mathbf{B}. \qquad (3.156)$$

Here μ_s is the intrinsic magnetic moment of the particle while μ_l is the magnetic moment caused by its orbital angular momentum $\mathbf{L}$. The second from last term vanishes in a uniform field.

3.14
Main Features

In the physical continuum, translations in position, time, velocity, and phase may occur. These would in general involve three orthogonal reference axes and three linearly independent spatial coordinates. As a result, the operators X, K, and P of Chapter 2 must be replaced by X_j, K_j, and P_j, with $j = 1$, 2, or 3. Commutators between components with the same j would appear as before; those between components with differing j would vanish.

For a one-particle system, we thus obtain the commutation relations

$$[H,\ P_j] = 0, \qquad (3.157)$$

$$[K_j,\ H] = -iP_j, \qquad (3.158)$$

$$[K_j,\ P_k] = -i\mu\delta_{jk}I, \qquad (3.159)$$

$$[K_j,\ K_k] = 0, \qquad (3.160)$$

$$[P_j,\ P_k] = 0. \qquad (3.161)$$

Because a person cannot discriminate between a free and an interacting particle over a small interval of time, we presume that these commutation relations also apply to the particle under interaction with a field. Then with

$$\mathbf{K} = -\mu\mathbf{X}, \qquad (3.162)$$

we get the 3-dimensional generalization of (2.164), a 3-dimensional Hamiltonian operator.

For an N-particle system there are $3N$ localization positions to be denoted. These may be plotted in a $3N$-dimensional Euclidean space and the argument repeated. We then obtain the Hamiltonian operator

$$H = \sum_{i=1}^{N} \frac{1}{2\mu_i}\ [\mathbf{P}_i - \mathbf{A}_i(\mathbf{X}_i,\ t)] \cdot (\mathbf{P}_i - \mathbf{A}_i(\mathbf{X}_i,\ t)]$$

$$+ U(\mathbf{X}_1, \mathbf{X}_2, \ldots, \mathbf{X}_N, t). \tag{3.163}$$

In a multidimensional space, rotations as well as translations can occur. Associated with each rotation is an antisymmetric infinitesimal operator. For physical space, one obtains the matrix representations:

$$\begin{pmatrix} 0 & 0 & 0 \\ 0 & 0 & 1 \\ 0 & -1 & 0 \end{pmatrix} = iJ_x, \text{ for rotation around the } x \text{ axis,} \tag{3.164}$$

$$\begin{pmatrix} 0 & 0 & -1 \\ 0 & 0 & 0 \\ 1 & 0 & 0 \end{pmatrix} = iJ_y, \text{ for rotation around the } y \text{ axis,} \tag{3.165}$$

$$\begin{pmatrix} 0 & 1 & 0 \\ -1 & 0 & 0 \\ 0 & 0 & 0 \end{pmatrix} = iJ_z, \text{ for rotation around the } z \text{ axis.} \tag{3.166}$$

These lead to the commutation rules

$$[J_x, J_y] = iJ_z, \tag{3.167}$$

$$[J_y, J_z] = iJ_x, \tag{3.168}$$

$$[J_z, J_x] = iJ_y, \tag{3.169}$$

which in index notation become

$$[J_j, J_k] = \sum_l i\,\epsilon_{jkl}J_l. \tag{3.170}$$

Similarly, we have

$$[J_j, K_k] = \sum_l i\,\epsilon_{jkl}K_l, \tag{3.171}$$

$$[J_j, P_k] = \sum_l i\,\epsilon_{jkl}P_l. \tag{3.172}$$

Also,

$$[H, J_k] = 0, \tag{3.173}$$

$$[H, J_k^2] = 0, \tag{3.174}$$

$$[J^2, J_k] = 0. \tag{3.175}$$

From (3.174), one obtains

$$[H, J^2] = 0. \tag{3.176}$$

Since J_z commutes with J^2 and H, while J^2 commutes with H, these operators have a common set of eigenkets. The pertinent eigenvalue equations are

$$J_z \,|\, j\, m\rangle = m\hbar \,|\, j\, m\rangle \, , \tag{3.177}$$

$$J^2 \,|\, j\, m\rangle = j(j + 1)\hbar^2 \,|\, j\, m\rangle \, , \tag{3.178}$$

$$H \,|\, j\, m\rangle = E_{jm} \,|\, j\, m\rangle \, , \tag{3.179}$$

with

$$m = -j, -j + 1, \ldots, j - 1, j. \tag{3.180}$$

Now, a submicroscopic particle is localizable to a point. When free, it is then subject to transformations about this point that are governed by (3.164), (3.165), (3.166), and by (3.177), (3.178), (3.179). It is observed to exist in a small number of angular momentum states called spin states. The corresponding operator is labeled **S**, with components S_x, S_y, S_z.

A submicroscopic particle may also travel in a closed orbital. When the field is spherically symmetric, transformations about the center of symmetry, without regard for spin, are governed by (3.164), (3.165), (3.166), and by (3.177), (3.178), (3.179). Furthermore, the particle is observed to exist in a limited number of angular momentum orbital states. The corresponding operator is labeled **L**, with components L_x, L_y, L_z.

The transformation equations may also govern the overall behavior. Here the operator is **J**. We consider that

$$\mathbf{J} = \mathbf{L} + \mathbf{S}, \tag{3.181}$$

since angular momenta add vectorially. The components of **L** and the components of **S** obey the same commutation rules as those of **J**.

Discussion Questions

3.1. Why are classical observables called properties? Why are quantum observables called propensities?

3.2. How are the kinematic possibilities affected on going from one to three spatial dimensions? How are the propensities of a point affected?

3.3. How can a rotation be modelled as a composite translation?

3.4. Explain the localizability requirement that influences this model rotation.

3.5. How does this localizability condition lead to the formulation

$$L_z = \frac{1}{i} \frac{\partial}{\partial \varphi}$$

when $\hbar = 1$?

3.6. Explain how the operator in Question 3.5 appears in Cartesian coordinates.

3.7. How are operators L_x and L_y obtained from operator L_z?

3.8. How are the standard angular momentum operators obtained from the form

$$\mathbf{L} = \mathbf{X} \times \mathbf{P}?$$

3.9. Explain the construction of the formula

$$[L_j, L_k] = -\frac{\hbar}{i} \sum_l \epsilon_{jkl} L_l.$$

3.10. How is the formula

$$[L_j, X_k] = -\frac{\hbar}{i} \sum_l \epsilon_{jkl} X_l$$

justified?

3.11. Explain

$$[L_j, P_k] = -\frac{\hbar}{i} \sum_l \epsilon_{jkl} P_l,$$

$$[H, L_j] = 0.$$

3.12. How is the relationship

$$[L_j, K_k] = -\frac{\hbar}{i} \sum_l \epsilon_{jkl} K_l$$

obtained from the above ones?

3.13. Formulate the key 3×3 matrices representing rotations in 3-dimensional space. From these obtain the basic infinitesimal generators.

3.14. How are the operations of the **SO**(3) group obtained from the infinitesimal generators?

3.15. What principles are used to obtain

$$[J_x, J_y] = iJ_z?$$

3.16. How is this equation rewritten for a general set of units?

3.17. Why can **J** be interpreted as an operator for angular momentum?

3.18. What operators do we need for describing the pertinent transformations of the physical continuum and the localizability of a particle?

3.19. Explain what determine the conditions on commutators of these operators.

3.20. How can the symmetry of the space-time continuum govern the motion of an interacting particle?

3.21. How do the standard commutation relations yield the form for the Hamiltonian operator?

3.22. How are the basic operators for an N-particle system chosen when the ket is represented by a function of the coordinates and time?

3.23. What condition must the Hamiltonian operator meet?

3.24. Show that the operator

$$H = -\frac{\hbar^2}{2\mu} \nabla^2 + U$$

is Hermitian.

3.25. Can k in Example 3.4 be interpreted as a local wavevector?

3.26. Under what conditions is the Hamiltonian invariant to rotations?

3.27. Explain why a system with a definite total angular momentum may exhibit a definite component of angular momentum around only one axis.

3.28. Why do we expect the lower limit on m to be the negative of the upper limit? Why do we expect η_j to be greater than the square of this upper limit?

3.29. Is operator $J_\pm$ Hermitian? Explain.

3.30. Why do we call J_+ a step-up operator and J_- a step-down operator?

3.31. Why may a particle possess an intrinsic angular momentum (spin)?

3.32. How is operator **S** defined?

3.33. On what choice of eigenkets are the conventional angular momentum matrices based?

3.34. What forms do these matrices assume when $j = \frac{1}{2}$?

3.35. How are these spin matrices normalized?

3.36. What criterion guides how spin is introduced into the Hamiltonian?

3.37. How is the operator equation

$$\nabla \times \mathbf{A} = (\nabla \times \mathbf{A}) - \mathbf{A} \times \nabla$$

justified?

Problems

3.1. Use the differential operator representation to establish that

$$[L_1, L_2] = -\frac{\hbar}{i} L_3.$$

3.2. Use the coordinate-time representation to show that

$$[L_1, X_2] = -\frac{\hbar}{i} X_3.$$

3.3. Employ the matrix representation to derive

$$[J_x, J_y] = iJ_z.$$

3.4. In an infinitesimal rotation of a body by angle $\delta\varphi$ about the z axis, the operator for coordinate z of a point on the body is unchanged. Show that

$$[J_z, Z] = 0.$$

3.5. In an infinitesimal rotation of a body by angle $\delta\varphi$ about the z axis, the operator for coordinate x on the body satisfies

$$X' = X - Y\delta\varphi.$$

Show that

$$[J_z, X] = iY.$$

3.6. By direct computation, derive the result in Example 3.7, for L^2, from the results in Example 3.6.

3.7. Show that

$$\sigma_x^2 = 1, \qquad \sigma_y^2 = 1, \qquad \sigma_z^2 = 1.$$

3.8. From (3.147), derive (3.145).
3.9. From (3.147), derive (3.146).

— — —

3.10. Use the coordinate representation to establish that

$$[L_x, Y] = -\frac{\hbar}{i} Z.$$

3.11. Employ the same method to derive

$$[L_1, P_2] = -\frac{\hbar}{i} P_3.$$

3.12. Use the coordinate-time representation to show that

$$[H, L_j] = 0.$$

3.13. With the established commutation relations, evaluate

$$\mathbf{J} \times \mathbf{J}.$$

3.14. Simplify

$$[J_j^2, X_j].$$

3.15. Evaluate

$$J_x J_y Z + Z J_y J_x - J_x Z J_y - J_y Z J_x.$$

3.16. Construct the operator for $L_\pm$ in spherical coordinates.
3.17. Justify the equations

$$\sigma_x \sigma_y = i \sigma_z, \qquad \sigma_y \sigma_z = i \sigma_x, \qquad \sigma_z \sigma_x = i \sigma_y.$$

3.18. Show that

$$\sigma_y \sigma_x = -i \sigma_z, \qquad \sigma_z \sigma_y = -i \sigma_x, \qquad \sigma_x \sigma_z = -i \sigma_y.$$

3.19. Show how (3.152) follows from (3.151).

References

Books

Edmonds, A. R.: 1957, *Angular Momentum in Quantum Mechanics*, Princeton University Press, Princeton, N.J., pp. 3-30. This book contains an extensive treatment of angular momentum based on the position-momentum commutation relations.
Kaempffer, F. A.: 1965, *Concepts in Quantum Mechanics*, Academic Press, New York, pp. 77-86. In the cited sections, an analysis of rotational transformations and invariance properties of the Hamiltonian appears. No group theory is employed.
Lipkin, H. J.: 1966, *Lie Groups for Pedestrians*, 2nd edn., North-Holland, Amsterdam, pp. 1-68. This thin book reviews the application of angular momentum algebra to elementary particle theory.

Rose, M. E.: 1957, *Elementary Theory of Angular Momentum*, Wiley, New York, pp. 3-75. We have here a detailed development of angular momentum theory based on transformation properties.

Tinkham, M.: 1964, *Group Theory and Quantum Mechanics*, McGraw-Hill, New York, pp. 94-106. A useful discussion of rotational transformation properties is presented.

Articles

Garcia Alvarez, E. T., and Gonzalez, A. D.: 1989. "Some General Commutation Rules and Identities for Three-Dimensional Quantum-Mechanical Operators." *Am. J. Phys.* **57**, 923-925.

Halprin, A.: 1978. "Pedagogy of Spin in Nonrelativistic Quantum Mechanics." *Am. J. Phys.* **46**, 768.

Kobe, D. H.: 1986. "Active and Passive Views of Gauge Invariance in Quantum Mechanics." *Am. J. Phys.* **54**, 77-80.

Levy-Leblond, J.-M.: 1963. "Galilei Group and Nonrelativistic Quantum Mechanics." *J. Math. Phys.* **4**, 776-788.

Levy-Leblond, J.-M.: 1974. "The Pedagogical Role and Epistemological Significance of Group Theory in Quantum Mechanics." *Riv. Nuovo Cimento* **4**, 99-143.

Swenson, R. J.: 1989. "The Correct Relation between Wavefunctions in Two Gauges." *Am. J. Phys.* **57**, 381-382.

Williams, B., and Robinson, T.: 1981. "Pedagogy of Spin in Nonrelativistic Quantum Mechanics." *Am. J. Phys.* **49**, 645-647.

4

KEY EIGENFUNCTIONS
AND KETS

4.1
The Nature of Eigenfunctions and Kets

From Section 1.6, each possible pure state of a quantum mechanical system is governed by an eigenvalue equation of the form

$$A \mid j \rangle = a_j \mid j \rangle. \tag{4.1}$$

Here A is the appropriate operator while $\mid j \rangle$ represents the state. Composite states are represented by superpositions

$$\mid g \rangle = \sum_j c_j \mid j \rangle, \tag{4.2}$$

while incoherent states are described by the density operator

$$\rho = \sum_j \mid j \rangle w_j \langle j \mid. \tag{4.3}$$

Measurements of a particular property, carried out on equivalent systems, generally yield a spectrum of values. The limiting average of these is called the expectation value. For a coherent state the theoretical expectation value is

$$\langle A \rangle = \langle g \mid A \mid g \rangle, \tag{4.4}$$

while for an incoherent state, it is

$$\langle A \rangle = \sum_k (A\rho)_{kk}. \tag{4.5}$$

Following axiom (I′), each function representing a ket, $|j\rangle$ or $|g\rangle$, must be a single-valued analytic expression satisfying the boundary conditions for the given system. From axiom (V′), a set of suitable functions for the $|j\rangle$'s make up an orthonormal set. Only for certain values of the a_j's do such functions exist. These are the allowed values, the eigenvalues.

In calculating a possible energy E_j for a given system, one employs the appropriate Hamiltonian operator H in the eigenvalue equation

$$H|j\rangle = E_j|j\rangle. \tag{4.6}$$

Spatial coordinates fitting the system are chosen and a representation of the eigenket is sought.

In general, the eigenfunctions are much more complicated than the potentials controlling them. Only for a few very simple potentials can the eigenfunctions be expressed in familiar closed form. These include a constant potential, a parabolic potential, a one-center Coulomb potential. Motions considered include free translation, confined translation, free rotation, harmonic vibration, and electron motion in the hydrogen atom.

For more complicated potentials one resorts to approximation methods. Nevertheless, the results developed here will in some cases serve as first approximations, in other cases as guides to how first approximations may be constructed.

4.2
Translational Eigenfunctions

The simplest kind of motion consists of a directed motion of particles in a constant potential field. Various such motions may be superposed to give the general motion within the region.

For such a field one may set the potentials equal to zero:

$$U = 0 \quad \text{and} \quad \mathbf{A} = 0. \tag{4.7}$$

With (3.68) and (3.69), expression (3.66) now reduces to

$$H = -\frac{\hbar^2}{2\mu} \sum \frac{\partial^2}{\partial x_j^2} = -\frac{\hbar^2}{2\mu} \nabla^2 \tag{4.8}$$

where μ is the particle mass. The corresponding eigenvalue equation is

$$-\frac{\hbar^2}{2\mu} \nabla^2 \Psi = E \Psi. \qquad (4.9)$$

A possible solution has the form

$$\Psi = Ne^{i(\mathbf{k}\cdot\mathbf{r} - \omega t)} \qquad (4.10)$$

with

$$\mathbf{k} \cdot \mathbf{r} = k_x x + k_y y + k_z z. \qquad (4.11)$$

This may be factored as

$$\Psi = X(x)Y(y)Z(z)e^{-i\omega t}. \qquad (4.12)$$

Indeed, substituting (4.10) into (4.9) yields

$$\frac{\hbar^2 k^2}{2\mu} = E. \qquad (4.13)$$

Since $\hbar k$ equals p, this merely states that the kinetic energy of a particle equals its total energy when (4.7) holds.

Solution (4.10) is appropriate when its symmetry fits that of the boundary conditions. Otherwise a person may construct superpositions that fit.

When the system exhibits rectangular symmetry, we place the x, y, z axes along the edges of a reference rectangular cell. This cell may be either the region in which the system is confined or a unit cell of a lattice that is independently periodic in the x, y, and z directions.

For the periodic conditions, formula (4.10) yields the factors in the first three lines of Table 4.1.

For movements between perfectly reflecting walls perpendicular to the x axis, one needs to superpose equivalent amounts of

$$X_1 = Ae^{i |k_x| x} \qquad (4.14)$$

and

$$X_2 = Ae^{-i |k_x| x} \qquad (4.15)$$

to get

$$X_3 = B \cos |k_x| x \qquad (4.16)$$

and

$$X_4 = B \sin |k_x| x. \qquad (4.17)$$

For movements between perfectly reflecting walls perpendicular to the y axis (the z axis), one replaces X by Y and x by y (X by Z and x by z) in (4.16) and (4.17).

Table 4.1

Normalized Translational Eigenfunctions

Boundary Condition	Factor	Possible Quantum Numbers	Formula
Periodic[a]	X	..., −1, 0, 1, ...	$\frac{1}{\sqrt{a}} \exp\left(i\,\frac{2\pi n_x}{a}\,x\right)$
Periodic[a]	Y	..., −1, 0, 1, ...	$\frac{1}{\sqrt{b}} \exp\left(i\,\frac{2\pi n_y}{b}\,y\right)$
Periodic[a]	Z	..., −1, 0, 1, ...	$\frac{1}{\sqrt{c}} \exp\left(i\,\frac{2\pi n_z}{c}\,z\right)$
Confining[b]	X	1, 2, 3, ...	$\sqrt{\frac{2}{a}} \sin\left(\frac{\pi n_x}{a}\,x\right)$
Confining[b]	Y	1, 2, 3, ...	$\sqrt{\frac{2}{b}} \sin\left(\frac{\pi n_y}{b}\,y\right)$
Confining[b]	Z	1, 2, 3, ...	$\sqrt{\frac{2}{c}} \sin\left(\frac{\pi n_z}{c}\,z\right)$

[a]The basic period along the x axis extends from $x = 0$ to $x = a$; that along the y axis, from $y = 0$ to $y = b$; that along the z axis, from $z = 0$ to $z = c$. The unit cell that is repeated periodically thus has edges a, b, c; and its size is $a \times b \times c$.
[b]The confining box is placed as the unit cell for periodic boundary conditions was; its size is $a \times b \times c$.

When the walls enclose a particle, as a box may, the probability density vanishes, so the state function Ψ is zero, outside the resulting pen. Consequently, Ψ goes to zero as each wall is approached from the inside. When the confining box is rectangular, of dimensions $a \times b \times c$, the Cartesian axes are placed along three intersecting edges, with the box in the first octant. The boundary conditions then lead to the results in the second three lines of Table 4.1.

Movement of the center of mass of a system of connected particles is considered to be translation. When this is free it is governed by the forms just obtained. In addition, such a system may rotate about its center of mass and vibrate in various modes.

Example 4.1

What is $\nabla^2\psi$ in spherical coordinates? If q_1, q_2, q_3 are generalized coordinates such that the line elements $h_1\,dq_1$, $h_2\,dq_2$, and $h_3\,dq_3$ are mutually perpendicular, then a standard derivation shows that

$$\nabla^2\psi = \frac{1}{h_1 h_2 h_3}\left[\frac{\partial}{\partial q_1}\left(\frac{h_2 h_3}{h_1}\frac{\partial\psi}{\partial q_1}\right) + \frac{\partial}{\partial q_2}\left(\frac{h_3 h_1}{h_2}\frac{\partial\psi}{\partial q_2}\right) + \frac{\partial}{\partial q_3}\left(\frac{h_1 h_2}{h_3}\frac{\partial\psi}{\partial q_3}\right)\right].$$

For the spherical coordinates described by Figure 3.1, we have

$$h_1 = 1, \qquad h_2 = r, \qquad h_3 = r\sin\vartheta.$$

Substituting into the above expression and reducing yields

$$\nabla^2\psi = \frac{1}{r^2}\frac{\partial}{\partial r}\left(r^2\frac{\partial\psi}{\partial r}\right) + \frac{1}{r^2\sin\vartheta}\frac{\partial}{\partial\vartheta}\left(\sin\vartheta\frac{\partial\psi}{\partial\vartheta}\right) + \frac{1}{r^2\sin^2\vartheta}\frac{\partial^2\psi}{\partial\varphi^2} =$$

$$\frac{1}{r^2}\left[\frac{\partial}{\partial r}\left(r^2\frac{\partial\psi}{\partial r}\right) - \frac{L^2}{\hbar^2}\psi\right].$$

In the last step, Example 3.7 was used to identify the explicit form for operator L^2.

4.3
Separating Angular Motion from the Other Motions

The next simplest kind of motion is that of free rotation by an essentially two particle system. The hydrogenlike atom and the diatomic molecule exhibit such behavior.

However, a two-particle system undergoes translation and oscillation as well as rotation. If m_1 and m_2 are the masses of the two particles, a single particle of mass

$$M = m_1 + m_2 \tag{4.18}$$

at the center of mass models the translatory motion. Furthermore, a particle of mass μ, where

$$\frac{1}{\mu} = \frac{1}{m_1} + \frac{1}{m_2}, \tag{4.19}$$

moving at the interparticle distance r from the center of mass, considered as a fixed point, models the rotation and oscillation.

The first motion was considered in Section 4.2. Let us here concentrate on the angular part of the second motion.

Let us suppose that the two particles are held together by a potential dependent only on the interparticle distance r:

$$U = U(r), \quad \mathbf{A} = 0. \tag{4.20}$$

For an electron moving about a nucleus, as in the hydrogenlike atom, this condition amounts to considering the Coulomb interaction alone. For a two-atom molecule, such as HC1, this amounts to considering the bonding as being a function of the internuclear distance alone. Presumably the electrons are so much lighter than the nuclei that their structures can adjust that rapidly.

From Example 3.4, the Schrödinger equation for the model particle is

$$\nabla^2\psi + k^2(r)\psi = 0. \tag{4.21}$$

In a pure energy state, we expect the radial motion of the representative particle to be independent of its angular motion. Then

$$\psi = R(r)Y(\vartheta, \varphi). \tag{4.22}$$

But because ∇R is perpendicular to ∇Y, we have

$$\nabla^2\psi = R\nabla^2 Y + Y\nabla^2 R. \tag{4.23}$$

The form constructed in Example 4.1 tells us that $\nabla^2 R$ does not contain ϑ and φ and that $r^2\nabla^2 Y$ does not contain r explicitly. So when a person substitutes (4.23) into (4.21) and separates r from the other variables, one obtains

$$r^2\left[\frac{1}{R}\nabla^2 R + k^2(r)\right] = -\frac{1}{Y}r^2\nabla^2 Y. \tag{4.24}$$

Substituting the last form in Example 4.1 into the right side of (4.24) yields

$$-\frac{1}{Y}r^2\nabla^2 Y = -\frac{1}{Y}r^2\frac{1}{r^2}\left(0 - \frac{L^2}{\hbar^2}Y\right)$$

$$= \frac{1}{Y}\frac{L^2 Y}{\hbar^2} = l(l+1). \tag{4.25}$$

In the last step, we take Y to be an eigenfunction of L^2. The result from (3.115) is then introduced.

4.4
Rotational Eigenfunctions

Equation (4.25) does not involve the wavevector k. As a consequence, the angular functions for $k = 0$ are those for any physical k. But setting k equal to zero in (4.21) gives us

$$\nabla^2 \psi = 0, \qquad (4.26)$$

Laplace's equation.

The simplest nontrivial solution to this equation is the reciprocal radius vector:

$$\psi = \frac{1}{r}. \qquad (4.27)$$

Furthermore, each spatial differentiating operator commutes with del squared. So every derivative of (4.27) is also a solution. And since Laplace's equation is linear, every linear combination of such derivatives is a solution.

Let us pass axes through the origin of our coordinate system pointing in l directions. Along the first one, we measure coordinate $x^{(1)}$, along the second one $x^{(2)}, \ldots ,$ along the lth one $x^{(l)}$. Now the derivative

$$\frac{\partial^l}{\partial x^{(1)} \partial x^{(2)} \ldots \partial x^{(l)}} \frac{1}{r} = \psi \qquad (4.28)$$

is a solution of Laplace's equation.

Form (4.28) can be expressed as a homogeneous polynomial in the Cartesian coordinates of degree l divided by r^{2l+1}. Multiplying it by r^{l+1} yields a function of x/r, y/r, and z/r. This is equivalent to a function of ϑ and φ alone. Multiplying it by a normalization factor N and by phase factor $(-1)^l$ gives us the appropriate angular factor.

Thus, we employ

$$Y(\vartheta, \varphi) = N(-1)^l r^{l+1} \frac{\partial^l}{\partial x^{(1)} \partial x^{(2)} \ldots \partial x^{(l)}} \frac{1}{r}. \qquad (4.29)$$

Results for small quantum numbers appear in Table 4.2. As these are real functions, they are standing-wave solutions. Corresponding traveling-wave solutions are the linear combinations appearing in Table 4.3.

Table 4.2
Normalized Standing-Wave
Rotational Eigenfunctions

Spherical Harmonic	Quantum Numbers l	$\lvert m \rvert$	Formula
Y_s	0	0	$\sqrt{\dfrac{1}{4\pi}}$
Y_{p_z}	1	0	$\sqrt{\dfrac{3}{4\pi}}\,\dfrac{z}{r}$
Y_{p_x}	1	1	$\sqrt{\dfrac{3}{4\pi}}\,\dfrac{x}{r}$
Y_{p_y}	1	1	$\sqrt{\dfrac{3}{4\pi}}\,\dfrac{y}{r}$
$Y_{d_{3z^2-r^2}}$	2	0	$\sqrt{\dfrac{5}{16\pi}}\,\dfrac{3z^2-r^2}{r^2}$
$Y_{d_{zx}}$	2	1	$\sqrt{\dfrac{15}{4\pi}}\,\dfrac{zx}{r^2}$
$Y_{d_{yz}}$	2	1	$\sqrt{\dfrac{15}{4\pi}}\,\dfrac{yz}{r^2}$
$Y_{d_{xy}}$	2	2	$\sqrt{\dfrac{15}{4\pi}}\,\dfrac{xy}{r^2}$
$Y_{d_{x^2-y^2}}$	2	2	$\sqrt{\dfrac{15}{16\pi}}\,\dfrac{x^2-y^2}{r^2}$

4.5
Resolving the Angular Motion

When a rotation involves a definite angular momentum along an axis, motion around this axis is independent of the wobbling orthogonal to it. In considering such a rotation let us place the origin of the reference Cartesian axes at the center of mass and point the z axis in the direction of the angular momentum vector. Spherical coordinates are then introduced as in Figure 3.1.

The system under consideration is still the one modeled by the particle of mass μ. We employ the result from Example 4.1 to represent the $\nabla^2 Y$ in (4.25). Dropping the vanishing terms and rearranging leads to

Table 4.3
Normalized Traveling-Wave
Rotational Eigenfunctions

Spherical Harmonic	Quantum Numbers l	m	Formula
$Y_0^{\,0}$	0	0	$\sqrt{\dfrac{1}{4\pi}}$
$Y_1^{\,0}$	1	0	$\sqrt{\dfrac{3}{4\pi}}\,\dfrac{z}{r}$
$Y_1^{\,\pm1}$	1	±1	$\mp\sqrt{\dfrac{3}{8\pi}}\,\dfrac{x\pm iy}{r}$
$Y_2^{\,0}$	2	1	$\sqrt{\dfrac{5}{16\pi}}\,\dfrac{3z^2-r^2}{r^2}$
$Y_2^{\,\pm1}$	2	±1	$\mp\sqrt{\dfrac{15}{8\pi}}\,\dfrac{(x\pm iy)z}{r^2}$
$Y_2^{\,\pm2}$	2	±2	$\sqrt{\dfrac{15}{32\pi}}\,\dfrac{(x\pm iy)^2}{r^2}$
$Y_3^{\,0}$	3	0	$\sqrt{\dfrac{7}{16\pi}}\,\dfrac{z(5z^2-3r^2)}{r^3}$
$Y_3^{\,\pm1}$	3	±1	$\mp\sqrt{\dfrac{21}{64\pi}}\,\dfrac{(x\pm iy)(5z^2-r^2)}{r^3}$
$Y_3^{\,\pm2}$	3	±2	$\sqrt{\dfrac{105}{32\pi}}\,\dfrac{(x\pm iy)^2z}{r^3}$
$Y_3^{\,\pm3}$	3	±3	$\mp\sqrt{\dfrac{35}{64\pi}}\,\dfrac{(x\pm iy)^3}{r^3}$

$$\frac{1}{\sin\vartheta}\frac{\partial}{\partial\vartheta}\left(\sin\vartheta\frac{\partial Y}{\partial\vartheta}\right)+\frac{1}{\sin^2\vartheta}\frac{\partial^2 Y}{\partial\varphi^2}+l(l+1)Y=0. \qquad (4.30)$$

Now, independence of the motion over φ is expressed by the factoring,

$$Y=\Theta(\vartheta)\Phi(\varphi), \qquad (4.31)$$

of the angular part of the state function. Substituting (4.31) into (4.30) and then separating the variables produces

$$\left[\frac{1}{\Theta\sin\vartheta}\frac{d}{d\vartheta}\left(\sin\vartheta\frac{d\Theta}{d\vartheta}\right)+l(l+1)\right]\sin^2\vartheta=-\frac{1}{\Phi}\frac{d^2\Phi}{d\varphi^2}=m^2, \qquad (4.32)$$

if m^2 is introduced as the separation constant.

The last equality yields

$$\frac{d^2\Phi}{d\varphi^2} + m^2\Phi = 0, \qquad (4.33)$$

whence

$$\Phi = e^{im\varphi}. \qquad (4.34)$$

For factor Φ to be single valued, we must have

$$m = \ldots -1, 0, 1, \ldots . \qquad (4.35)$$

The overall equality in (4.32) yields

$$\frac{1}{\sin\vartheta} \frac{d}{d\vartheta}\left(\sin\vartheta \frac{d\Theta}{d\vartheta}\right) + \left[l(l+1) - \frac{m^2}{\sin^2\vartheta}\right]\Theta = 0. \qquad (4.36)$$

But letting

$$\cos\vartheta = w \qquad (4.37)$$

and

$$\Theta(\vartheta) = P(w) \qquad (4.38)$$

changes (4.36) to

$$\frac{d}{dw}\left[(1 - w^2)\frac{dP}{dw}\right] + \left[l(l+1) - \frac{m^2}{1-w^2}\right]P = 0. \qquad (4.39)$$

The single-valued, normalizable solutions to this equation are the *associated Legendre functions*. They are given by the *Rodrigues formula*

$$P_l^{|m|}(w) = \frac{(-1)^l}{2^l l!}(1 - w^2)^{|m|/2}\frac{d^{l+|m|}}{dw^{l+|m|}}(1 - w^2)^l \qquad (4.40)$$

in which

$$l = 0, 1, 2, \ldots \qquad (4.41)$$

and

$$|m| \le l. \qquad (4.42)$$

Using (4.38) and (4.37) to return to the original variables, multiplying the result by (4.34), and normalizing leads to the forms listed in Table 4.3. These solutions can be superposed to represent any possible rotational movement.

Example 4.2

Show that the *m* introduced after the separation of variables, in Equation (4.32), is the same as the *m* introduced in Section 3.9.

From formula (3.9), the operator for angular momentum around the *z* axis is

$$L_z = \frac{\hbar}{i} \frac{\partial}{\partial \varphi}.$$

Let this act on the Φ in (4.34):

$$L_z \Phi = \frac{\hbar}{i} \frac{\partial}{\partial \varphi} e^{im\varphi} = \frac{\hbar}{i} im \, e^{im\varphi} = m\hbar\Phi.$$

We see that the *m* introduced in Equation (4.32) gives the number of $\hbar$ units in the angular momentum. So this *m* is the same as the *m* in (3.118).

4.6
Simple Vibrational Eigenfunctions

A molecule consists of atomic cores held together by valence electrons. In the classical picture, the small relative mass of each electron would allow it to complete many orbits while the nuclei with their cores would move only a short distance. A single cycle executed by an electron would be practically the same as it would be if the nuclei were fixed.

In the corresponding quantum picture, we argue that the state functions for the electrons at any stage are practically the same as they would be if the nuclei were fixed. Because they replace the classical orbits, these functions are generally called orbitals.

As a consequence the electronic energies are calculated for representative fixed positions of the nuclei. The energies for all other configurations are obtained by interpolation and extrapolation. The results are considered to make up the potential energy function for the vibrational motions of the molecule. Since this manner of proceeding was developed by Max Born and J. Robert Oppenheimer, it is called the *Born-Oppenheimer approximation*.

Any relative motion of the atoms causes this potential energy to vary. Furthermore, such a motion can be resolved into vibrational modes, which are only roughly independent.

An approximate model for a mode is a particle of reduced mass μ in a parabolic potential

$$U = \tfrac{1}{2} fx^2. \tag{4.43}$$

Here x is the appropriate generalized coordinate and f is the force constant. For a diatomic molecule, μ is given by (4.19) and x is the displacement from the equilibrium position,

$$x = r - r_e. \tag{4.44}$$

Substituting into the 1-dimensional Schrödinger equation, from Example 3.4, yields

$$\frac{d^2\psi}{dx^2} + \frac{2\mu}{\hbar^2}(E - \frac{1}{2} fx^2)\psi = 0. \tag{4.45}$$

To get a common form, we introduce parameters a and b with

$$a^2 = \frac{f\mu}{\hbar^2}, \qquad ab = \frac{2\mu E}{\hbar^2}. \tag{4.46}$$

Also, we set

$$w = a^{\frac{1}{2}}x. \tag{4.47}$$

Then (4.45) reduces to

$$\frac{d^2\psi}{dw^2} + (b - w^2)\psi = 0. \tag{4.48}$$

Only when

$$b = 2v + 1 \tag{4.49}$$

with

$$v = 0, 1, 2, \ldots \tag{4.50}$$

does a person get analytic solutions that are normalizable. These are given by the formula

$$\psi = Ne^{-w^2/2}(-1)^v e^{w^2} \frac{d^v}{dw^v} e^{-w^2}$$

$$= Ne^{-w^2/2}H_v(w). \tag{4.51}$$

Here $H_v(w)$ is the *Hermite polynomial* of the vth degree in w.

The results one obtains for the low levels appear in Table 4.4. The corresponding energies, from (4.46) and (4.49), are

$$E = \frac{\hbar^2}{2\mu} \frac{(f\mu)^{1/2}}{\hbar} (2v + 1)$$
$$= (v + \tfrac{1}{2})\hbar\left(\frac{f}{\mu}\right)^{1/2}. \tag{4.52}$$

Table 4.4
Normalized Eigenfunctions for a Harmonic Oscillator[a]

Quantum Number v	Formula
0	$\left(\dfrac{a}{\pi}\right)^{1/4} e^{-ax^2/2}$
1	$\sqrt{2}\left(\dfrac{a}{\pi}\right)^{1/4} a^{1/2}xe^{-ax^2/2}$
2	$\dfrac{1}{\sqrt{2}}\left(\dfrac{a}{\pi}\right)^{1/4} (2ax^2 - 1)e^{-ax^2/2}$
3	$\dfrac{1}{\sqrt{3}}\left(\dfrac{a}{\pi}\right)^{1/4} (2a^{3/2}x^3 - 3a^{1/2}x)e^{-ax^2/2}$
4	$\dfrac{1}{2\sqrt{6}}\left(\dfrac{a}{\pi}\right)^{1/4} (4a^2x^4 - 12ax^2 + 3)e^{-ax^2/2}$
5	$\dfrac{1}{2\sqrt{15}}\left(\dfrac{a}{\pi}\right)^{1/4} (4a^{5/2}x^5 - 20a^{3/2}x^3 + 15a^{1/2}x)e^{-ax^2/2}$
6	$\dfrac{1}{4\sqrt{45}}\left(\dfrac{a}{\pi}\right)^{1/4} (8a^3x^6 - 60a^2x^4 + 90ax^2 - 15)e^{-ax^2/2}$

[a]Here $a = (f\mu)^{1/2}/\hbar$, with f = force constant and μ = reduced mass.

4.7
Radial Oscillatory Motion in a Coulomb Field

The simplest atom consists of a single electron spread around a nucleus. The angular structure of each bound state of this atom is described by a rotational eigenfunction, as given by (4.29), or by (4.34) and (4.40).

The radial structure, on the other hand, involves oscillations of the model particle of mass μ in the Coulomb potential

$$U = -\frac{Ze^2}{4\pi\epsilon_0 r}. \tag{4.53}$$

Here Ze is the charge on the nucleus, $-e$ the charge on the electron, ϵ_0 the permittivity of space, and r the interparticle distance.

The energy states of this hydrogenlike atom are governed by its Schrödinger equation. But potential (4.53) makes the k^2 of Example 3.4 take on the form

$$k^2 = \frac{2\mu}{\hbar^2}\left(E + \frac{Ze^2}{4\pi\epsilon_0 r}\right). \tag{4.54}$$

The radial motion separates from the angular motion as described in Section 4.3. Substituting the last expression in (4.25) into the right side of (4.24) and the next-to-last form of Example 4.1 into the left side gives us

$$\frac{1}{R}\frac{d}{dr}\left(r^2\frac{dR}{dr}\right) + k^2(r)r^2 = l(l + 1) \tag{4.55}$$

whence

$$\frac{1}{r^2}\frac{d}{dr}\left(r^2\frac{dR}{dr}\right) + \left[k^2(r) - \frac{l(l+1)}{r^2}\right]R = 0. \tag{4.56}$$

A convenient unit for length is the *Bohr radius*

$$a = \frac{4\pi\epsilon_0\hbar^2}{\mu Ze^2}, \tag{4.57}$$

while a dimensionless radial distance is

$$\rho = \frac{r}{a}. \tag{4.58}$$

If we also let

$$E = \mp\frac{b}{n^2} \tag{4.59}$$

where

$$b = \frac{\mu Z^2 e^4}{32\pi^2\epsilon_0^2\hbar^2} \tag{4.60}$$

and set

$$r = \tfrac{1}{2}\,anx, \tag{4.61}$$

then (4.56) and (4.54) yield

$$\frac{d^2R}{dx^2} + \frac{2}{x}\frac{dR}{dx} + \left[\mp\frac{1}{4} + \frac{n}{x} - \frac{l(l+1)}{x^2}\right]R = 0. \tag{4.62}$$

When the energy is positive, the electron is not bound to the nucleus. Out at large distance, the electron then behaves as a free translator. Correspondingly, a continuum of energy levels exists.

For negative energies the electron is bound. Then only when

$$n = 1, 2, 3, \ldots \tag{4.63}$$

with

$$n \geq l + 1 \tag{4.64}$$

does a person get analytic solutions that are normalizable. These are given by the formula

$$R = Nx^l e^{-x/2} L_{n+l}^{2l+1}(x) \tag{4.65}$$

in which the *associated Laguerre polynomial* is

$$L_j^k(x) = \frac{d^k}{dx^k}\left[e^x \frac{d^j}{dx^j}(x^j e^{-x})\right]. \tag{4.66}$$

Explicit forms for the low lying states appear in Table 4.5. The corresponding energies are

$$E = -\frac{Z^2}{2n^2}\alpha^2\mu c^2 \tag{4.67}$$

with

$$\alpha = \frac{e^2}{4\pi\epsilon_0 \hbar c}. \tag{4.68}$$

Parameter α is called the *fine structure constant* while c is the velocity of light.

4.8
The Hydrogenlike Atom

An atom consisting of a single electron and a nucleus is modeled by a particle of reduced mass μ moving at the interparticle distance r from the center of mass, considered as an inertial point.

The interparticle potential U is given by (4.53). Since this fits form (4.20), the Schrödinger equation reduces to form (4.21). When the system is in a definite energy state the radial motion is independent of the angular motion, so the radial coordinate separates from the angular coordinates in the corresponding state function.

Table 4.5
Normalized Eigenfunctions for
the Radial Motion in a Coulomb Field[a]

Quantum Numbers		Formula
n	l	
1	0	$a^{-3/2}2e^{-\rho}$
2	0	$\dfrac{a^{-3/2}}{2\sqrt{2}}(2-\rho)e^{-\rho/2}$
2	1	$\dfrac{a^{-3/2}}{2\sqrt{6}}\rho e^{-\rho/2}$
3	0	$\dfrac{2a^{-3/2}}{81\sqrt{3}}(27-18\rho+2\rho^2)e^{-\rho/3}$
3	1	$\dfrac{2\sqrt{2}a^{-3/2}}{81\sqrt{3}}(6-\rho)\rho e^{-\rho/3}$
3	2	$\dfrac{4a^{-3/2}}{81\sqrt{30}}\rho^2 e^{-\rho/3}$

[a]Here $\rho = r/a$ with a = Bohr radius. Note that $a = a_0/Z$ where a_0 is the Bohr radius of the hydrogen atom calculated using the same reduced mass.

Orthogonal solutions for the angular factor appear in Tables 4.2 and 4.3; orthogonal solutions for the radial factor appear in Table 4.5. Employing the traveling-wave rotational solutions and converting to spherical coordinates using

$$x \pm iy = r \sin \vartheta\, e^{\pm i\varphi} \qquad (4.69)$$

and

$$z = r \cos \vartheta \qquad (4.70)$$

yields the results in Table 4.6.

From (4.67), the energy when magnetic and Einstein-relativistic effects are neglected depends only on the principal quantum number n. So a definite energy state need not have a definite l or m. It could be described as a superposition of different contributing l and m states, all with the same n.

Table 4.6
Eigenfunctions for Hydrogenlike Atoms[a]

Quantum Numbers			Formula
n	l	m	
1	0	0	$\left(\dfrac{1}{\pi}\right)^{\frac{1}{2}} a^{-3/2} e^{-\rho}$
2	0	0	$\dfrac{1}{4}\left(\dfrac{1}{2\pi}\right)^{\frac{1}{2}} a^{-3/2}(2-\rho)e^{-\rho/2}$
2	1	0	$\dfrac{1}{4}\left(\dfrac{1}{2\pi}\right)^{\frac{1}{2}} a^{-3/2}\rho e^{-\rho/2}\cos\vartheta$
2	1	± 1	$\dfrac{1}{8}\left(\dfrac{1}{\pi}\right)^{\frac{1}{2}} a^{-3/2}\rho e^{-\rho/2}\sin\vartheta\, e^{\pm i\varphi}$
3	0	0	$\dfrac{1}{81}\left(\dfrac{1}{3\pi}\right)^{\frac{1}{2}} a^{-3/2}(27-18\rho+2\rho^2)e^{-\rho/3}$
3	1	0	$\dfrac{1}{81}\left(\dfrac{2}{\pi}\right)^{\frac{1}{2}} a^{-3/2}(6-\rho)\rho e^{-\rho/3}\cos\vartheta$
3	1	± 1	$\dfrac{1}{81}\left(\dfrac{1}{\pi}\right)^{\frac{1}{2}} a^{-3/2}(6-\rho)\rho e^{-\rho/3}\sin\vartheta\, e^{\pm i\varphi}$
3	2	0	$\dfrac{1}{81}\left(\dfrac{1}{6\pi}\right)^{\frac{1}{2}} a^{-3/2}\rho^2 e^{-\rho/3}(3\cos^2\vartheta-1)$
3	2	± 1	$\dfrac{1}{81}\left(\dfrac{1}{\pi}\right)^{\frac{1}{2}} a^{-3/2}\rho^2 e^{-\rho/3}\sin\vartheta\cos\vartheta\, e^{\pm i\varphi}$
3	2	± 2	$\dfrac{1}{162}\left(\dfrac{1}{\pi}\right)^{\frac{1}{2}} a^{-3/2}\rho^2 e^{-\rho/3}\sin^2\vartheta\, e^{\pm 2i\varphi}$

[a]The spherical coordinates "locating" the electron with respect to the nucleus are r, ϑ, φ. We let $r/a = \rho$, where a = Bohr radius.

4.9
Approximate Orbitals for Multielectron Atoms

When an atom contains more than a single electron, interactions exist among the electrons as well as with the nucleus. The electric interactions involve the position propensities through Coulomb's law. The magnetic interactions involve the orbital and spin angular momenta propensities through Ampere's law. Accurately determining the effects of all of these taxes one's ingenuity. So one proceeds by a method of successive approximations.

A person may first neglect the interelectron forces. Each electron would then interact with the nucleus as in a hydrogenlike atom. For its orbital, one would employ a radial factor from Table 4.5 and an angular factor from Tables 4.2 or 4.3. So that the conditions of Sections 1.11 and 1.12 can be satisified, each electron is placed in a different spin-orbital state. Thus, it is labeled by a different set of hydrogenlike quantum numbers.

In the *independent-particle model*, one considers that each electron moves in the average field of all the rest. Allowance is made for the antisymmetry over identical particles by requiring each electron to have a different set of quantum numbers n, l, m, s_z. Magnetic effects are ignored. The electric interaction is taken to be spherically symmetric. Then, only the average distribution of charge at smaller radial distances than that being considered for the given electron influences it. These act to reduce effectively the charge on the nucleus. This part of the electron distribution is said to shield or screen the nucleus, insofar as the electron under consideration is concerned.

Since the field seen by each electron is taken to be spherically symmetric, the orbital for each electron has the form

$$\psi_j = R_{n,l}(r) Y_l^m(\vartheta, \varphi). \tag{4.71}$$

With the electrons independent, the complete state function would be

$$\psi = \psi_1 \psi_2 \ldots \psi_N \tag{4.72}$$

where N is the number of orbital electrons.

A person might guess what the orbital should be for each electron. The average field that this electron senses from the others can then be calculated as a function of the radial distance r. The result can be put into a Schrödinger equation and this can be solved numerically. One then has a basis for revising each orbital and repeating the calculation. One iterates until the field is consistent with the newly calculated orbitals.

A crude analytic representation of the results is provided by the radial factor

$$R = N\rho^{n'-1} e^{-(Z-s)\rho/n'} \tag{4.73}$$

where

$$\rho = \frac{r}{a_0} \qquad (4.74)$$

and N is a normalization constant.

Effective principal quantum number n' and screening constant s are approximated by the rules of John C. Slater:

(1) The conventional quantum number n determines the effective n' as follows:

n	1	2	3	4	5	6
n'	1	2	3	3.7	4.0	4.2

(2) For determining constant s, the electrons are divided into the groups

$$1s;\ 2s,\ 2p;\ 3s,\ 3p;\ 3d;\ 4s,\ 4p;\ 4d,\ 4f;\ \ldots .$$

(3) Then one sums the contributions:
   (a) nothing from any group to the right (higher) than the one under consideration;
   (b) for each other electron in the same group, amount 0.35 (except when this electron is the other $1s$ electron, then 0.30);
   (c) for an s or p orbital, 0.85 for each electron with 1 unit lower principal quantum number, and 1.00 for each electron further in; for a d electron, 1.00 for every electron further in.

Better approximations are available in the literature, but these are more complicated. See, for instance, the work by E. Clementi and E. L. Raimondi listed at the end of this chapter.

In the simplest calculations on molecules, one employs only those Slater orbitals that correspond to occupied atomic orbitals in the separated atom limit. The orbital exponents may be varied to minimize the energy. The basis set may be improved by adding additional Slater orbitals to each atom.

Some matrix elements encountered with Slater orbitals are difficult to compute. Consequently, an alternative basis-set class if often used. This consists of Gaussian functions.

The exponential factor in a Gaussian orbital has the form

$$e^{-\alpha r^2}, \qquad (4.75)$$

as in harmonic oscillator eigenfunctions. However, each Slater orbital has to be replaced by a number of Gaussian orbitals with different α's. Multiplication by the standard ϑ and φ dependences generates the $p, d, f, \ldots$ orbitals.

4.10
Elements of Photon Transmission, Production,
and Absorption

The photon structure of the electromagnetic field is described in a statistical manner by the electric and magnetic vectors of Maxwell's equations. Since the equations are linear in free space, these vectors obey the superposition principle. And, a complicated field can be resolved into a sum of monochromatic, polarized components traveling in mutually perpendicular directions. The state of each of these is determined by the behavior of its electric vector $\mathbf{E}$, since the perpendicular magnetic vector $\mathbf{H}$ is proportional in magnitude to the $\mathbf{E}$.

Because the governing Maxwell equations are linear, a person can add to an $\mathbf{E}$ and $\mathbf{H}$ similar imaginary parts without affecting the real functions. For convenience, the amplitude of the imaginary part is set equal to that of the real part, the phase, 90° retarded. The conventional trigonometric functions for a sinusoidal variation are thus replaced by exponential functions.

A constitutent with angular frequency ω and wavevector $\mathbf{k}$ in the z direction can be expressed as a linear combination of

$$\mathbf{E}_1 = \mathbf{e}_x \exp i(kz - \omega t) \tag{4.76}$$

and

$$\mathbf{E}_2 = \mathbf{e}_y \exp i(kz - \omega t), \tag{4.77}$$

where $\mathbf{e}_x$ and $\mathbf{e}_y$ are unit vectors in the x and y directions, respectively. The wave being transverse, no component in the z direction exists. On examination, however, we find that (4.76) and (4.77) are not eigenstates for the spin angular momentum.

From Table 4.3, a set of eigenfunctions for the angular momentum $l = 1$ is

$$Y_1^{+1} = -\sqrt{\frac{3}{8\pi}} \frac{x+iy}{r}, \tag{4.78}$$

$$Y_1^0 = \sqrt{\frac{3}{4\pi}} \frac{z}{r}, \tag{4.79}$$

$$Y_1^{-1} = \sqrt{\frac{3}{8\pi}} \frac{x-iy}{r}. \tag{4.80}$$

Corresponding to these are the unit vectors

$$\mathbf{e}_{+1} = -\frac{1}{\sqrt{2}} (\mathbf{e}_x + i\mathbf{e}_y), \tag{4.81}$$

$$\mathbf{e}_0 = \mathbf{e}_z, \tag{4.82}$$

$$\mathbf{e}_{-1} = \frac{1}{\sqrt{2}} (\mathbf{e}_x - i\mathbf{e}_y), \tag{4.83}$$

So to get eigenfunctions for spin, we combine (4.76) and (4.77) in the forms

$$\mathbf{E}_{+1} = \mathbf{e}_{+1} \exp i(kz - \omega t), \tag{4.84}$$

$$\mathbf{E}_{-1} = \mathbf{e}_{-1} \exp i(kz - \omega t), \tag{4.85}$$

Since no longitudinal field is associated with real photons, an $\mathbf{e}_0$ component does not exist. Only components with spin about the axis of propagation equal to $+1$ unit, from (4.84), and -1 unit, from (4.85), exist. These are called *helicity* states.

A photon may thus appear, or disappear, with quantum number s equal to 1 and quantum number s_z equal to $+1$ or -1. The state with $s_z = +1$ is right circularly polarized; that with $s_z = -1$, left circularly polarized. The linear polarization states arise from superposing right and left polarization states with appropriate relative phasing.

The emission, or absorption, of a real photon involves a relative angular momentum between the photon and the interacting system, measured by an l. The total angular momentum that the photon carries off, or adds, equals a vector sum of this relative angular momentum and the photon spin. Since these are perpendicular, the photon requires at least one unit total angular momentum, which is measured by j. Consequently, an emitting, or absorbing, system $J = 0$ to $J = 0$ transition is absolutely forbidden. The addition of s and any l must give a j of 1 or greater. Corresponding to each j and $j_{z'}$, we may have $l = j - 1, j, j + 1$.

For the changing system, we have

$$\Delta J = 0 \text{ (except when } J = 0), \pm 1, \pm 2, \pm 3, \dots \tag{4.86}$$

The electric dipole transitions with

$$\Delta J = \pm 1 \tag{4.87}$$

are the most rapid by far. These are determined by the matrix element $\langle n \mid q\mathbf{r} \mid m \rangle$, as we will find later. So each is connected to a shift in position of charge.

Both electric and magnetic changes in the source or target system produce or alter propagating electromagnetic fields. The transition induced by a magnetic change at a given ΔJ is much slower than one induced by an electric change with the same ΔJ, other things being equal. For here, the determining matrix element is connected to a change in circulation of charge.

4.11
Photon Kets

From the Compton effect, from the Raman effect, from the photoelectric effect, and from the absorption or emission of radiation in other contexts, we know that interaction of the electromagnetic field with a localized system is quantized according to the Einstein equation

$$E = \hbar\omega \tag{4.88}$$

Measurements with discriminating detectors (counters, photographic emulsions, etc.) show that the interaction is localized in space and time. So we consider it to occur through particles called *photons*.

However, there is no way to distinguish between equivalent interactions, so we consider equivalent photons to be indistinguishable. Since there can be any number in a given state, they obey Bose-Einstein statistics.

Furthermore, only the probability of an interaction occurring can be determined. Because the density of this probability travels at the speed of light in all inertial frames, a photon cannot be observed at rest. But through equation (4.88), the magnitude of the probability density can be obtained from the energy density in the wave. Consequently kets describing the possible states can be constructed.

Now, each translational state involves an amplitude, a wavevector, an angular frequency, and a polarization. Since the phase velocity is c, the angular frequency is related to the wavevector by the equation

$$\frac{\omega}{k} = c. \tag{4.89}$$

The independent polarization states may be taken as the ones with s_k (the spin-component quantum number for the direction of $\mathbf{k}$) equal to $+1$ and to -1. These are right handed (R) and left handed (L), respectively.

For a photon traveling with wavevector $\mathbf{k}$ from a source, we have the independent kets

$$| R, + \mathbf{k} \rangle \tag{4.90}$$

and

$$| L, + \mathbf{k} \rangle. \tag{4.91}$$

Reflection through the origin, which is called *parity operation P*, reverses **k** and the helicity:

$$P \, | \, R, + \mathbf{k} \rangle = | \, L, - \mathbf{k} \rangle, \tag{4.92}$$

$$P \, | \, L, + \mathbf{k} \rangle = | \, R, - \mathbf{k} \rangle. \tag{4.93}$$

In a one-photon process, the photon does not carry the parity symmetry assigned to it. However, the state function governing the spatial distribution of **k**'s about the source does.

Since different photons are independent of each other, the kets for different photon states combine multiplicatively. Let us suppose that two photons are formed at the same source with equivalent energies at the same time. From symmetry considerations, conservation of momentum, they travel in opposite directions.

The state with +2 units of spin in the +**k** direction is then represented by

$$| R, + \mathbf{k} \rangle \, | L, - \mathbf{k} \rangle = | R, + \mathbf{k}; L, - \mathbf{k} \rangle, \tag{4.94}$$

while the state with −2 units of spin in the +**k** direction is represented by

$$| L, + \mathbf{k} \rangle \, | R, - \mathbf{k} \rangle = | L, + \mathbf{k}; R, - \mathbf{k} \rangle, \tag{4.95}$$

With both (4.94) and (4.95), the parity operation changes the first one-photon ket into the second and the second into the first. Operation *P* has the same effect on the products as multiplication by +1; so we say that their parity equals +1.

For two oppositely traveling photons from a single source, we also have the products

$$| R, + \mathbf{k} \rangle \, | R, - \mathbf{k} \rangle = | R, + \mathbf{k}; R, - \mathbf{k} \rangle, \tag{4.96}$$

and

$$| L, + \mathbf{k} \rangle \, | L, - \mathbf{k} \rangle = | L, + \mathbf{k}; L, - \mathbf{k} \rangle, \tag{4.97}$$

These have a resultant spin of zero. However, the parity operation converts (4.96) into (4.97) and (4.97) into (4.96). So these are not suitable as they stand. But, the superpositions

$$\frac{1}{\sqrt{2}} \left(| R, + \mathbf{k}; R, -\mathbf{k} \rangle + | L, +\mathbf{k}; L, -\mathbf{k} \rangle \right) \tag{4.98}$$

and

$$\frac{1}{\sqrt{2}} \left(| R, + \mathbf{k}; R, -\mathbf{k} \rangle - | L, +\mathbf{k}; L, -\mathbf{k} \rangle \right) \tag{4.99}$$

are suitable, exhibiting parities of $+1$ and -1, respectively.

The circular polarization kets can be resolved into planar polarized kets. Let a photon traveling with wavevector $\pm\mathbf{k}$ while polarized in the x direction be represented by

$$| X, \pm\mathbf{k} \rangle \qquad (4.100)$$

and when polarized in the y direction by

$$| Y, \pm\mathbf{k} \rangle \qquad (4.101)$$

Then use (4.81) to analyze the $+1$ spin kets:

$$| R, +\mathbf{k} \rangle = -\frac{1}{\sqrt{2}} (| X, +\mathbf{k} \rangle + i | Y, +\mathbf{k} \rangle), \qquad (4.102)$$

$$| L, -\mathbf{k} \rangle = -\frac{1}{\sqrt{2}} (| X, -\mathbf{k} \rangle + i | Y, -\mathbf{k} \rangle). \qquad (4.103)$$

Similarly employ (4.83) to resolve the -1 spin kets:

$$| R, -\mathbf{k} \rangle = \frac{1}{\sqrt{2}} (| X, -\mathbf{k} \rangle - i | Y, -\mathbf{k} \rangle), \qquad (4.104)$$

$$| L, +\mathbf{k} \rangle = \frac{1}{\sqrt{2}} (| X, +\mathbf{k} \rangle - i | Y, +\mathbf{k} \rangle). \qquad (4.105)$$

Substituting into (4.98) and (4.99) yields

$$-\frac{1}{\sqrt{2}} (| X, +\mathbf{k} \rangle | X, -\mathbf{k} \rangle + | Y, +\mathbf{k} \rangle |Y, -\mathbf{k} \rangle) \qquad (4.106)$$

for parity $P = 1$ and

$$\frac{i}{\sqrt{2}} (| X, +\mathbf{k} \rangle | Y, -\mathbf{k} \rangle - | Y, +\mathbf{k} \rangle |X, -\mathbf{k} \rangle) \qquad (4.107)$$

for parity $P = -1$.

In the even parity configuration, the two photons appear with the same plane of polarization. In the odd parity configuration, they appear with their planes of polarization perpendicular. Thus, a person can determine which configuration is involved in a particular case by determining the relative planes of polarization.

When the two photons do not have the same energy, formulas (4.106) and (4.107) may be replaced by

$$-\frac{1}{\sqrt{2}} (| X, \mathbf{k}_1 \rangle | X, \mathbf{k}_2 \rangle + | Y, \mathbf{k}_1 \rangle |Y, \mathbf{k}_2 \rangle) \qquad (4.108)$$

and

$$\frac{i}{\sqrt{2}} \, (\mid X, \, \mathbf{k}_1 \rangle \mid Y, \, \mathbf{k}_2 \rangle - \mid Y, \, \mathbf{k}_1 \rangle \mid X, \, \mathbf{k}_2 \rangle \,). \qquad\qquad (4.109)$$

4.12
Molecular Structure

In Section 1.10, we noted that under certain conditions a composite of particles may exhibit a particular average form or shape. The resulting chemical structure was described as a classical observable. In Section 4.6, however, the form did not come out of the theory. Rather, one considered the atomic nuclei to be at various relatively fixed positions and determined the electronic structure for each arrangement. Corresponding energies were obtained. The excess above the minimum was then considered to be the potential energy, for the configuration, to be used in determining the vibrational behavior of the molecule.

Some authors, such as R. G. Woolley, have consequently come to the following conclusions:

(1) Molecular structure has to be put into quantum mechanics.
(2) Consequently, there is no fundamental significance to this structure.

They have supported these statements by the observation that the Hamiltonian operator is the same for all possible chemical structures that the given particles can yield. However, the same remark applies to the quantized states for any quantum mechanical system. The Hamiltonian operator is the same for them all.

A given molecular structure corresponds to a certain subset of the possible pure states for the electrons and nuclei involved. It represents a part of the Hilbert space. It is in a sense a kind of quantization for the given system.

4.13
Summary

The energy states of a particle translating freely are governed by Equation (4.9). The resulting eigenfunctions are exponentials in the form (4.10). With rectangularly periodic boundary conditions, the factors

$$X = \frac{1}{\sqrt{a}} e^{ik_x x}, \qquad (4.110)$$

$$Y = \frac{1}{\sqrt{b}} e^{ik_y y}, \qquad (4.111)$$

$$Z = \frac{1}{\sqrt{c}} e^{ik_z z}, \qquad (4.112)$$

$$T = e^{-i\omega t} \qquad (4.113)$$

are obtained. Here k_x, k_y, and k_z are components of the wavevector $\mathbf{k}$ while ω is the de Broglie angular frequency.

A free linear rotator involves two angular coordinates, ϑ and φ. For it, we found the angular eigenfunction

$$Y = NP_l^{|m|} (\cos \vartheta) e^{im\varphi}, \qquad (4.114)$$

where l is the azimuthal quantum number and m the magnetic quantum number. Explicit forms for the low lying levels appear in Tables 4.3 and 4.2.

An approximate model for a normal vibrational mode of a molecule is the one-dimensional harmonic oscillator, a mass μ in the potential

$$U = \frac{1}{2} fx^2. \qquad (4.115)$$

Its eigenfunctions have the form

$$\psi = Ne^{-w^2/2} H_v(w), \qquad (4.116)$$

where

$$w = \frac{(f\mu)^{1/4}}{\hbar^{1/2}} x, \qquad (4.117)$$

while v is the vibrational quantum number. Explicit forms appear in Table 4.4.

Radial motion in a Coulomb field is governed by the eigenfunction

$$R = Nx^l e^{-x/2} L_{n+l}^{2l+1}(x), \qquad (4.118)$$

with

$$x = \frac{2r}{an}, \qquad (4.119)$$

where a is the Bohr radius and n is the principal quantum number. Explicit forms appear in Table 4.5.

Eigenfunctions for a hydrogenlike atom are formed by multiplying particular radial factors, from Table 4.5, by pertinent angular factors, from Table 4.3 or Table 4.2. Results for the standard quantum numbers appear in Table 4.6.

A multiparticle structure may be constructed by multiplying the appropriate single particle orbitals in the various allowed ways and combining the terms so the result has the proper permutational symmetry. In an atom, the single electron orbitals would be perturbed forms of the hydrogenlike orbitals.

The unique direction for a photon is the direction of propagation of the corresponding electromagnetic wave. About this direction, the photon may exhibit ±1 unit of spin. The zero component of spin is ruled out by the limiting transverse nature of the wave.

Discussion Questions

4.1 What conditions must a ket $| j \rangle$ describing a system satisfy?

4.2 On changing levels, may the eigenvalues vary continuously?

4.3 How does a function satisfying an eigenvalue equation behave when the parameter in the equation is not an eigenvalue?

4.4 Why are state functions more complicated than the governing potentials?

4.5 How are the state functions describing unidirectional free motion obtained?

4.6 Explain how (a) periodic boundary conditions, (b) rectangular confining boundary conditions, restrict these functions.

4.7 How can one obtain the orthogonal coordinate expression for $\nabla^2\psi$?

4.8 Why can the angular motion of a linear system be represented by an independent factor in the state function?

4.9 The Schrödinger equation for the model particle representing a two-particle system has the form

$$\nabla^2\psi + k^2(r)\psi = 0.$$

Why is the angular dependence of ψ independent of the nature of $k^2(r)$?

4.10 How can the angular dependence be generated from $1/r$? Explain.

4.11 When does the dependence on the azimuth angle φ separate from the dependence on the colatitude angle ϑ?

4.12 What can a person conclude when a function of one independent variable, such as φ, equals a function of another independent variable, such as ϑ?

4.13 Why can the Born-Oppenheimer approximation be applied in analyzing the vibrations of a molecule?

4.14 Why are the vibrational modes approximately independent?

4.15 In the harmonic oscillator state function, what role is played by (a) the transcendental factor, (b) the polynomial factor?

4.16 What boundary conditions must the state function for radial motion in a Coulomb field satisfy?

4.17 What roles are played by the different factors in the radial function for the Coulomb field?

4.18 What interactions are important in a multielectron atom?

4.19 How are these approximated in the independent-particle model?

4.20 What equations describe photon behavior as the Schrödinger equation describes electron behavior?

4.21 What spin states can photons exhibit? How are these described?

4.22 What evidence do we have that the electromagnetic field acts by means of photons?

4.23 What statistics do photons obey? Explain.

4.24 Explain how a parity operation affects a photon.

4.25 Why do the kets for different photons combine multiplicatively?

4.26 What objection may be raised to the ket

$$| R, +\mathbf{k} \rangle | R, -\mathbf{k} \rangle?$$

How is the defect corrected?

4.27 How does the structure of a molecule come out of quantum mechanics?

Problems

4.1 Consider the translation of independent equivalent particles with rectangularly symmetric periodic boundary conditions. Substitute the pertinent solutions from Table 4.1 into the eigenvalue equation to obtain an expression for the particle energy.

4.2 Construct $\nabla^2 \psi$ in cylindrical coordinates.

4.3 Show how one obtains

$$Y_{p_z} = N \frac{z}{r}$$

from formula (4.28) or (4.29).

4.4 Calculate the normalization constant for Y_{p_z}.

4.5 Construct

$$Y_{d_{x^2-y^2}} = N \frac{x^2 - y^2}{r^2}$$

using formula (4.29).

4.6 Show how the ψ in Equation (4.45) is related to the radial factor R from the corresponding Equation (4.55), with $l = 0$.

4.7 Obtain the asymptotic solution to (4.62), valid when x is large.

4.8 Construct the Slater radial factors for the $1s$, $2s$, and $2p$ orbitals of nitrogen.

4.9 Substitute the solution

$$\psi = Ae^{-ar}$$

into the Schrödinger equation for the hydrogen atom and determine a and the corresponding energy.

4.10 Resolve the photon ket

$$| X, +\mathbf{k} \rangle$$

into kets with definite spins.

4.11 Show how (4.16) and (4.17) are obtained from (4.14) and (4.15).

4.12 Obtain $\nabla^2\psi$ in the parabolic coordinates for which

$$x = \sqrt{\xi\eta} \cos \varphi, \qquad y = \sqrt{\xi\eta} \sin \varphi, \qquad z = \tfrac{1}{2} (\xi - \eta).$$

4.13 Show how one obtains

$$Y_{d_{3z^2-r^2}} = N \frac{3z^2-r^2}{r^2}$$

from formula (4.28) or (4.29).

4.14 Normalize the orbital in Problem 4.13.

4.15 Construct

$$Y_{d_{xy}} = N \frac{xy}{r^2}$$

using formula (4.29).

4.16 Explain what form (4.29) assumes when $| m | = 0$. So what can one say about $P_l^0 (\cos \vartheta)$?

4.17 Show that w in formula (4.47) is dimensionless.

4.18 Construct the Slater radial factors for the $1s$, $2s$, and $2p$ orbitals of carbon.

4.19 A particle is in a 1-dimensional confining potential with energy E. Furthermore, both $\psi_1(x)$ and $\psi_2(x)$ are state functions at this energy. Construct the corresponding Schrödinger equations. From these, eliminate both U and E. Integrate the resulting equation to get an expression for $\psi_1 \, d\psi_2/dx$. Consider the particle to be bound and determine ψ_2 as a function of ψ_1. Can any bound level in a 1-dimensional one-particle system be degenerate?

References

Books

Buckingham, R. A.: 1961. "Exactly Soluble Bound State Problems," in Bates, D. R. (editor), *Quantum Theory I. Elements*, Academic Press, New York, pp. 81-145. A solution in the form of a well-known function is said to be exact. Buckingham presents such solutions of the Schrödinger equations for a rectangular potential, a harmonic oscillator, a rigid rotator, a spherical well, a one-center Coulomb potential, and a two-center Coulomb potential.

Duffey, G. H.: 1984. *A Development of Quantum Mechanics Based on Symmetry Considerations*, Reidel, Dordrecht, pp. 1-142, 212-241. In this introductory text, Duffey gives step-by-step solutions for a translator, a rotator, a harmonic oscillator, and a hydrogenlike atom.

Lowe, J. P.: 1978, *Quantum Chemistry*, Academic Press, New York, pp. 27-134. This introductory text contains detailed solutions for a translating particle, a harmonic oscillator, and a hydrogenlike atom.

Slater, J. C.: 1960. *Quantum Theory of Atomic Structure, Volume I*, McGraw-Hill Book Company, New York, pp. 51-85, 166-233. In this comprehensive monograph, Slater presents methods for solving the Schrödinger equation in one dimension. Then the hydrogenlike atom and the central field model for atoms in general are considered.

Articles

Amus'ya, M. Ya, and Ivanov, V. K.: 1987. "Intershell Interaction in Atoms," *Soviet Phys. Uspekhi* **30**, 449-474.

Blukis, U., and Howell, J. M.: 1983. "Numerical Solution of the One-Dimensional Schrödinger Equation," *J. Chem. Educ.* **60**, 207-212.

Clementi, E., and Raimondi, D. L.: 1963. "Atomic Screening Constants from SCF Functions," *J. Chem. Phys.* **38**, 2686-2689.

David. C. W.: 1981. "On Orbital Drawings," *J. Chem. Educ.* **58**, 377-380.

Gershenfeld, N.: 1984. "Recurring Textbook Error: Graphs of the Hydrogen Radial Probability Density," *Am. J. Phys.* **52**, 81-82.

Jiang, H. X., and Lin, J. Y.: 1987. "Band Structure of a Periodic Potential with Two Wells and Two Barriers per Period," *Am. J. Phys.* **55**, 462-465.

Killingbeck, J.: 1987. "Microcomputer Numerical Experiments in Quantum Mechanics. 2: A Recurrence-Relation Approach for Eigenvalues," *Eur. J. Phys.* **8**, 263-267.

Kjaergaard, H. G., and Mortensen, O. S.: 1990. "The Quantum Mechanical Hamiltonian in Curvilinear Coordinates: A Simple Derivation," *Am. J. Phys.* **58**, 344-347.

Lee, S.-Y.: 1985, "The Hydrogen Atom as a Morse Oscillator," *Am. J. Phys.* **53**, 753-757.

Li, W.-K., and Blinder, S. M.: 1987. "Particle in an Equilateral Triangle: Exact Solution of a Nonseparable Problem," *J. Chem. Educ.* **64**, 130-132.

Marin, J. L., and Cruz, S. A.: 1988. "On the Harmonic Oscillator Inside an Infinite Potential Well," *Am. J. Phys.* **56**, 1134-1136.

Neethiulagarajan, A., and Balasubramanian, S.: 1989. "On Numerical Solutions of the Radial Schrödinger Equation," *Eur. J. Phys.* **10**, 93-95.

Nieto, N. M.: 1979. "Hydrogen Atom and Relativistic Pi-Mesic Atom in *N* Space Dimensions," *Am. J. Phys.* **47**, 1067-1072.

Norbury, J. W.: 1989. "The Quantum Mechanical Few-Body Problem," *Am. J. Phys.* **57**, 264-266.

Solt, G., and Amiet, J.-P.: 1985. "On the 'Series Method' to Solve Schrödinger's Equation for the Linear Oscillator," *Am. J. Phys.* **53**, 180-181.

Snygg, J.: 1977. "Wave Functions Rotated in Phase Space," *Am. J. Phys.* **45**, 58-60.

5

STEP OPERATORS

5.1
Transformation Operators in a Given Hilbert Space

We have seen how the possible states of a given system are represented by normalized kets, unit vectors, in a particular Hilbert space. An observable is represented by an operator that does not alter the direction of each ket in a set that spans the Hilbert space. The possible values are given by the corresponding eigenvalues.

Since measurements yield real numbers these eigenvalues must be real. The reality is ensured by requiring the operator to be Hermitian, as axiom (X) notes.

An operator A is *Hermitian* when Equation (1.82) holds for any $|j\rangle$ and $|k\rangle$ in the Hilbert space for the system:

$$\langle j \,|\, A \,|\, k \rangle = \langle A \,|\, j \,|\, k \rangle = \langle k \,|\, A \,|\, j \rangle^*. \tag{5.1}$$

Multiplying each integral of (5.1) by i and bringing each i inside leads to

$$\langle j \,|\, iA \,|\, k \rangle = \langle -iA \,|\, j \,|\, k \rangle = - \langle k \,|\, iA \,|\, j \rangle^*. \tag{5.2}$$

Now, an operator B that equals i times a Hermitian operator is said to be *anti-Hermitian*. From (5.2), we find that

$$\langle j \,|\, B \,|\, k \rangle = - \langle B \,|\, j \,|\, k \rangle = - \langle k \,|\, B \,|\, j \rangle^*. \tag{5.3}$$

A general transformation operator can be expressed as the sum of a Hermitian operator A and an anti-Hermitian operator B. See Example 5.1.

In principle, an operator exists that converts each eigenket of a given physical operator into another eigenket. Such an operator is called a *step operator*. We will find that, when the eigenkets are not degenerate, the step operator must satisfy a pair of commutation relations.

For a single type or mode of motion in a given system, the eigenkets form a simple progression without degeneracy. Recall Problem 4.19. A step operator then acts by introducing or eliminating nodes in the wave function. When this operator employs differentiating operators no higher than first order, it can introduce or eliminate only one node and the shift is to the next wave function up or down.

When the eigenvalues are equally spaced, a step operator linking successive eigenkets can be constructed. As illustrations, we have the angular momentum problem and will consider the harmonic oscillator system.

For most one-dimensional motions the eigenvalues are not equally spaced. But if the spacing increases or decreases with the observable, in a certain way, first-order step operators can still be constructed. As illustrations, we will consider the Morse oscillator and the radial mode in a Coulombic field.

Example 5.1

Show that a general transformation operator for an arbitrary ket in a given Hilbert space may be expressed as the sum of a Hermitian operator and an anti-Hermitian operator.

For any two kets $|j\rangle$ and $|k\rangle$ in the Hilbert space and any transformation operator C, one can construct the matrix element C_{jk}. But the corresponding adjoint of C also exists:

$$\langle j \mid C \mid k \rangle = \langle C^\dagger \mid j \mid \mid k \rangle = \langle k \mid C^\dagger \mid j \rangle^*.$$

Let us form the sum and the difference of C and $C^\dagger$:

$$C + C^\dagger = 2A, \qquad C - C^\dagger = 2B.$$

We thus obtain two independent operators A and B. Their sum equals C; their difference, $C^\dagger$:

$$C = A + B, \qquad C^\dagger = A - B.$$

Substitute into the matrix-element equation to get

$$\langle j \mid A + B \mid k \rangle = \langle k \mid A - B \mid j \rangle^*.$$

Since A and B are independent, this equation implies that

$$\langle j \mid A \mid k \rangle = \langle k \mid A \mid j \rangle^*$$

and

$$\langle j \mid B \mid k \rangle = -\langle k \mid B \mid j \rangle^*.$$

We see that part A is Hermitian while part B is anti-Hermitian. The anti-Hermitian part of C can be expressed as i times a Hermitian operator.

Example 5.2

Show that the commutator of two Hermitian operators is anti-Hermitian.

Again consider $\mid j \rangle$ and $\mid k \rangle$ to be any two kets in the given Hilbert space. Also consider A and B to be two Hermitian operators for that space. Then both A and B are self-adjoint:

$$\langle j \mid A \mid k \rangle = \langle A \mid j \mid \mid k \rangle = \langle k \mid A \mid j \rangle^*,$$
$$\langle j \mid B \mid k \rangle = \langle B \mid j \mid \mid k \rangle = \langle k \mid B \mid j \rangle^*.$$

Consequently, we find that

$$\langle j \mid AB \mid k \rangle = \langle A \mid j \mid \mid B \mid k \rangle = \langle BA \mid j \mid \mid k \rangle = \langle k \mid BA \mid j \rangle^*$$

and

$$\langle j \mid BA \mid k \rangle = \langle k \mid AB \mid j \rangle^*.$$

The differences of these two equations yields

$$\langle j \mid AB - BA \mid k \rangle = \langle k \mid BA - AB \mid j \rangle^* = -\langle k \mid AB - BA \mid j \rangle^*.$$

Thus, operator

$$[A, B] = AB - BA$$

is anti-Hermitian. Similarly, the commutator of a Hermitian operator and an anti-Hermitian operator is Hermitian.

5.2
Commutation Relations for a Step Operator

As a guide to constructing step operators, we now derive certain conditions. Let us consider a given physical system or an independent part of a physical system with its concomitant Hilbert space. Let us also consider an observable, for which A is the pertinent Hermitian operator.

Suppose that $|i\rangle$ and $|j\rangle$ are two eigenkets for the operator,

$$A|i\rangle = a_i|i\rangle, \tag{5.4}$$

$$A|j\rangle = a_j|j\rangle, \tag{5.5}$$

and that B is the operator that converts $|i\rangle$ to $|j\rangle$:

$$B|i\rangle = |j\rangle. \tag{5.6}$$

Operating on (5.6) with A yields

$$AB|i\rangle = A|j\rangle = a_j|j\rangle = a_j B|i\rangle, \tag{5.7}$$

while operating on (5.4) with B yields

$$BA|i\rangle = Ba_i|i\rangle = a_i B|i\rangle. \tag{5.8}$$

Subtracting (5.8) from (5.7) gives us

$$(AB - BA)|i\rangle = (a_j - a_i)B|i\rangle = kB|i\rangle \tag{5.9}$$

or

$$[A, B]|i\rangle = kB|i\rangle \tag{5.10}$$

with

$$a_j = a_i + k. \tag{5.11}$$

If the ith and jth eigenkets are successive ones and if the spacing between the eigenvalues is constant, then k is constant.

From (5.10), we obtain the operator condition

$$[A, B] = kB. \tag{5.12}$$

According to Example 5.2, the commutator of two Hermitian operators is anti-Hermitian, while the commutator of a Hermitian operator and an anti-Hermitian operator is Hermitian. So B must have a Hermitian part B_1 and an anti-Hermitian part iB_2, where B_2 is also Hermitian. We write

$$B = B_1 + iB_2. \tag{5.13}$$

From (5.13), we obtain both

$$[A, B] = [A, B_1] + i[A, B_2] \tag{5.14}$$

and

$$kB = kB_1 + ikB_2. \tag{5.15}$$

Substituting into (5.12) yields

$$[A, B_1] + i[A, B_2] = kB_1 + ikB_2. \tag{5.16}$$

Transformation operators have only Hermitian and anti-Hermitian parts. In an operator equation, these must be separately equal. Here

$$[A, B_1] = ikB_2, \tag{5.17}$$

$$[A, B_2] = -ikB_1. \tag{5.18}$$

Now, multiplying B by a coefficient c produces

$$cB = cB_1 + icB_2 \tag{5.19}$$

and

$$[A, cB_1] = ikcB_2, \tag{5.20}$$

$$[A, cB_2] = -ikcB_1. \tag{5.21}$$

If we let

$$cB = C = C_1 + iC_2, \tag{5.22}$$

then (5.20) and (5.21) become

$$[A, C_1] = ikC_2, \tag{5.23}$$

$$[A, C_2] = -ikC_1. \tag{5.24}$$

Conversely, if Equations (5.23) and (5.24) hold for a given operator A, then (5.22) can be constructed. Constant c can be determined from the normalization condition on kets $|i\rangle$ and $|j\rangle$ and so step operator B constructed.

Example 5.3

Show that multiplying a step operator by a complex number of magnitude 1 does not destroy its nature.

The complex number of magnitude 1 can be written as $e^{i\alpha}$. Multiplying (5.6) by this yields

$$e^{i\alpha}B|i\rangle = e^{i\alpha}|j\rangle.$$

If we let

$$e^{i\alpha}B = C,$$

then

$$C|i\rangle = e^{i\alpha}|j\rangle.$$

Going from operator B to operator C merely shifts the phase of ket $|j\rangle$ by angle α. This does not alter what is represented by the ket.

5.3
Angular-Momentum Step Operators

A simple application of this theory is provided by the angular-momentum operators. The procedure is similar for the orbital angular momentum, the spin, and the total angular momentum.

Here let us consider operator A to be J_z, the operator for the z component of total angular momentum. From (3.44) and (3.43), we have

$$[J_z, J_x] = iJ_y, \tag{5.25}$$

$$[J_z, J_y] = -iJ_x. \tag{5.26}$$

These have the same form as (5.23) and (5.24) with $k = \pm 1$.

When $k = +1$, C_1 is J_x, C_2 is J_y, and (5.22) yields

$$C = J_x + iJ_y = J_+. \tag{5.27}$$

Thus, we obtain the step-up operator J_+. When $k = -1$, C_1 is J_y, C_2 is J_x, and

$$C = J_y + iJ_x \tag{5.28}$$

Multiplying this operator by $-i$ yields

$$-iC = J_x - iJ_y = J_-, \tag{5.29}$$

the step-down operator J_-.

From (3.109), (3.93), (3.92), and (3.115), we obtain

$$J_-J_+|jm\rangle = [j(j + 1) - m(m + 1)]|jm\rangle. \tag{5.30}$$

Multiplying by bra $\langle jm|$ then yields

$$\langle jm|J_-J_+|jm\rangle = [j(j + 1) - m(m + 1)]. \tag{5.31}$$

Recall that J_x and J_y are Hermitian. So iJ_y is anti-Hermitian and the adjoint of $J_\pm$ is $J_\mp$. Consequently,

$$\langle jm|J_-J_+|jm\rangle = \langle J_+|jm||J_+|jm\rangle. \tag{5.32}$$

Since J_+ is the step-up operator, we have

$$|J_+|jm\rangle = c_{jm}^+ |jm + 1\rangle \qquad (5.33)$$

and

$$\langle J_+|jm| = (c_{jm}^+)^* \langle jm + 1|. \qquad (5.34)$$

If the new ket is normalized as the original ket, we obtain

$$\langle J_+|jm||J_+|jm\rangle = (c_{jm}^+)^* c_{jm}^+ . \qquad (5.35)$$

Combining (5.35), (5.32), (5.31) then leads to the condition

$$(c_{jm}^+)^* c_{jm}^+ = [j(j + 1) - m(m + 1)]. \qquad (5.36)$$

We may choose the relative phasing so that c_{jm}^+ is real. Then

$$c_{jm}^+ = [j(j + 1) - m(m + 1)]^{\frac{1}{2}} \qquad (5.37)$$

and Equation (5.33) becomes

$$J_+|jm\rangle = [j(j + 1) - m(m + 1)]^{\frac{1}{2}}|jm + 1\rangle. \qquad (5.38)$$

Similarly working with

$$J_+ J_-|jm\rangle = [j(j + 1) - m(m - 1)]|jm\rangle \qquad (5.39)$$

leads to

$$J_-|jm\rangle = [j(j + 1) - m(m - 1)]^{\frac{1}{2}}|jm - 1\rangle. \qquad (5.40)$$

Table 5.1.
Commutation Relations Involving the z-Component
of Angular Momentum and the Corresponding Step Operators

$$[J_z, J_x] = \pm i(\pm 1)J_y$$
$$[J_z, J_y] = \mp i(\pm 1)J_x$$

$\Delta m = +1$	$\Delta m = -1$
	$iJ_- = J_y + iJ_x$
$J_+ = J_z + iJ_y$	$J_- = J_x - iJ_y$

$$[J_+, J_-] \equiv J_+ J_- - J_- J_+ = 2J_z$$
$$\{J_+. J_-\} \equiv J_+ J_- + J_- J_+ = 2J_x^2 + 2J_y^2$$

Remember that in these results, units in which $\hbar = 1$ are being used. In a general set of units, the right sides of (5.38) and (5.40) would be multiplied by $\hbar$.

Important formulas involving the step operators for angular momentum are listed in Table 5.1. Corresponding to formula (2.67) defining the commutator of two operators is the equation

$$AB + BA = \{A, B\} \tag{5.41}$$

defining the *anticommutator* of operators A and B.

5.4
Commutation Relations for the Harmonic Oscillator

Since the energy levels of the harmonic oscillator are equispaced, this system provides another simple application of the theory.

A model for a vibrational mode with a definite classical frequency is a particle of mass μ in the potential

$$U = \tfrac{1}{2} fx^2. \tag{5.42}$$

The classical angular frequency is

$$\omega = \sqrt{\frac{f}{\mu}}. \tag{5.43}$$

Furthermore, the classical form for the kinetic energy is

$$T = \frac{p^2}{2\mu}; \tag{5.44}$$

so the Hamiltonian operator has the form

$$H = \frac{P^2}{2\mu} + \tfrac{1}{2} fX^2 = \frac{P^2}{2\mu} + \tfrac{1}{2}\mu\omega^2 X^2. \tag{5.45}$$

For simplicity, let us employ the natural units in which

$$\mu = 1, \qquad f = 1, \qquad \hbar = 1. \tag{5.46}$$

Then the Hamiltonian reduces to

$$H = \tfrac{1}{2}P^2 + \tfrac{1}{2}X^2. \tag{5.47}$$

From the localizability requirement for a particle we obtained (2.68), which we rearrange to

$$PX = XP - i(1). \tag{5.48}$$

Combining this with operator X from the right yields

$$PX^2 = XPX - iX = X[XP - i(1)] - iX$$
$$= X^2P - 2iX, \tag{5.49}$$

while combining it with operator P from the left leads to

$$P^2X = PXP - iP = [XP - i(1)]P - iP$$
$$= XP^2 - 2iP. \tag{5.50}$$

Now, the commutator of the Hamiltonian with the momentum operator is

$$[H, P] = HP - PH = (\tfrac{1}{2}P^2 + \tfrac{1}{2}X^2)P - P(\tfrac{1}{2}P^2 + \tfrac{1}{2}X^2)$$
$$= \tfrac{1}{2}X^2P - \tfrac{1}{2}PX^2 = \tfrac{1}{2}X^2P - \tfrac{1}{2}X^2P + \tfrac{1}{2}(2iX)$$
$$= iX. \tag{5.51}$$

while the commutator of the Hamiltonian with the coordinate operator is

$$[H, X] = HX - XH = (\tfrac{1}{2}P^2 + \tfrac{1}{2}X^2)X - X(\tfrac{1}{2}P^2 + \tfrac{1}{2}X^2)$$
$$= \tfrac{1}{2}P^2X - \tfrac{1}{2}XP^2 = \tfrac{1}{2}XP^2 - \tfrac{1}{2}(2iP) - \tfrac{1}{2}XP^2$$
$$= -iP. \tag{5.52}$$

5.5
Harmonic Oscillator Step Operators

Overall Equations (5.51) and (5.52) have the same form as Equations (5.23) and (5.24). So the corresponding step operators can be readily constructed. Again the two values of k are ± 1.

When $k = +1$, C_1 is P, C_2 is X, and (5.22) yields

$$C = P + iX. \tag{5.53}$$

Multiplying this operator by $-i$ gives us

$$-iC = X - iP = \sqrt{2}\, a^+. \tag{5.54}$$

Here a^+ has been introduced by definition. Solving for it leads to the standard form

$$a^+ = \frac{1}{\sqrt{2}}(X - iP) \tag{5.55}$$

for the step-up operator.

When $k = -1$, C_1 is X, C_2 is P, and (5.22) produces

$$C = X + iP = \sqrt{2}\, a^-. \tag{5.56}$$

Solving for a^- gives us the standard form

$$a^- = \frac{1}{\sqrt{2}} (X + iP) \qquad (5.57)$$

for the step-down operator.

Combining these step operators leads to

$$a^- a^+ = \frac{1}{2}(X + iP)(X - iP) = \frac{1}{2} X^2 + \frac{1}{2} P^2 + \frac{1}{2} iPX - \frac{1}{2} iXP$$

$$= H + \frac{1}{2} iXP - \frac{1}{2} i^2(1) - \frac{1}{2} iXP = H + \frac{1}{2}(1). \qquad (5.58)$$

and

$$a^+ a^- = \frac{1}{2}(X - iP)(X + iP) = \frac{1}{2} X^2 + \frac{1}{2} P^2 - \frac{1}{2} iPX + \frac{1}{2} iXP$$

$$= H - \frac{1}{2} iXP + \frac{1}{2} i^2(1) + \frac{1}{2} iXP = H - \frac{1}{2}(1). \qquad (5.59)$$

In the reductions, (5.47) and (5.48) have been used.

The Hamiltonian is the operator for energy E. If we label each of its states with a number v, we have

$$H| v \rangle = E| v \rangle. \qquad (5.60)$$

Successive energies obey Equation (5.11). For $a^{\pm}$ acting on $| v \rangle$, we have $k = \pm 1$ and

$$\Delta E = \pm 1. \qquad (5.61)$$

Let us complete the definition of v by taking

$$v = E - \frac{1}{2} \qquad (5.62)$$

or

$$v + 1 = E + \frac{1}{2}, \qquad (5.63)$$

so

$$\Delta v = \pm 1. \qquad (5.64)$$

Let the combined operator $a^- a^+$ act on ket $| v \rangle$ and employ (5.58), (5.60) to reduce:

$$a^- a^+ | v \rangle = [H + \frac{1}{2}(1)]| v \rangle = (E + \frac{1}{2})| v \rangle. \qquad (5.65)$$

Then multiply by bra $\langle v |$:

$$\langle v | a^- a^+ | v \rangle = (E + \frac{1}{2}) \langle v | v \rangle. \qquad (5.66)$$

Similarly, let the combined operator $a^+ a^-$ act on ket $| v \rangle$ and use (5.59), (5.60) to reduce:

$$a^+ a^- \mid v \rangle = [H - \tfrac{1}{2}(1)] \mid v \rangle = (E - \tfrac{1}{2}) \mid v \rangle. \tag{5.67}$$

Then multiply by bra $\langle v \mid$:

$$\langle v \mid a^+ a^- \mid v \rangle = (E - \tfrac{1}{2}) \langle v \mid v \rangle. \tag{5.68}$$

Now, operator a^+ acting on $\mid v \rangle$ yields a constant times the next higher ket:

$$a^+ \mid v \rangle = c_v^+ \mid v + 1 \rangle, \tag{5.69}$$

while the action of a^- on $\mid v \rangle$ is

$$a^- \mid v \rangle = c_v^- \mid v - 1 \rangle. \tag{5.70}$$

The coefficients are determined by requiring all the kets to be normalized to 1. We recall that both X and P are Hermitian. As a consequence, iP is anti-Hermitian and

$$\langle v \mid a^- a^+ \mid v \rangle = \langle a^+ \mid v \mid \mid a^+ \mid v \rangle$$
$$= (c_v^+)^* c_v^+ \langle v + 1 \mid v + 1 \rangle. \tag{5.71}$$

From (5.66) and (5.71) with

$$\langle v \mid v \rangle = \langle v + 1 \mid v + 1 \rangle, \tag{5.72}$$

we get

$$(c_v^+)^* c_v^+ = E + \tfrac{1}{2} = v + 1 \tag{5.73}$$

If we take the coefficient to be real, we obtain

$$c_v^+ = \sqrt{v + 1} . \tag{5.74}$$

Similarly with iP anti-Hermitian, we have

$$\langle v \mid a^+ a^- \mid v \rangle = \langle a^- \mid v \mid \mid a^- \mid v \rangle$$
$$= (c_v^-)^* c_v^- \langle v - 1 \mid v - 1 \rangle. \tag{5.75}$$

From (5.68) and (5.75) with (5.72), we get

$$(c_v^-)^* c_v^- = E - \tfrac{1}{2} = v. \tag{5.76}$$

Choosing the coefficient to be real leads to

$$c_v^- = \sqrt{v}. \tag{5.77}$$

We know that the harmonic oscillator has a lowest level. Let us label this v_0 and note that for it

$$a^- \mid v_0 \rangle = 0 \tag{5.78}$$

and

$$a^+ a^- \mid v_0\rangle = 0. \tag{5.79}$$

On comparing with (5.68), we see that

$$E = \tfrac{1}{2} \quad \text{when} \quad v = v_0. \tag{5.80}$$

Substituting this result into (5.62) yields

$$v_0 = 0. \tag{5.81}$$

With (5.64), we obtain the possible values

$$v = 0, 1, 2, 3, \dots \tag{5.82}$$

Important results from this section are listed in Table 5.2.

Table 5.2
Commutation Relations and the Corresponding Step Operators for the Harmonic-Oscillator Kets

$$[H, P] = \pm i(\pm 1)X$$
$$[H, X] = \mp i(\pm 1)P$$

$\Delta v = +1$	$\Delta v = -1$
$a^+ = \dfrac{1}{\sqrt{2}}\,(X - iP)$	$a^- = \dfrac{1}{\sqrt{2}}\,(X + iP)$
$a^+ \mid v\rangle = \sqrt{v+1}\,\mid v + 1\rangle$	$a^- \mid v\rangle = \sqrt{v}\,\mid v - 1\rangle$

$$[a^-, a^+] \equiv a^- a^+ - a^+ a^- = (1)$$
$$\{a^-, a^+\} \equiv a^- a^+ + a^+ a^- = 2H$$

Example 5.4

Find a representation of the ground state ket for the harmonic oscillator. Combine (5.78) and (5.81). Then introduce $\sqrt{2}$ times (5.57) to get

$$(X + iP)\mid 0\rangle = 0.$$

In the coordinate representation, (2.128), (2.129) hold and this equation becomes

$$\left(x + \frac{d}{dx}\right)\psi = 0.$$

Integration and normalization then leads to

$$\psi_0 = \left(\frac{1}{\pi}\right)^{1/4} e^{-x^2/2}.$$

In a general system of units, x^2 is replaced with ax^2 and coefficient $(1/\pi)^{1/4}$ with $(a/\pi)^{1/4}$, as in Table 4.4.

Example 5.5

Explain how representations of higher level kets for the harmonic oscillator are found.

From (5.69) and (5.74), the action of the step-up operator on a harmonic oscillator eigenket is

$$a^+ | v - 1 \rangle = \sqrt{v} \, | v \rangle,$$

whence

$$| v \rangle = \frac{1}{\sqrt{v}} a^+ | v - 1 \rangle = \frac{1}{\sqrt{2v}} (X - iP) | v - 1 \rangle.$$

In the coordinate representation, the overall equation becomes

$$\psi_v = \frac{1}{\sqrt{2v}} \left(x - \frac{d}{dx} \right) \psi_{v-1}.$$

Introducing the result from Example 5.4 and iterating up to the vth level then gives us

$$\psi_v = \frac{1}{(2^v v!)^{1/2}} \left(x - \frac{d}{dx} \right)^v \left(\frac{1}{\pi} \right)^{1/4} e^{-x^2/2}.$$

Since

$$\left(x - \frac{d}{dx} \right) e^{x^2/2} \frac{d^{u-1}}{dx^{u-1}} e^{-x^2} = -e^{x^2/2} \frac{d^u}{dx^u} e^{-x^2}$$

for any u, we have

$$\left(x - \frac{d}{dx} \right)^v e^{-x^2/2} = (-1)^v e^{-x^2/2} e^{x^2} \frac{d^v}{dx^v} e^{-x^2}$$

by induction. Introducing the definition of a *Hermite polynomial*

$$H_v(x) = (-1)^v e^{x^2} \frac{d^v}{dx^v} e^{-x^2},$$

in addition, we obtain

$$\psi_v = \frac{1}{(2^v v!)^{1/2}} \left(\frac{1}{\pi} \right)^{1/4} e^{-x^2/2} H_v(x).$$

5.6
More General Step Operators

With most one-dimensional systems the eigenvalues for a pertinent physical operator are not equally spaced. Furthermore, at a certain level, transition to a continuum of levels may appear. Nevertheless, for the simpler of such systems one can construct step operators without trouble. Here we will consider a generalization of a^+ and a^- which still yields the form of the Schrödinger equation.

Let us replace operator X in (5.55) and (5.57) by a function of X and a parameter k:

$$a_k^{\pm} = \frac{1}{\sqrt{2}} [q(X, k) \mp iP]. \tag{5.83}$$

In the coordinate representation, operator X reduces to multiplication by variable x, while i times the momentum operator P is

$$iP \equiv D = \frac{d}{dx}. \tag{5.84}$$

The step operator now becomes

$$a_k^{\pm} = \frac{1}{\sqrt{2}} [q(x, k) \mp D]. \tag{5.85}$$

Next, construct $a^- a^+$ for the parameter equalling $k + 1$ and $a^+ a^-$ for it equalling just k:

$$a_{k+1}^- a_{k+1}^+ = \frac{1}{2}[q(x, k + 1) + D] [q(x, k + 1) - D]$$
$$= \frac{1}{2}[q^2(x, k + 1) + q'(x, k + 1) - D^2], \tag{5.86}$$

$$a_k^+ a_k^- = \frac{1}{2}[q(x, k) - D] [q(x, k) + D]$$
$$= \frac{1}{2}[q^2(x, k) - q'(x, k) - D^2]. \tag{5.87}$$

By analogy with (5.65) and (5.67), we consider a system for which action of operator (5.86) on ket $|jk\rangle$ yields a number times $|jk\rangle$, thus

$$[q^2(x, k + 1) + q'(x, k + 1) - D^2] |jk\rangle = 2[j - K(k + 1)] |jk\rangle, \tag{5.88}$$

and for which action of operator (5.87) on ket $|jk\rangle$ yields a different number times $|jk\rangle$, thus

$$[q^2(x, k) - q'(x, k) - D^2] |jk\rangle = 2[j - K(k)] |jk\rangle. \tag{5.89}$$

Rearranging (5.88) leads to

$$[D^2 + g(x, k) + 2j] |jk\rangle = 0 \tag{5.90}$$

if

$$q^2(x, k + 1) + q'(x, k + 1) + 2K(k + 1) = -g(x, k) \tag{5.91}$$

or

$$q^2(x, k) + q'(x, k) + 2K(k) = -g(x, k - 1). \tag{5.92}$$

Rearranging (5.89) also leads to (5.90) if

$$q^2(x, k) - q'(x, k) + 2K(k) = -g(x, k). \tag{5.93}$$

Adding (5.92) and (5.93), dividing by 2,

$$q^2(x, k) + 2K(k) = -\tfrac{1}{2}[g(x, k - 1) + g(x, k)], \tag{5.94}$$

then differentiating with respect to x gives us

$$2q(x, k)q'(x, k) = -\tfrac{1}{2}[g'(x, k - 1) + g'(x, k)]. \tag{5.95}$$

Subtracting (5.93) from (5.92) and dividing by 2 gives us

$$q'(x, k) = -\tfrac{1}{2}[g(x, k - 1) - g(x, k)]. \tag{5.96}$$

Now, substituting (5.96) into (5.95) and solving for $q(x, k)$ yields

$$q(x, k) = \frac{1}{2} \frac{g'(x, k - 1) + g'(x, k)}{g(x, k - 1) - g(x, k)}, \tag{5.97}$$

while solving for $2K(k)$ in (5.94) yields

$$2K(k) = -\tfrac{1}{2}[g(x, k - 1) + g(x, k)] - q^2(x, k). \tag{5.98}$$

For the procedure to work, the combination of functions on the right of (5.98) must yield a function of k alone; coordinate x must drop out. Then the (q, k) constructed as in (5.97) makes

$$a_k^{\pm} = \frac{1}{\sqrt{2}} [q(x, k) \mp D] \tag{5.99}$$

a step-up or step-down operator.

Now, substitute (5.88) into (5.86) and let the result act on ket $|jk\rangle$:

$$a_{k+1}^{-} a_{k+1}^{+} |jk\rangle = [j - K(k + 1)]|jk\rangle . \tag{5.100}$$

Similarly, substitute (5.89) into (5.87) and let the result act on ket $|jk\rangle$:

$$a_k^{+} a_k^{-} |jk\rangle = [j - K(k)]|jk\rangle. \tag{5.101}$$

Then operate on (5.100) with a_{k+1}^{+} and on (5.101) with a_k^{-} :

$$a_{k+1}^{+} a_{k+1}^{-} (a_{k+1}^{+} |jk\rangle) = [j - K(k + 1)](a_{k+1}^{+} |jk\rangle, \tag{5.102}$$

$$a_k^{-} a_k^{+} (a_k^{-} |jk\rangle) = [j - K(k)](a_k^{-} |j\rangle. \tag{5.103}$$

On comparing (5.102) with (5.101), (5.103) with (5.100), one is led to write

$$a_{k+1}^+ \mid jk\rangle = c_1 \mid jk{+}1\rangle \tag{5.104}$$

and

$$a_k^- \mid jk\rangle = c_2 \mid jk{-}1\rangle, \tag{5.105}$$

as long as j and k determine the initial state. The normalization condition fixes the magnitudes of c_1 and c_2.

5.7
The Uniform Normalization Condition

Important eigenvalue equations can be put into form (5.90) with the function $g(x, k)$ satisfying restriction (5.98). In some of the situations, parameter k would be chosen so that $K(k)$ increases monotonically with k; in others, so that $K(k)$ decreases monotonically as k increases.

First, consider a suitable equation in which K increases as k increases, with

$$j \geqslant K(k + 1). \tag{5.106}$$

Let a typical eigenket, $\mid jk\rangle$, for the system be normalized to 1:

$$\langle jk \mid jk\rangle = 1. \tag{5.107}$$

When the eigenket one step up the ladder also exists, formulas (5.104) and (5.100) tell us that

$$
\begin{aligned}
\mid c_1 \mid^2 \langle jk{+}1 \mid jk{+}1\rangle &= \langle a_{k+1}^+ \mid jk \mid\mid a_{k+1}^+ \mid jk\rangle \\
&= \langle jk \mid\mid a_{k+1}^- a_{k+1}^+ \mid jk\rangle = [j - K(k + 1)] \langle jk \mid jk\rangle. \tag{5.108}
\end{aligned}
$$

For this ket to be normalizable the expression in brackets must be positive. But the increase in K with k causes this expression to become negative above a valid k which we label k_{max}. To avoid a contradiction we must cut off the ladder at this stage by setting

$$j = K(k_{max} + 1). \tag{5.109}$$

Now from (5.108), successive kets have the same normalizations when

$$\mid c_1 \mid^2 = K(k_{max} + 1) - K(k + 1). \tag{5.110}$$

We may choose the relative phasing so

$$c_1 = [K(k_{max} + 1) - K(k + 1)]^{1/2} \tag{5.111}$$

and (5.104) becomes

$$a_{k+1}^+ \mid jk\rangle = [K(k_{\max} + 1) - K(k + 1)]^{\frac{1}{2}} \mid jk{+}1\rangle. \qquad (5.112)$$

Next, consider a suitable eigenvalue equation in which K increases as k decreases, with

$$j \geqslant K(k). \qquad (5.113)$$

Let a typical eigenket, $\mid jk\rangle$, for the system be normalized to 1:

$$\langle jk \mid jk\rangle = 1. \qquad (5.114)$$

When the eigenket one step down the ladder also exists, formulas (5.105) and (5.101) allow us to construct

$$\mid c_2 \mid^2 \langle jk{-}1 \mid jk{-}1\rangle = \langle a_k^- \mid jk \mid\mid a_k^- \mid jk\rangle = \langle jk \mid\mid a_k^+ a_k^- \mid jk\rangle$$

$$= [j - K(k)] \langle jk \mid jk\rangle. \qquad (5.115)$$

Again, the factor in brackets must be positive for the scalar product on the left to be positive. This condition is met as long as inequality (5.113) holds, but the increase in K with decreasing k causes this factor to become negative below a valid k labeled $k_{\min}$. Since the scalar product of a ket with itself must be positive, we have a contradiction unless the ladder is truncated by the equality

$$j = K(k_{\min}). \qquad (5.116)$$

Substituting (5.116) and the generalization of (5.114) into (5.115) yields

$$\mid c_2 \mid^2 = K(k_{\min}) - K(k). \qquad (5.117)$$

We may choose the relative phasing so

$$c_2 = [K(k_{\min}) - K(k)]^{\frac{1}{2}} \qquad (5.118)$$

and (5.105) becomes

$$a_k^- \mid jk\rangle = [K(k_{\min}) - K(k)]^{\frac{1}{2}} \mid jk{-}1\rangle. \qquad (5.119)$$

A summary of the restrictions and results appears in Table 5.3.

5.8
Application to the Radial Motion in a Coulomb Field

The energy levels for a hydrogenlike atom are determined by the radial part of its Schrödinger equation. In a natural set of units, this part assumes form (4.62). But as it stands, (4.62) does not fit (5.90).

Table 5.3
The Parameterized Eigenvalue Equation
and Properties of Step Operators Therefor

$$[D^2 + g(x, k) + 2j] \mid jk\rangle = 0$$

$$a_k^{\pm} = \frac{1}{\sqrt{2}} [q(x, k) \mp D]$$

$$q(x, k) = \frac{1}{2} \frac{g'(x, k - 1) + g'(x, k)}{g(x, k - 1) - g(x, k)}$$

$$2K(k) = -\tfrac{1}{2}[g(x, k - 1) + g(x, k)] - q^2(x, k)$$

$$\mid jk+1\rangle = \frac{a_{k+1}^{+}}{[j - K(k + 1)]^{\frac{1}{2}}} \mid jk\rangle$$

$$\mid jk-1\rangle = \frac{a_k^{-}}{[j - K(k)]^{\frac{1}{2}}} \mid jk\rangle$$

K increases as k decreases	K increases as k increases
$j = K(k_{min})$	$j = K(k_{max} + 1)$

$$[a^-, a^+] = K(k) - K(k + 1)$$
$$\{a^-, a^+\} = 2j - [K(k) + K(k + 1)]$$

To remove the first order derivative, one introduces the transformation

$$y = xR. \tag{5.120}$$

Then the radial differential equation for bound states, from (4.62), reduces to

$$\frac{d^2y}{dx^2} + \left[\frac{n}{x} - \frac{l(l + 1)}{x^2} - \frac{1}{4} \right] y = 0. \tag{5.121}$$

On comparing (5.121) with (5.90), we see that

$$j = -\frac{1}{8} \tag{5.122}$$

and

$$g(x, l) = \frac{n}{x} - \frac{l(l + 1)}{x^2}, \tag{5.123}$$

whence

$$g(x, l - 1) = \frac{n}{x} - \frac{(l - 1)l}{x^2} \tag{5.124}$$

and

$$g'(x, l) = -\frac{n}{x^2} + \frac{2l(l + 1)}{x^3}, \tag{5.125}$$

$$g'(x, l - 1) = -\frac{n}{x^2} + \frac{2(l - 1)l}{x^3}. \tag{5.126}$$

We let l play the role of k. Then substituting (5.123), (5.124), (5.125), (5.126) into (5.98) yields

$$2K = -\frac{n^2}{4l^2}. \tag{5.127}$$

Since this result is independent of x, the method is applicable.

We note that K increases as parameter l increases; so the ladder must be cut off where condition (5.109) is met. Here

$$j = K(l_{\max} + 1). \tag{5.128}$$

Introducing (5.122) and (5.127) changes (5.128) to

$$-\frac{1}{8} = -\frac{n^2}{8(l_{\max} + 1)^2}. \tag{5.129}$$

This equation is satisfied when

$$l_{\max} = n - 1. \tag{5.130}$$

Next, substituting (5.123), (5.124), (5.125), (5.126) into (5.97) yields

$$q = \frac{l}{x} - \frac{n}{2l}. \tag{5.131}$$

So the unnormalized step operators are given by

$$a_l^{\ddagger} = \frac{1}{\sqrt{2}} \left[\frac{l}{x} - \frac{n}{2l} \mp D \right]. \tag{5.132}$$

Substituting into the formulas in Table 5.3 and reducing leads to

$$| j l+1 \rangle = \frac{2(l + 1)}{[(n + l + 1)(n - l - 1)]^{1/2}} \left[\frac{l + 1}{x} - \frac{n}{2(l + 1)} - D \right] | j l \rangle, \tag{5.133}$$

$$| j l - 1 \rangle = \frac{2l}{[(n + l)(n - l)]^{1/2}} \left[\frac{l}{x} - \frac{n}{2l} + D \right] | j l \rangle. \tag{5.134}$$

At the upper cutoff, the step-up operator $a_{l+1}^{\ddagger}$ acting on the coordinate representation of $| j l \rangle$ produces a vanishing result; and, the orbital angular momentum is a maximum for the given energy.

Since the orbital angular momentum also has a minimum, there is a lower cutoff. According to (5.134), this can only occur with l equal to zero. And since the parameter l increases by integral steps, to the value given by (5.130), we have

$$l = 0, 1, 2, \ldots, n - 1. \qquad (5.135)$$

Because l_{max} is an integer, n must also be an integer and

$$n = 1, 2, 3, \ldots . \qquad (5.136)$$

Equations (4.59) and (4.60) now determine the energy levels.

Example 5.6

Construct the radial factor for the hydrogenlike atom at the maximum in l. At the upper cutoff, we have

$$l + 1 = n$$

from (5.130), and

$$d^+_{l+1} \, y = a^+_n \, y = 0$$

from (5.104). Employing (5.132) and canceling the $\sqrt{2}$ leads to

$$\left[\frac{n}{x} - \frac{1}{2} - \frac{d}{dx} \right] y = 0.$$

Separate the variables,

$$\left(\frac{n}{x} - \frac{1}{2} \right) dx = \frac{dy}{y}$$

and integrate,

$$\ln A + n \ln x - \frac{x}{2} = \ln y.$$

Then take the antilogarithm,

$$y = A x^n e^{-x/2},$$

substitute into a rearranged (5.120), and eliminate x with (4.61):

$$R_{n,n-1} = A x^{n-1} e^{-x/2} = A \left(\frac{2r}{an} \right)^{n-1} e^{-r/an}.$$

Example 5.7

Evaluate

$$\int_0^\infty x^n e^{-\alpha x} dx$$

for a positive α and an integral n.

We know that

$$\int_0^\infty e^{-\alpha x} dx = - \left. \frac{e^{-\alpha x}}{\alpha} \right|_0^\infty = \frac{1}{\alpha} .$$

Now, differentiate the overall equation with respect to α, repeatedly, to get

$$\int_0^\infty -x e^{-\alpha x} dx = -\frac{1}{\alpha^2} \quad \text{or} \quad \int_0^\infty x e^{-\alpha x} dx = \frac{1}{\alpha^2} ,$$

$$\int_0^\infty x^2 e^{-\alpha x} dx = \frac{2}{\alpha^3} ,$$

$$\int_0^\infty x^3 e^{-\alpha x} dx = \frac{3!}{\alpha^4} .$$

By induction, we obtain

$$\int_0^\infty x^n e^{-\alpha x} dx = \frac{n!}{\alpha^4} .$$

Example 5.8

Normalize the radial factor obtained in Example 5.6, with A real and positive. The normalization factor is chosen so that

$$\int_0^\infty R^2 r^2 \, dr = \left(\frac{an}{2} \right)^3 \int_0^\infty R^2 x^2 \, dx = 1.$$

Insert the radial factor from Example 5.6, factor out the A^2, and apply the result from Example 5.7:

$$A^2 \left(\frac{an}{2} \right)^3 \int_0^\infty x^{2n} e^{-x} dx = A^2 \left(\frac{an}{2} \right)^3 (2n)! = 1.$$

Then solve for A:

$$A = \left(\frac{2}{an} \right)^{3/2} \frac{1}{[(2n)!]^{1/2}} .$$

With $R_{n,l\,=\,n-1}$ normalized, application of (5.134) yields the other radial factors for the given n in normalized form. The results check those in Table 4.5.

5.9
The Morse Potential

A mode of vibratory motion in a molecule or a solid behaves approximately as a harmonic oscillator when it possesses little excess energy. But with further excitation, its apparent frequency falls. At a certain level this frequency reaches zero and dissociation may occur.

As a model, consider the internal motion of a diatomic molecule. Let the distance between the nuclear centers be r. Let this distance, when the potential U is a minimum, be r_e. Also, let the depth of the potential below its value at infinite separation, at the minimum point r_e, be U_0. When the mass of the first atom is m_1 and the mass of the second atom is m_2, the reduced mass μ is given by

$$\frac{1}{\mu} = \frac{1}{m_1} + \frac{1}{m_2}. \tag{5.137}$$

If the interaction between the two atoms were governed by the parabolic potential

$$U = \tfrac{1}{2} f(r - r_e)^2, \tag{5.138}$$

with f the force constant, the molecule would oscillate at the angular frequency

$$\omega = \sqrt{\frac{f}{\mu}}, \tag{5.139}$$

according to classical theory. Then it would behave as a harmonic oscillator. Differentiating (5.138) twice yields the relationship

$$\frac{\partial^2 U}{\partial r^2} = f. \tag{5.140}$$

But a parabolic potential increases without limit as r increases. The simplest potential that approaches a constant with increasing r, approximates the parabolic U around r_e, and increases rapidly as r decreases is the one,

$$U = U_0[1 - e^{-a(r-r_e)}]^2$$

$$= U_0 - 2U_0 e^{-a(r-r_e)} + U_0 e^{-2a(r-r_e)}, \tag{5.141}$$

proposed by Philip M. Morse. Parameter U_0 is the extreme depth of the potential below its asymptotic limit; parameter a, on the other hand, is related to the force constant.

Indeed, differentiating (5.141) twice yields

$$\left(\frac{\partial^2 U}{\partial r^2} \right)_{r=r_e} = 2a^2 U_0. \tag{5.142}$$

For the parabolic potential to fit the Morse potential in the neighborhood of the minimum, as Figure 5.1 shows, this second derivative must equal the one in (5.140); then

$$f = 2a^2 U_0. \tag{5.143}$$

The corresponding angular frequency, from (5.139), is

$$\omega = a \left(\frac{2U_0}{\mu} \right)^{\frac{1}{2}}. \tag{5.144}$$

Now, the Schrödinger equation with the Morse potential does not appear to fit the key equations in Table 5.3. Some transformations are needed. First, let us make the substitution

$$a(r - r_e) = x + \ln 2U_0^{\frac{1}{2}} \tag{5.145}$$

in (5.141) to get

$$U = U_0 - U_0^{\frac{1}{2}} e^{-x} + \frac{1}{4} e^{-2x}. \tag{5.146}$$

Then in the middle term on the right, we set

$$U_0^{\frac{1}{2}} = k + \frac{1}{2}. \tag{5.147}$$

We choose natural units for which

$$\hbar = 1, \qquad \mu = \frac{1}{2}, \qquad a = 1. \tag{5.148}$$

We also note that

$$\frac{d}{dr} = \frac{d}{dx} \qquad \text{and} \qquad \frac{d^2}{dr^2} = \frac{d^2}{dx^2}. \tag{5.149}$$

So the Schrödinger equation assumes the form

$$\frac{d^2\psi}{dx^2} + (E - U)\psi = 0. \tag{5.150}$$

Employing potential (5.146) with parameterization (5.147) leads to

$$\left[\frac{d^2}{dx^2} - \frac{1}{4} e^{-2x} + (k + \frac{1}{2})e^{-x} + (E - U_0) \right] \psi = 0. \tag{5.151}$$

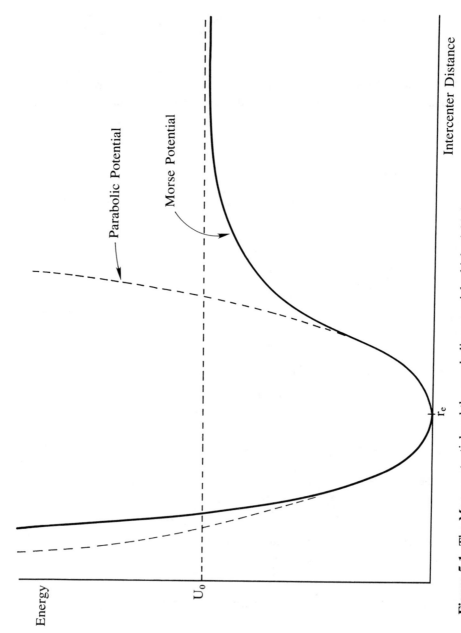

Figure 5.1. The Morse potential and the parabolic potential which yield the same small-amplitude classical frequency.

5.10
Energy Levels for the Morse Oscillator

Equation (5.90) with condition (5.98) does not depend on the existence of a ladder of levels at the given j, even though a visualization of such a structure led us to the formulas. However, the two together do require the different values of k to differ by integers. Some or all of the k's may correspond to different j's, as we will now see.

The system under consideration is the Morse oscillator described by (5.151). On comparing this equation with (5.90), or with the first equation in Table 5.3, we see that

$$g(x, k) = -\tfrac{1}{4}e^{-2x} + (k + \tfrac{1}{2})e^{-x}, \qquad (5.152)$$

$$g'(x, k) = \tfrac{1}{2}e^{-2x} - (k + \tfrac{1}{2})e^{-x}, \qquad (5.153)$$

and

$$2j = E - U_0. \qquad (5.154)$$

Substituting into (5.97), or into the third equation in Table 5.3, gives us

$$q = -\frac{e^{-x}}{2} + k. \qquad (5.155)$$

Most importantly, substituting into the right side of (5.98), or into the fourth equation in Table 5.3, leads to

$$2K = -k^2, \qquad (5.156)$$

a function of k alone.

Furthermore, this function increases as k decreases. Following the result in Section 5.7, summarized in Table 5.3, we see that for the lowest level,

$$2j = 2K(k_{min}) = -k_{min}^2. \qquad (5.157)$$

Substituting this into a rearranged (5.154) and employing (5.147) leads to

$$E = U_0 + 2j = U_0 - k_{min}^2 = U_0 - (U_0^{1/2} - \tfrac{1}{2})^2$$
$$= \tfrac{1}{2}(2U_0^{1/2}) - (\tfrac{1}{2})^2. \qquad (5.158)$$

For the higher levels, we set

$$k = k_{min} + v \qquad \text{with} \qquad v = 1, 2, 3, \ldots . \qquad (5.159)$$

So now

$$E = U_0 - k_{min}^2 = U_0 - (k - v)^2 = U_0 - [U_0^{1/2} - (v + 1/2)]^2$$
$$= (v + 1/2)2U_0^{1/2} - (v + 1/2)^2. \tag{5.160}$$

With (5.144), we identify $2U_0^{1/2}$ in our natural units as $\hbar\omega$. Then the 1 multiplying the last term is identified as $b\hbar\omega$. Thus, we obtain

$$E = (v + 1/2)\hbar\omega - (v + 1/2)^2 b\hbar\omega \quad \text{with} \quad v = 0, 1, 2, \ldots . \tag{5.161}$$

The Schrödinger equation for the Morse oscillator, with all parameters explicitly present, has the form

$$\left[\frac{a^2\hbar^2}{2\mu} \frac{d^2}{d(ar)^2} + \ldots + (E - U_0) \right] \psi = 0. \tag{5.162}$$

But ar is dimensionless, while E and U_0 have the dimensions of energy. So a dimensionless quantity has to be multiplied by the coefficient

$$\frac{a^2\hbar^2}{2\mu} \tag{5.163}$$

to obtain its value in energy units.

From (5.160), we obtain

$$E = (v + 1/2)2U_0^{1/2} \frac{a\hbar}{(2\mu)^{1/2}} - (v + 1/2)^2 \frac{a^2\hbar^2}{2\mu}$$
$$= (v + 1/2)\hbar\omega - (v + 1/2)^2 b\hbar\omega. \tag{5.164}$$

In the last step, (5.144) has been introduced, together with

$$b = \frac{\hbar a}{2} \frac{1}{(2U_0\mu)^{1/2}}. \tag{5.165}$$

Number b is called the *anharmonicity constant*. Combining (5.165) with (5.144) yields

$$b = \frac{\hbar\omega}{4U_0}. \tag{5.166}$$

Example 5.9

Removing the rotational shifts from the vibrational spectrum of HCl leaves a fundamental at 2886 cm^{-1} and an overtone at 5668 cm^{-1}. Calculate the harmonic oscillator energy $\hbar\omega$ and the anharmonicity constant b for HCl.

Let the vibrational quantum number for the higher level be v_2, for the lower level, v_1. Then the energy of the photon absorbed (or emitted) is

$$\Delta E = [(v_2 + 1/2) - (v_2 + 1/2)^2 b - (v_1 + 1/2) + (v_1 + 1/2)^2 b]\hbar\omega$$

from Equation (5.161).

The fundamental results when $v_2 = 1$ and $v_1 = 0$; the first overtone when $v_2 = 2$ and $v_1 = 0$. Putting the given data into the equation for E gives us

$$2886 \text{ cm}^{-1} = \left[\frac{3}{2} - \left(\frac{3}{2} \right)^2 b - \frac{1}{2} + \left(\frac{1}{2} \right)^2 b \right] \hbar\omega = (1 - 2b)\hbar\omega,$$

$$5668 \text{ cm}^{-1} = \left[\frac{5}{2} - \left(\frac{5}{2} \right)^2 b - \frac{1}{2} + \left(\frac{1}{2} \right)^2 b \right] \hbar\omega = (2 - 6b)\hbar\omega.$$

Solving these equations simultaneously yields

$$\hbar\omega = 2990 \text{ cm}^{-1}$$

and

$$b = 0.0174.$$

Substituting these values into (5.166) would yield a dissociation energy that is too large, because of the approximate nature of the Morse potential. The RKR method described by Castano et al. (*J. Chem. Ed.* **60**, 91-93) does yield an accurate potential.

5.11
Creation and Annihilation Operators for Bosons

We have noted that interaction of a charged particle with the ambient electromagnetic field is quantized following the Einstein equation

$$E = \hbar\omega. \tag{5.167}$$

Here ω is the classical angular frequency of the field and, by assignment, of the photon involved.

But (5.167) is the energy associated with the transition

$$\Delta v = \pm 1 \tag{5.168}$$

of a harmonic oscillator. So the production of a photon is analogous to raising the energy of a harmonic oscillator one step; absorption of a photon is analogous to lowering the energy of a harmonic oscillator one step. Furthermore, these transitions are effected by the operators a^+ and a^- of Table 5.2. So we expect similar operators to exist for the production and absorption of a photon of given energy.

Let us consider the electromagnetic radiation in a box of convenient size. Let n_j be the number of photons of angular frequency ω_j within the box. Then addition of one of these photons to the box is represented by

$$b_j^+ \mid n_j \rangle = \sqrt{n_j + 1} \mid n_j + 1 \rangle, \tag{5.169}$$

while subtraction of one of these photons from the box is represented by

$$b_j^- \mid n_j \rangle = \sqrt{n_j} \mid n_j - 1 \rangle. \tag{5.170}$$

Note how quantum number v is replaced by the photon number n_j, while b_j^+ is the step-up operator and b_j^- the step-down operator for the given angular frequency. We also have the commutation rule

$$b_j^- \, b_j^+ - b_j^+ \, b_j^- = (1) \tag{5.171}$$

as before.

We suppose that in the box, the radiation angular frequencies

$$\omega_1, \ \omega_2, \ \ldots, \ \omega_j, \ \ldots \tag{5.172}$$

can be induced. These are described by the kets

$$\mid n_1 \rangle, \ \mid n_2 \rangle, \ \ldots, \ \mid n_j \rangle, \ \ldots \ . \tag{5.173}$$

Since the different frequencies of radiation are independent, these combine multiplicatively to yield the overall ket

$$\mid n_1 \rangle \mid n_2 \rangle \ldots \mid n_j \rangle \ldots = \mid n_1, n_2, \ldots, n_j \ldots \rangle. \tag{5.174}$$

Furthermore, we have

$$b_j^+ \mid n_1, n_2, \ldots, n_j, \ldots \rangle = \sqrt{n_j + 1} \mid n_1, n_2, \ldots, n_j + 1, \ldots \rangle \tag{5.175}$$

and

$$b_j^- \mid n_1, n_2, \ldots, n_j, \ldots \rangle = \sqrt{n_j} \mid n_1, n_2, \ldots, n_j - 1, \ldots \rangle. \tag{5.176}$$

Also because of the independence, we find that

$$b_j^+ \, b_k^+ \mid n_1, n_2, \ldots, n_j, \ldots, n_k, \ldots \rangle$$
$$= \sqrt{n_k + 1} \, b_j^+ \mid n_1, n_2, \ldots, n_j, \ldots, n_k + 1, \ldots \rangle$$
$$= \sqrt{n_j + 1} \, \sqrt{n_k + 1} \mid n_1, n_2, \ldots, n_j + 1, \ldots, n_k + 1, \ldots \rangle \tag{5.177}$$

and

$$b_k^+ \, b_j^+ \mid n_1, n_2, \ldots, n_j, \ldots, n_k, \ldots \rangle$$
$$= \sqrt{n_j + 1} \, b_k^+ \mid n_1, n_2, \ldots, n_j + 1, \ldots, n_k, \ldots \rangle$$
$$= \sqrt{n_k + 1} \, \sqrt{n_j + 1} \mid n_1, n_2, \ldots, n_j + 1, \ldots, n_k + 1, \ldots \rangle \tag{5.178}$$

Consequently, we can say that

$$b_j^+ \, b_k^+ - b_k^+ \, b_j^+ = 0. \tag{5.179}$$

Similarly, we find that

$$b_j^- b_k^- - b_k^- b_j^- = 0 \qquad (5.180)$$

and that

$$b_j^- b_k^+ - b_k^+ b_j^- = \delta_{jk}(1). \qquad (5.181)$$

Since b_j^+ adds a photon of angular frequency ω_j to the radiation system, it is called the jth *creation operator*. Since b_j^- removes a photon of angular frequency ω_j from the radiation system, it is called the jth *annihilation operator*.

These operator relations may be applied to other bosons also. The basic properties are listed in Table 5.4. The Hilbert space for each system is extensive enough to allow for any occupancy of individual states.

Table 5.4
Relationships Involving Boson Creation and Annihilation Operators

$$b_j^+ \, | \, n_j \rangle = \sqrt{n_j + 1} \, | \, n_j + 1 \rangle$$

$$b_j^- \, | \, n_j \rangle = \sqrt{n_j} \, | \, n_j - 1 \rangle$$

$$[b_j^+ , b_k^+] \equiv b_j^+ b_k^+ - b_k^+ b_j^+ = 0$$

$$[b_j^- , b_k^-] \equiv b_j^- b_k^- - b_k^- b_j^- = 0$$

$$[b_j^- , b_k^+] \equiv b_j^- b_k^+ - b_k^+ b_j^- = \delta_{jk}$$

5.12
Creation and Annihilation Operators for Fermions

Fermions differ from bosons in that any state function for them must be antisymmetric in the exchange of identical particles, rather than symmetric. This property restricts the occupancy of any individual state or cell to one. However, the foregoing procedure can be readily altered to satisfy this condition.

For simplicity, we limit our consideration to p orthonormal orbitals which a fermion of the given species may occupy with $s_z = \pm 1$. Spin $+1$ may be indicated by $\uparrow$ (or by factor α); spin -1 may be indicated by $\downarrow$ (or by factor β). An n-fermion state is then represented by the ket

$$| \, p \rangle = | \, n_{1\uparrow}, n_{1\downarrow}, n_{2\uparrow}, \ldots, n_{js}, \ldots, n_{p\downarrow} \rangle, \qquad (5.182)$$

where n_{js} is the number of fermions of the given species in the js-th state.

Since n_{js} may be only 0 or 1, we change (5.175) to the form

$$f_{js}^+ \mid \ldots n_{js} \ldots \rangle = (-1)^{N_{js}} \sqrt{1 - n_{js}} \mid \ldots n_{js} + 1 \ldots \rangle, \qquad (5.183)$$

where

$$N_{js} = n_{1\uparrow} + n_{1\downarrow} + \ldots + n_{js}, \qquad (5.184)$$

and (5.176) to the form

$$f_{js}^- \mid \ldots n_{js} \ldots \rangle = (-1)^{N_{js}-1} \sqrt{n_{js}} \mid \ldots n_{js} - 1 \ldots \rangle. \qquad (5.185)$$

The signs are chosen so that the required antisymmetry is maintained. Applying the step-down and the step-up operators in succession yields

$$f_{js}^+ f_{js}^- \mid \ldots n_{js} \ldots \rangle = f_{js}^+ (-1)^{N_{js}-1} \sqrt{n_{js}} \mid \ldots n_{js}-1 \ldots \rangle$$

$$= (-1)^{N_{js}-1} \sqrt{n_{js}} \, (-1)^{N_{js}-1} \sqrt{1 - (n_{js} - 1)} \mid \ldots n_{js} \ldots \rangle. \qquad (5.186)$$

Applying them in reverse order leads to

$$f_{js}^- f_{js}^+ \mid \ldots n_{js} \ldots \rangle = f_{js}^- (-1)^{N_{js}} \sqrt{1 - n_{js}} \mid \ldots n_{js} + 1 \ldots \rangle$$

$$= (-1)^{N_{js}} \sqrt{1 - n_{js}} \, (-1)^{N_{js}} \sqrt{n_{js} + 1} \mid \ldots n_{js} \ldots \rangle. \qquad (5.187)$$

We see that when $n_{js} = 1$, the combinations reduce as follows:

$$f_{js}^+ f_{js}^- = 1 \quad \text{and} \quad f_{js}^- f_{js}^+ = 0. \qquad (5.188)$$

But when $n_{js} = 0$, the reduction is

$$f_{js}^+ f_{js}^- = 0 \quad \text{and} \quad f_{js}^- f_{js}^+ = 1. \qquad (5.189)$$

In either case, the anticommutator of f_{js}^+ and f_{js}^- is

$$\{f_{js}^+ f_{js}^-\} \equiv f_{js}^+ f_{js}^- + f_{js}^- f_{js}^+ = 1. \qquad (5.190)$$

Furthermore, the operator

$$n_{js} = f_{js}^+ f_{js}^- \qquad (5.191)$$

yields the number of fermions occupying the js-th spin orbital. So it is called a *number operator*.

When either operator

$$f_{js}^+ f_{ks'}^-, \quad \text{or} \quad f_{js}^- f_{ks'}^-, \qquad (5.192)$$

acts on a ket with an empty ks'-th spin orbital, the state is destroyed; nothing is obtained. When either operator

$$f_{js}^- f_{ks'}^+, \quad \text{or} \quad f_{js}^+ f_{ks'}^+, \qquad (5.193)$$

acts on a ket with a filled ks'-th spin orbital, the state is also destroyed; nothing is obtained.

But when a shift of a fermion from the ks'-th spin orbital to the js-th one is possible, then operators

$$f_{js}^+ \, f_{ks'}^-, \quad \text{and} \quad f_{ks'}^- \, f_{js}^+ \quad (5.194)$$

both effect this shift. But because of the choice of signs in (5.183) and (5.185), the resulting kets possess opposite signs; they cancel each other. This outcome, together with (5.190), can be summarized with the operator equation

$$f_{js}^+ \, f_{ks'}^- + f_{ks'}^- \, f_{js}^+ = \delta_{jk} \, \delta_{ss'}. \quad (5.195)$$

When addition of fermions to the js-th and the ks'-th spin orbitals is possible, the operators

$$f_{js}^+ \, f_{ks'}^+ \quad \text{and} \quad f_{ks'}^+ \, f_{js}^+ \quad (5.196)$$

both effect this process. But because of the choice of signs in (5.183) and (5.184), the resulting kets cancel each other. Thus, we write

$$f_{js}^+ \, f_{ks'}^+ + f_{ks'}^+ \, f_{js}^+ = 0. \quad (5.197)$$

Similarly, we find that

$$f_{js}^- \, f_{ks_3}^- + f_{ks'}^- \, f_{js}^- = 0. \quad (5.198)$$

In various discussions, the two indices may be represented by a single one. Then $1\uparrow, 1\downarrow, 2\uparrow, 2\downarrow, \ldots, p\downarrow$ would be replaced by $1, 2, 3, 4, \ldots, j, \ldots, 2p$, with j the jth spin orbital in the series. The key relationships for the fermion creation and annihilation operators are thus listed in Table 5.5. These operators can be used to generate all the n-fermion states associated with a given set of spin orbitals from a single reference state. This reference state may be the vacuum state, in which all the spin orbitals are empty.

Example 5.10

Show that the choice of signs in Table 5.5 leads to the correct antisymmetry with respect to exchange of particles.

Start with the jth and kth spin orbitals filled. The operator that takes a particle from the kth spin orbital is f_k^-. Then the operator that moves a particle from the jth to the kth spin orbital is $f_k^+ \, f_j^-$. Finally, the operator that places the previously removed particle into the jth spin orbital is f_j^+.

From Table 5.5, we have

Table 5.5
Relationships Involving Fermion Creation and Annihilation Operators

n_j = number in jth spin orbital $\qquad\qquad N_j = \sum_{i=1}^{j} n_i$

$$f_j^+ \mid \ldots n_j \ldots \rangle = (-1)^{N_j} \sqrt{1 - n_j} \mid \ldots n_j + 1 \ldots \rangle$$

$$f_j^- \mid \ldots n_j \ldots \rangle = (-1)^{N_j - 1} \sqrt{n_j} \mid \ldots n_j - 1 \ldots \rangle$$

$$\{f_j^+, f_k^+\} \equiv f_j^+ f_k^+ + f_k^+ f_j^+ = 0$$

$$\{f_j^-, f_k^-\} \equiv f_j^- f_k^- + f_k^- f_j^- = 0$$

$$\{f_j^+, f_k^-\} \equiv f_j^+ f_k^- + f_k^- f_j^+ = \delta_{jk}$$

$$f_j^+ (f_k^+ f_j^-) f_k^- = (-1)^{N_j - 1}(-1)^{N_k - 2}(-1)^{N_j - 1}(-1)^{N_k - 1}$$

$$= (-1)^{2N_j - 2}(-1)^{2N_k - 3}$$

$$= -1.$$

Thus, the sign conventions in (5.183) and (5.185), and in Table 5.5, cause the sign of a ket, and of the corresponding state function, to change when two identical fermions in the ket are interchanged.

5.13
Boson to Fermion and Fermion to Boson Operators

In Section 1.11 we saw how the Hilbert space for indistinguishable particles of a given kind is split into a boson sector, a mixed statistics sector, and a fermion sector. We also saw that a transition between the extreme sectors was forbidden by a superselection rule. However, this restriction does not preclude the existence of step operators linking these sectors.

Indeed, these can be constructed from the creation and annihilation operators for bosons and fermions. Thus, the binary operator

$$Q_+ = q b^- f^+, \qquad (5.199)$$

acting on the appropriate ket, converts a boson to a fermion. Here q is a number that ensures normalization is not destroyed. Similarly, the binary operator

$$Q_- = q b^+ f^-, \qquad (5.200)$$

acting on the appropriate ket, converts a fermion to a boson.

5.14
Recapitulation

The nature and behavior of particles is circumscribed by the commutation relations that derive from the symmetries of space, time, phase, and from the localizability of the particles. Indeed, these relations lead to forms for the physical operators that are employed in key eigenvalue equations, as we have seen.

For certain important systems, commutation relations also determine the nature of step operators that transform a typical eigenket into a neighboring eigenket.

Suppose that A is a physical operator governing a complete set of states for the given system by the equations

$$A|i\rangle = a_i|i\rangle, \tag{5.201}$$

$$A|j\rangle = a_j|j\rangle. \tag{5.202}$$

Also, suppose that we have operators C_1 and C_2 for which

$$[A, C_1] = ikC_2 \tag{5.203}$$

and

$$[A, C_2] = -ikC_1. \tag{5.204}$$

Then construct

$$C = C_1 + iC_2 \tag{5.205}$$

and note that

$$[A, C] = [A, C_1] + i[A, C_2] = kiC_2 + kC_1 = kC. \tag{5.206}$$

Letting the initial and final operators in (5.206) act on ket $|i\rangle$ yields

$$(AC - CA)|i\rangle = kC|i\rangle, \tag{5.207}$$

whence

$$AC|i\rangle = (a_i + k)C|i\rangle. \tag{5.208}$$

We see that $C|i\rangle$ is an eigenket (unnormalized) of A with the eigenvalue $a_j + k$. Whenever C introduces (or removes) a single node into (or from) a wave function, the shift is to the neighboring wave function. For functions in space, such a shift is accomplished by having C involve differentiating operators no higher than first order.

For angular momenta, the resulting step operators are

$$J_+ = J_x + iJ_y \tag{5.209}$$

and

$$J_- = J_x - iJ_y. \tag{5.210}$$

For harmonic oscillators, the standard step operators are

$$a^+ = \frac{1}{\sqrt{2}} (X - iP) \tag{5.211}$$

and

$$a^- = \frac{1}{\sqrt{2}} (X + iP) \tag{5.212}$$

in natural units. With normalized kets, we find that

$$a^+ \mid v\rangle = \sqrt{v + 1} \mid v + 1\rangle, \tag{5.213}$$

$$a^- \mid v\rangle = \sqrt{v} \mid v - 1\rangle. \tag{5.214}$$

Electromagnetic radiation of a given frequency confined in a cavity behaves as a harmonic oscillator; energy is absorbed and evolved in units of $\hbar\omega$. Each unit can be considered to form a particle called a photon. The state is then labeled by the number of photons of the given frequency that it contains. The quantization into photons is called second quantization.

When the given system contains n_1 photons of angular frequency $\omega_1, \ldots, n_j$ photons of angular frequency $\omega_j, \ldots$, adding one photon of angular frequency ω_j is represented by

$$b_j^+ \mid n_1, n_2, \ldots, n_j, \ldots\rangle = \sqrt{n_j + 1} \mid n_1, n_2, \ldots, n_j + 1, \ldots\rangle, \tag{5.215}$$

while removing one photon of this frequency is represented by

$$b_j^- \mid n_1, n_2, \ldots, n_j, \ldots\rangle = \sqrt{n_j} \mid n_1, n_2, \ldots, n_j - 1, \ldots\rangle. \tag{5.216}$$

We call b_j^+ the creation operator for a photon of energy $\hbar\omega_j$, b_j^- the annihilation operator for a similar photon. We find that these obey the commutation relations

$$[b_j^\pm, b_k^\pm] = 0, \tag{5.217}$$

$$[b_j^-, b_k^+] = \delta_{jk}. \tag{5.218}$$

The states of other bosons are also related by these formulas. But for fermions, the formulas need to be altered so that the state functions become antisymmetric in the exchange of like particles.

This antisymmetry is ensured by replacing (5.215) with

$$f_j^+ \mid \ldots n_j \ldots\rangle = (-1)^{N_j} \sqrt{1 - n_j} \mid \ldots n_j + 1 \ldots\rangle, \tag{5.219}$$

and by replacing (5.216) with

$$f_j^- \mid \ldots n_j \ldots\rangle = (-1)^{N_j - 1} \sqrt{n_j} \mid \ldots n_j - 1 \ldots\rangle. \tag{5.220}$$

Here f_j^+ is the creation operator for a fermion in the jth state while f_j^- is the annihilation operator for a fermion in the jth state.

These changes cause the commutation relations to be replaced by the anticommutation relations

$$\{f_j^{\pm}, f_k^{\pm}\} = 0, \tag{5.221}$$

$$\{f_j^{+}, f_k^{-}\} = \delta_{jk}. \tag{5.222}$$

Furthermore, the operator

$$n = \Sigma f_j^{+} f_j^{-} \tag{5.223}$$

acts to yield the number of fermions in the states over which the summation proceeds.

Discussion Questions

5.1. What is a step operator?

5.2. How can a general transformation operator be expressed as the sum of a Hermitian operator and an anti-Hermitian operator?

5.3. Why is the commutator of two Hermitian operators anti-Hermitian?

5.4. Explain what relation exists among different eigenkets of A when

$$[A, B] \, |i\rangle = kB \, |i\rangle$$

(a) if $k \neq 0$, (b) if $k = 0$.

5.5. When is the k in Question 5.4 a constant for any eigenket of A in the given Hilbert space?

5.6. How do we obtain two linked commutation relations from

$$[A, B] = kB?$$

5.7. What do the simultaneous equations

$$[A, C_1] = ikC_2,$$
$$[A, C_2] = -ikC_1$$

imply? Explain.

5.8. What commutation relations are the basis for the step-up and step-down operators J_+ and J_-?

5.9. What natural units does one employ for (a) a rotator, (b) a harmonic oscillator?

5.10. How are the commutation relations of H with P and with X, for the harmonic oscillator, constructed?

5.11. How do these commutation relations lead to the standard harmonic oscillator step operators?

5.12. How do the commutation relations yield the spacing between successive harmonic oscillator energy levels?

5.13. Show that among the natural units for the harmonic oscillator the unit of distance is

$$\frac{\hbar^{1/2}}{(f\mu)^{1/4}} \; .$$

5.14. How may the operator relationship

$$a^+ a^- = H - \tfrac{1}{2}(1)$$

be used to obtain the energy E for the lowest level of the harmonic oscillator?

5.15. How is step-operator theory employed in constructing a representation of the ground state ket for the harmonic oscillator?

5.16. How are normalized state functions for higher level kets obtained?

5.17. How may the harmonic oscillator step functions be generalized?

5.18. Why is the step-operator method often called the factorization method?

5.19. What condition had to be satisfied for both $a_k^+ a_k^- \psi_k$ and $a_{k+1}^- a_{k+1}^+ \psi_k$ to yield numbers times ψ_k?

5.20. What conditions truncate a ladder of states?

5.21. What role is played by the normalization condition here?

5.22. What transformation allowed us to construct step operators for the eigenkets for radial motion in a Coulomb field?

5.23. Why is the Morse potential more realistic than the parabolic potential?

5.24. What natural units did we use in applying the Morse potential?

5.25. How does one obtain an energy in general units from its value in these natural units?

5.26. Compare the bound states for the Morse oscillator qualitatively with those for the radial motion in a Coulomb field.

5.27. Why does the method used not depend on the existence of a ladder of levels at each j?

5.28. How is the electromagnetic field analogous to an infinite set of harmonic oscillators?

5.29. Why are the step operators for each of these modes called creation and annihilation operators?

5.30. Need the bosons be photons for such creation and annihilation operators to work? Explain.

5.31. How does the Hilbert space for fermions differ from that for bosons?

5.32. How are the creation and annihilation operators altered to allow for this difference?

5.33. Why are the commutator relations replaced by anticommutator relations in going to the fermion operators?

5.34. How can the fermion operators be used to generate all the n-fermion states associated with a given set of spin orbitals?

5.35. When can one construct operators that change fermions to bosons and the bosons to the fermions?

Problems

5.1. Show that the commutator of a Hermitian operator and an anti-Hermitian operator is Hermitian.

5.2. From the results in Example 3.6, construct

$$L_\pm = L_x \pm iL_y$$

in spherical coordinates.

5.3. Let the operator constructed in Problem 5.2 act on the extreme

$$Y = \Theta(\vartheta)\, e^{\pm il\varphi}$$

so that zero is obtained. Integrate the resulting differential equation.

5.4. Show that for the harmonic oscillator we have the relationship

$$2H = -\tfrac{1}{2}(a^- - a^+)^2 + \tfrac{1}{2}(a^- + a^+)^2.$$

5.5. Act on the function

$$\psi_o = \left(\frac{1}{\pi}\right)^{1/4} e^{-x^2/2}$$

with a harmonic-oscillator step operator to obtain the normalized eigenfunction for the first excited state of the oscillator.

5.6. Fit the Schrödinger equation for the harmonic oscillator to (5.90) with $j = 0$ and $E = E(k)$. What form must $E(k)$ assume for condition (5.98) to be met? How is condition (5.116) satisfied? What are the resulting energy levels?

5.7. Show how (5.127) is obtained from (5.121).

5.8. From Examples 5.6 and 5.8, write down the factor y for $n = 3, l = 2$. Apply the appropriate step operator to obtain the y for $n = 3, l = 1$. Then convert it to the corresponding $R(r)$.

5.9. Derive formula (5.155) from (5.151).

5.10. For HCl, energies $\hbar\omega$ and U_0 are 2999 cm^{-1} and 37,200 cm^{-1}, respectively. Calculate its anharmonicity constant b.

— — —

5.11. Show that the eigenvalues of an anti-Hermitian operator are imaginary.

5.12. Let the operator that was constructed in Problem 5.2, that reduces the $|m|$ by 1, act on

$$Y = A(\sin \vartheta)^l e^{\pm il\varphi}.$$

5.13. Show that

$$iPX^n = iX^n P + nX^{n-1}.$$

5.14. The Hamiltonian H for an anharmonic oscillator obeys

$$2H = P^2 + X^2 + 2bX.$$

Rewrite this in terms of the a^- and a^+ for a harmonic oscillator.

5.15. Act on the function

$$\psi_1 = \sqrt{2}\left(\frac{1}{\pi}\right)^{1/4} x e^{-x^2/2}$$

with a harmonic oscillator step operator to obtain the normalized eigenfunction for the second excited state of the oscillator.

5.16. Can the theory in Table 5.3 apply to

$$[D^2 - \epsilon x^2 + f(k) + j]\psi = 0$$

in the limit when

$$\epsilon \to 0?$$

Explain.

5.17. Take the y of (5.120) for a hydrogenlike atom with $n = 3$, $l = 1$, from the answer to Problem 5.8. Then apply a step operator to generate $R(r)$ for $n = 3$ and $l = 0$.

5.18. Show how (5.156) is obtained from preceding equations.

5.19. How is the vibrational quantum number v related to the anharmonicity constant b when the energy of the level equals U_0, the dissociation energy plus the ground state energy?

5.20. From the data in Problem 5.10 and the result of Problem 5.19, calculate the number of discrete vibrational levels which the HCl molecule exhibits.

5.21. How small must U_0 be for the Morse oscillator to have no bound states?

References

Books

Dirac, P. A. M.: 1947, *The Principles of Quantum Mechanics*, 3rd edn., Oxford University Press, London, pp. 136-149. In Chapter VI, Dirac applies his abstract methods to the harmonic oscillator and to angular momentum.

Duffey, G. H.: 1984, *A Development of Quantum Mechanics Based on Symmetry Considerations*, Reidel, Dordrecht, pp. 157-166, 247-250. For the harmonic oscillator and for radial motion in a Coulomb field, a differential operator in the Schrödinger equation is here factored in two different ways. The results then lead to forms for the step operators. With angular motion, the commutation relations form the basis.

Harris, L., and Loeb, A. L.: 1963, *Introduction to Wave Mechanics*, McGraw-Hill, New York, pp. 131-195, 224-229. Harris and Loeb develop an operator algebra for the harmonic oscillator and for motion in a central field. Angular momentum and electron spin are also considered in some detail.

Kaempffer, F. A.: 1965, *Concepts in Quantum Mechanics*, Academic Press, New York, pp. 119-150. Here creation and annihilation operators are employed in describing fermion and boson states.

Merzbacher, E.: 1970, *Quantum Mechanics*, 2nd edn., Wiley, New York, pp. 356-359, 375-378. Merzbacher's treatment of step operators for the harmonic oscillator and for angular momentum is based on the pertinent commutation relations.

Park, D.: 1974, *Introduction to the Quantum Theory*, 2nd edn., McGraw-Hill, New York, pp. 127-133, 163-167, 618-623. Besides developing step operators for the harmonic oscillator and for angular momentum in a similar way, Park discusses the completeness question.

Articles

Bose, A. K.: 1988. "A Note on the Angular Momentum Operators." *Am. J. Phys.* **56**, 945-946.

Boya, L. J.: 1988. "Supersymmetric Quantum Mechancis: Two Simple Examples," *Eur. J. Phys.* **9**, 139-144.

Castano, F., Juan, J. de, and Martinez, E.: 1983. "The Calculation of Potential Energy Curves of Diatomic Molecules: The RKR Method," *J. Chem. Educ.* **60**, 91-93.

Das, R., and Sannigrahi, A. B.: 1981. "The Factorization Method and Its Applications in Quantum Chemistry," *J. Chem. Educ.* **58**, 383-388.

Fernandez, F. M., and Castro, E. A.: 1984. "Resolution of the Schrödinger Equation through a Simple Algebraic Technique," *Am. J. Phys.* **52**, 344-346.

Henry, R. W., and Glotzer, S. C.: 1988. "A Squeezed-State Primer," *Am. J. Phys.* **56**, 318-328.

Infeld, L., and Hull, T. E.: 1951. "The Factorization Method," *Rev. Mod. Phys.* **23**, 21-68.

Lange, O. L. de, and Raab, R. E.: 1986. "Ladder Operators for Orbital Angular Momentum," *Am. J. Phys.* **54**, 372-375.

Lange, O. L. de, and Raab, R. E.: 1987. "Factorizations of Two Vector Operators for the Coulomb Problem," *Phys. Rev. A* **35**, 951-956.

Lange, O. L. de, and Raab, R. E.: 1987. "An Operator Solution for the Hydrogen Atom with Application to the Momentum Representation," *Am. J. Phys.* **55**, 913-917.

Leach, P. G. L.: 1984. "Exact Solutions of the Schrödinger Equation for a Class of Three-Dimensional Isotropic Anharmonic Oscillators," *J. Math. Phys.* **25**, 2974-2978.

Milner, J. D., and Peterson, C.: 1985. "Notes on the Factorization Method for Quantum Chemistry," *J. Chem. Educ.* **62**, 567-568.

Montemayor, R., and Urrutia, L.: 1983. "Ladder Operators in the Morse Potential Obtained from a Related Harmonic Oscillator," *Am. J. Phys.* **51**, 641-644.

Newmarch, J. D., and Golding, R. M.: 1978. "Ladder Operators for Some Spherically Symmetric Potentials in Quantum Mechanics," *Am. J. Phys.* **46**, 658-660.

Pena, L. de la, and Montemayor, R.: 1980. "Raising and Lowering Operators and Spectral Structure: A Concise Algebraic Technique," *Am. J. Phys.* **48**, 855-860.

Peterson, C.: 1975. "The Radial Equation for Hydrogen-Like Atoms," *J. Chem. Educ.* **52**, 92-94.

Sannigrahi, A. B.: 1985. "On the Genesis of Ladder Operators," *J. Chem. Educ.* **62**, 205.

Silverman, M. P.: 1981. "An Exactly Soluble Quantum Model of an Atom in an Arbitrarily Strong Uniform Magnetic Field," *Am. J. Phys.* **49**, 546-551.

Stahlhofen, A., and Bleuler, K.: 1989. "An Algebraic Form of the Factorization Method." *Nuovo Cimento* **104B**, 447-465.

6

VARIATIONAL PROCEDURES

6.1
Using Parameters to Introduce Variations

We have seen how operators for position, linear momentum, angular momentum, and energy can be constructed. We have also described representations of the eigenkets for free translation, free rotation, harmonic vibration, and motion in a Coulomb field. Eigenenergies have been obtained for these modes and also for the Morse oscillator.

Most systems, however, have more complicated potentials. To make progress with these we need to develop additional methods.

In this chapter we will consider how the kinetic energy expectation value can be separated from the total energy. A theorem relating this expectation value to the potential energy expectation value will be developed. The calculation will be facilitated by introducing a rescaling parameter into the Hamiltonian operator in pertinent ways.

We will also see how the calculus of variations can be applied in evaluating representations of the eigenkets for a given observable for a system. The parameters that are varied may act anywhere in the constructed functions.

Of particular interest, however, are representations that are linear combinations of basis functions, with the parameters acting as the coefficients. These representations lead to the secular equation that is widely employed in investigations.

One or more parameters may enter into the construction of the operator for any physical property. In perturbation procedures, complicating terms are introduced by a multiplier that increases from zero to a limiting value. Such calculations will be considered in later chapters.

6.2
Rate of Change of the State Energy with a Parameter

The Hamiltonian operator H for an N-particle system is made up of operators for the particle positions and for the particle momenta. The positions are represented by a $3N$-dimensional $\mathbf{X}$, the momenta by a $3N$-dimensional $\mathbf{P}$. A person may also introduce a real parameter λ that varies the action of some part of the expression.

For the parameterized system, we have

$$H = H(\mathbf{X}, \mathbf{P}, \lambda). \tag{6.1}$$

The jth energy state, described by ket $|j\rangle$, is governed by the eigenvalue equation

$$H |j\rangle = E_j(\lambda) |j\rangle, \tag{6.2}$$

in which the eigenenergy E_j depends on λ. We presume that $|j\rangle$ is normalized to 1. When both sides of (6.2) are multiplied by bra $\langle j|$, there results

$$\langle j | H | j \rangle = E_j(\lambda) \langle j | j \rangle = E_j(\lambda). \tag{6.3}$$

Let us differentiate (6.3) with respect to the parameter:

$$\frac{\partial E_j}{\partial \lambda} = \langle j | \frac{\partial H}{\partial \lambda} | j \rangle + \langle \frac{\partial}{\partial \lambda} | j \,\|\, H | j \rangle$$
$$+ \langle j \,\|\, H | \frac{\partial}{\partial \lambda} | j \rangle. \tag{6.4}$$

But since H is Hermitian, the last two terms become

$$\langle \frac{\partial}{\partial \lambda} | j \,\|\, H | j \rangle + \langle H | j \,\|\, \frac{\partial}{\partial \lambda} | j \rangle = E_j \langle \frac{\partial}{\partial \lambda} | j \,\|\, j \rangle$$

$$+ E_j \langle j \,\|\, \frac{\partial}{\partial \lambda} | j \rangle = E_j \frac{\partial}{\partial \lambda} \langle j | j \rangle = 0. \tag{6.5}$$

Also, the first term on the right of (6.4) is the expectation value for $\partial H / \partial \lambda$. So we obtain

$$\frac{\partial E_j}{\partial \lambda} = \left\langle \frac{\partial H}{\partial \lambda} \right\rangle. \tag{6.6}$$

This result is known as the *Feynman-Hellmann theorem.*
Next, we rewrite Hamiltonian (3.77) in the form

$$H = \sum_{i=1}^{N} \frac{\lambda}{2\mu_i} (\mathbf{P}_i - \mathbf{A}_i) \cdot (\mathbf{P}_i - \mathbf{A}_i) + U. \tag{6.7}$$

Then

$$\lambda \frac{\partial H}{\partial \lambda} = \sum \frac{\lambda}{2\mu_i} (\mathbf{P}_i - \mathbf{A}_i) \cdot (\mathbf{P}_i - \mathbf{A}_i) = T \tag{6.8}$$

where

$$T = H - U. \tag{6.9}$$

Let us multiply (6.6) by λ and bring in (6.8) to construct

$$\lambda \frac{\partial E_j}{\partial \lambda} = \left\langle \lambda \frac{\partial H}{\partial \lambda} \right\rangle = \langle T \rangle. \tag{6.10}$$

For the last equality to be true, parameter λ must be introduced as in (6.7). One can say that in this Hamiltonian each mass μ_i has been replaced by

$$\frac{\mu_i}{\lambda}. \tag{6.11}$$

Thus, the parameterization corresponds to a simultaneous rescaling of all particle masses in the system.

6.3
Relating the Kinetic Energy Expectation Value to the Potential Energy Expectation Value

A person can introduce the parameter λ so that it alters the spread of the potential functions for the given system symmetricly. Here we will do this, neglecting magnetic effects. The introduction will be found equivalent to varying the kinetic energy. A relation between the mean kinetic energy and the mean potential energy will thus be obtained.

Consider an N-particle system for which the mass and coordinates of the ith particle are μ_i and $\mathbf{x}_i$, respectively. Let us consider only those interactions that a scalar potential can express. The Hamiltonian operator is then

$$H = \sum_{i=1}^{N} -\frac{\hbar^2}{2\mu_i} \nabla_{\mathbf{x}_i}^2 + U(\mathbf{x}_1, \mathbf{x}_2, \ldots, \mathbf{x}_N). \tag{6.12}$$

Now, let us replace each $\mathbf{x}_i$ in the potential with $\lambda \mathbf{x}_i$, thus altering the spread of the potential U:

$$H' = \sum_{i=1}^{N} -\frac{\hbar^2}{2\mu_i} \nabla_{x_i}^2 + U(\lambda\mathbf{x}_1, \lambda\mathbf{x}_2, \ldots, \lambda\mathbf{x}_N). \tag{6.13}$$

But with the transformation

$$\mathbf{y}_i = \lambda\mathbf{x}_i, \tag{6.14}$$

the altered Hamiltonian becomes

$$H' = \sum_{i=1}^{N} -\frac{\lambda^2\hbar^2}{2\mu_i} \nabla_{y_i}^2 + U(\mathbf{y}_1, \mathbf{y}_2, \ldots, \mathbf{y}_N). \tag{6.15}$$

Let us apply the Feynman-Hellmann theorem to (6.15) to get

$$\lambda \frac{\partial E_j}{\partial \lambda} = \lambda \left\langle \frac{\partial H'}{\partial \lambda} \right\rangle = \lambda \left\langle -2 \sum \frac{\lambda \hbar^2}{2\mu_i} \nabla_{y_i}^2 \right\rangle = 2 \langle T' \rangle. \tag{6.16}$$

Also, we apply (6.6) to the altered Hamiltonian in form (6.13) to construct

$$\lambda \frac{\partial E_j}{\partial \lambda} = \lambda \left\langle \frac{\partial H'}{\partial \lambda} \right\rangle = \left\langle \lambda \frac{\partial U}{\partial \lambda} \right\rangle. \tag{6.17}$$

With the indicated dependence of U in (6.13), the right side of (6.17) can be rewritten as

$$\left\langle \lambda \sum_{i=1}^{N} \sum_{j=1}^{3} x_{ij} \frac{\partial U}{\partial(\lambda x_{ij})} \right\rangle = \left\langle \sum_i \sum_j x_{ij} \frac{\partial U}{\partial x_{ij}} \right\rangle. \tag{6.18}$$

Since the left sides of (6.16) and (6.17) are the same, the final right sides of (6.16) and (6.18) are equal. If we then set $\lambda = 1$, we obtain

$$2 \langle T \rangle = \left\langle \sum_i \sum_j x_{ij} \frac{\partial U}{\partial x_{ij}} \right\rangle. \tag{6.19}$$

When the potential in the given system is a homogeneous function of the coordinates of degree n, we have

$$U(\lambda\mathbf{x}_1, \lambda\mathbf{x}_2, \ldots, \lambda\mathbf{x}_N) = \lambda^n U(\mathbf{x}_1, \mathbf{x}_2, \ldots, \mathbf{x}_N). \tag{6.20}$$

Then

$$\lambda \frac{\partial U}{\partial \lambda} = nU \tag{6.21}$$

and (6.17) becomes

$$\lambda \frac{\partial E_j}{\partial \lambda} = n \langle U \rangle. \tag{6.22}$$

Combining (6.16), (6.22), and setting $\lambda = 1$ yields

$$2 \langle T \rangle = n \langle U \rangle, \qquad (6.23)$$

a quantum mechanical form of the *virial theorem*.

Example 6.1

How are the expectation values for kinetic energy and for potential energy related in the harmonic oscillator?

The potential

$$U = \tfrac{1}{2} \, fx^2$$

for the harmonic oscillator is homogeneous of degree 2. So in Equation (6.23)

$$n = 2.$$

Canceling the 2 on both sides gives us

$$\langle T \rangle = \langle U \rangle.$$

6.4
The Variation Theorem

Representations of kets are much more complicated than the potentials that generate them; so usually, only approximate descriptions of them can be constructed. Flexibility in these can be afforded by making them vary with parameters, but then, one has the problem of choosing the best values for the parameters. A useful answer, however, is provided by the variation theorem.

We consider an observable of a given system represented by a Hermitian operator A. In each pure state with respect to A, we have

$$A \, | \, j \, \rangle = a_j \, | \, j \, \rangle. \qquad (6.24)$$

Let us also suppose the kets are ordered so that

$$a_{j+1} \geq a_j \quad \text{with} \quad j = 1, 2, 3, \ldots . \qquad (6.25)$$

Without loss of generality, we may consider the eigenkets to be normalized and orthogonal,

$$\langle \, j \, | \, k \, \rangle = \delta_{jk}. \qquad (6.26)$$

Since the eigenkets span the Hilbert space for the given system, any suitable approximate ket can be expressed as a superposition,

$$| g \rangle = | \Sigma c_j | j \rangle = \Sigma c_j | j \rangle. \tag{6.27}$$

The bra form of this is

$$\langle g | = \langle \Sigma c_j | j | = \Sigma c_j^* \langle j |. \tag{6.28}$$

Now let operator A act on the trial ket $| g \rangle$,

$$A | g \rangle = \Sigma c_j A | j \rangle = \Sigma c_j a_j | j \rangle. \tag{6.29}$$

Then multiply overall (6.29) by bra (6.28),

$$\langle g | A | g \rangle = \Sigma c_j^* \langle j | \Sigma c_k a_k | k \rangle = \sum_{j,k} c_j^* c_k a_k \langle j | k \rangle$$

$$= \Sigma c_j^* c_j a_j, \tag{6.30}$$

and divide by the normalization integral,

$$\frac{\langle g | A | g \rangle}{\langle g | g \rangle} = \frac{\Sigma c_j^* c_j a_j}{\langle g | g \rangle} = \Sigma w_j a_j. \tag{6.31}$$

Here w_j is the weight of the jth state in the composite state $| g \rangle$.
    Since

$$a_j \geq a_1 \tag{6.32}$$

and

$$\Sigma w_j = 1, \tag{6.33}$$

we find that

$$\frac{\langle g | A | g \rangle}{\langle g | g \rangle} \geq a_1. \tag{6.34}$$

When the candidate ket $| g \rangle$ is normalized to 1, this reduces to

$$\langle g | A | g \rangle \geq a_1. \tag{6.35}$$

The best approximation to the lowest eigenvalue is obtained when the left side of (6.34), or of (6.35), is minimized. The corresponding values of the parameters may be substituted into the representation of the ket.

One may then construct an approximate ket orthogonal to the ket just obtained and repeat the procedure to get an approximation to the next level, and so on.

Formula (6.34) and the deduction that the best approximation to a_1 that a trial ket $| g \rangle$ yields is obtained on minimizing the left side constitute the *variation theorem*.

Example 6.2

As an approximate representation of the ground-state ket for a harmonic oscillator, consider the function in Figure 6.1, for which

$$\psi = \frac{b}{a}x + b \qquad \text{when} \qquad -a < x < 0,$$

$$\psi = -\frac{b}{a}x + b \qquad \text{when} \qquad 0 < x < a.$$

and

$$\psi = 0 \qquad \text{elsewhere.}$$

Determine the a that minimizes E. Then calculate the resulting energy E and the slope b/a.

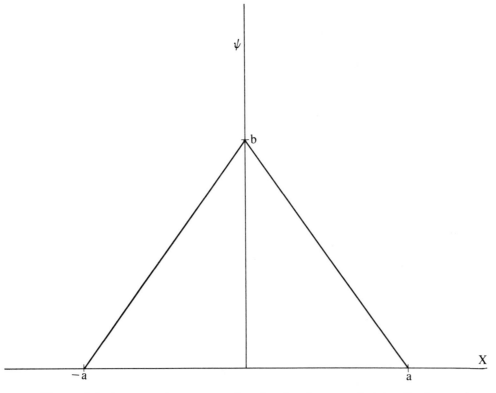

Figure 6.1. An approximate wave function for the ground state of a harmonic oscillator.

The variation theorem in form (6.34) will be employed. Since a coordinate representation of the trial ket is given, we need the Hamiltonian in terms of x also. For simplicity, we employ natural units for the harmonic oscillator, such that

$$2H = x^2 - \frac{d^2}{dx^2},$$

as in (5.47).

Since the slope of ψ changes abruptly at $x = -a$, at $x = 0$, and at $x = a$, singularities in $d^2\psi/dx^2$ occur at these points. An evaluation of their contributions to the numerator of (6.34) can be made by going to an increment formulation.

Consider a singularity at x_3 with ψ known at equally spaced points x_1, x_2, x_3, x_4, x_5 such that

$$x_{j+1} - x_j = \epsilon.$$

Here ϵ is small and in the limit will be allowed to vanish. Then

$$\frac{d^2\psi}{dx^2} \simeq \frac{\psi_5 - 2\psi_3 + \psi_1}{(x_4 - x_2)(x_3 - x_1)}.$$

The contribution to the numerator of (6.34) from the neighborhood of the maximum is

$$\int_{x_2}^{x_4} \psi \left(-\frac{d^2}{dx^2} \right) \psi \, dx \simeq -\psi_3 \frac{\psi_5 - 2\psi_3 + \psi_1}{(x_4 - x_2)(x_3 - x_1)} (x_4 - x_2)$$

$$= -b \frac{\dfrac{b}{a}(x_1 - x_5) + 2b - 2b}{x_3 - x_1} = \frac{2b^2}{a}.$$

Since ψ is zero at $x = -a$ and at $x = a$, the contributions from the singularities there vanish.

The other contributions to the numerator of (6.34) are

$$\int_{-a}^{0} \left(\frac{b}{a}x + b \right) \left(x^2 - \frac{d^2}{dx^2} \right) \left(\frac{b}{a}x + b \right) dx$$

$$+ \int_{0}^{a} \left(-\frac{b}{a}x + b \right) \left(x^2 - \frac{d^2}{dx^2} \right) \left(-\frac{b}{a}x + b \right) dx$$

$$= \frac{2b^2}{5a^2} a^5 - \frac{4b^2}{4a} a^4 + \frac{2b^2}{3} a^3 = 2b^2 \frac{1}{30} a^3.$$

Furthermore, the integral in the denominator is

$$\int_{-a}^{0} \left(\frac{b}{a} x + b \right)^2 dx + \int_{0}^{a} \left(-\frac{b}{a} x + b \right)^2 dx$$

$$= \frac{2b^2}{3a^2} a^3 - \frac{2b^2}{a} a^2 + 2b^2 a = 2b^2 \frac{1}{3} a.$$

Substituting into (6.34) and canceling $2b^2$ yields

$$2E = \frac{\dfrac{1}{30} a^3 + \dfrac{1}{a}}{\dfrac{1}{3} a} = \frac{1}{10} a^2 + \frac{3}{a^2} .$$

The minimum occurs where

$$\frac{d(2E)}{da} = \frac{1}{5} a - \frac{6}{a^3} = 0,$$

or

$$a = (30)^{1/4} = 2.3403.$$

Substituting into the formula for energy yields

$$2E = 0.5477 + 0.5477 = 1.0954 \text{ unit.}$$

Since the correct ground state energy is ½ unit, this result is 9.54% high.
When the state function is normalized to 1, we have

$$\langle g \mid g \rangle = \frac{2}{3} b^2 a = 1.$$

Then

$$b = \left(\frac{3}{2a} \right)^{1/2} = 0.8006$$

and

$$\frac{b}{a} = \frac{0.8006}{2.3403} = 0.3421.$$

6.5
The Ritz Theorem

More generally, we find that, for an observable of a given system, the expectation value is stationary in the neighborhood of each discrete eigenvalue.

We consider a system in the state described by the unnormalized ket $|f\rangle$. The expectation value for the observable of operator A is then

$$\langle A \rangle = \frac{\langle f | A | f \rangle}{\langle f | f \rangle} \qquad (6.36)$$

whence

$$\langle A \rangle \langle f | f \rangle = \langle f | A | f \rangle. \qquad (6.37)$$

Let us alter the ket and the bra continuously. When the alteration is infinitesimal, it produces the variations

$$\langle \delta f |, \qquad | \delta f \rangle, \qquad \text{and} \qquad \delta \langle A \rangle. \qquad (6.38)$$

The corresponding variation of (6.37) is

$$\langle f | f \rangle \delta \langle A \rangle + \langle A \rangle (\langle \delta f | f \rangle + \langle f | \delta f \rangle)$$
$$= \langle \delta f | A | f \rangle + \langle f | A | \delta f \rangle, \qquad (6.39)$$

for the higher order infinitesimals are negligible.

Wherever the expectation value is stationary, as at a minimum, we have

$$\delta \langle A \rangle = 0 \qquad (6.40)$$

and (6.39) rearranges to

$$\langle \delta f | A - \langle A \rangle | f \rangle + \langle f | A - \langle A \rangle | \delta f \rangle = 0. \qquad (6.41)$$

Since $A - \langle A \rangle$ is Hermitian, (6.41) can be rewritten as

$$\langle \delta f | A - \langle A \rangle | f \rangle + \langle \delta f | A - \langle A \rangle | f \rangle * = 0. \qquad (6.42)$$

This relationship must be satisfied for any infinitesimal bra $\langle \delta f |$. So we may choose

$$\langle \delta f | = \epsilon \langle A - \langle A \rangle | f \|, \qquad (6.43)$$

where ϵ is an infinitesimal real number. Then (6.42) becomes

$$\epsilon \langle A - \langle A \rangle | f \| A - \langle A \rangle | f \rangle$$
$$+ \epsilon \langle A - \langle A \rangle | f \| A - \langle A \rangle | f \rangle * = 0 \qquad (6.44)$$

or

$$2 \epsilon \langle A - \langle A \rangle | f \| A - \langle A \rangle | f \rangle = 0. \qquad (6.45)$$

Since ϵ may differ from zero, (6.45) implies that

$$A | f \rangle = \langle A \rangle | f \rangle. \qquad (6.46)$$

whenever $\langle A \rangle$ satisfies condition (6.40). Thus, each stationary expectation value, when there is no constraint on $| f \rangle$ in the Hilbert space, is an eigenvalue of the operator. This is known as the *Ritz theorem*.

In many calculations, constraint is imposed by the form of the representation used. Each stationary $\langle A \rangle$ then is an approximation to an eigenvalue. By the variation theorem, when the eigenvalues have a lower limit as (6.25) specifies, the approximate values run high.

A constraint may be viewed as restricting the trial kets to a subspace of the complete Hilbert space for the given system. Then the variational method reduces to constructing the eigenvalues for the operator in this subspace.

6.6
Linear Variation Functions

For all except the simplest systems and operators, representations of the eigenkets are complicated unfamiliar transcendental functions. However, explicit forms for the most important contributions to such functions are available in many cases. A criterion for constructing the best combinations of these is provided by the theorems just introduced.

From (1.62) and (6.36), the expectation value for the observable corresponding to operator A is

$$a = \frac{\langle g \mid A \mid g \rangle}{\langle g \mid g \rangle} \tag{6.47}$$

when the ket $| g \rangle$ describing the state of the given system is not normalized. From the Ritz theorem, each unconstrained ket that causes

$$\delta a = 0 \tag{6.48}$$

is an eigenket for the system.

But when the ket is limited by its representation, condition (6.48) yields the best approximation, consistent with the constraints, to the corresponding eigenket, insofar as the observable is concerned. A finite basis set generally does not include all possible contributions and so imposes constraints.

An example is the superposition

$$| g \rangle = \sum_{k=1}^{n} c_k | k \rangle \tag{6.49}$$

in which the known kets

$$| 1 \rangle, | 2 \rangle, \ldots, | n \rangle \tag{6.50}$$

describe limited aspects of the behavior of the system. While these kets should be linearly independent, they need not be mutually orthogonal. The bra form of (6.49) is

$$\langle g \mid = \sum_{j=1}^{n} c_j^* \langle j \mid. \tag{6.51}$$

Substituting (6.49) and (6.51) into the rearrangement

$$\langle g \mid A \mid g \rangle - a \langle g \mid g \rangle = 0 \tag{6.52}$$

of (6.47) yields the equation

$$\sum_{j=1}^{n} \sum_{k=1}^{n} c_j^* c_k \langle j \mid A \mid k \rangle - a \sum_{j=1}^{n} \sum_{k=1}^{n} c_j^* c_k \langle j \mid k \rangle = 0. \tag{6.53}$$

The scalar products may be considered as matrix elements. Thus, we let

$$\langle j \mid A \mid k \rangle = A_{jk}, \tag{6.54}$$

$$\langle j \mid k \rangle = S_{jk}, \tag{6.55}$$

and rewrite (6.53) as

$$\sum_{j=1}^{n} \sum_{k=1}^{n} c_j^* c_k A_{jk} - a \sum_{j=1}^{n} \sum_{k=1}^{n} c_j^* c_k S_{jk} = 0. \tag{6.56}$$

Now, we may vary both the real and the imaginary parts of each coefficient c_k independently. But with

$$c_j = Re\, c_j + i\, Im\, c_j, \tag{6.57}$$

$$c_j^* = Re\, c_j - i\, Im\, c_j, \tag{6.58}$$

we have

$$\frac{\partial c_j}{\partial (Re\, c_j)} = 1, \qquad \frac{\partial c_j^*}{\partial (Re\, c_j)} = 1, \tag{6.59}$$

$$\frac{\partial c_j}{\partial (Im\, c_j)} = i, \qquad \frac{\partial c_j^*}{\partial (Im\, c_j)} = -i, \tag{6.60}$$

At the points where condition (6.48) is satisfied, a acts as a constant and the derivative of (6.56) with respect to $Re\, c_l$ is

$$\sum_{j=1}^{n} c_j^* A_{jl} + \sum_{k=1}^{n} c_k A_{lk} - a \sum_{j=1}^{n} c_j^* S_{jl} - a \sum_{k=1}^{n} c_k S_{lk} = 0. \tag{6.61}$$

Differentiating (6.56) with respect to $Im\, c_l$, with condition (6.48), and canceling i similarly gives us

$$\sum_{j=1}^{n} c_j^* A_{jl} - \sum_{k=1}^{n} c_k A_{lk} - a \sum_{j=1}^{n} c_j^* S_{jl} + a \sum_{k=1}^{n} c_k S_{lk} = 0. \qquad (6.62)$$

Subtracting (6.62) from (6.61) and dividing by 2 produces

$$\sum_{k=1}^{n} c_k(A_{lk} - aS_{lk}) = 0 \qquad (6.63)$$

or

$$(A_{11} - aS_{11})c_1 + (A_{12} - aS_{12})c_2 + \ldots + (A_{1n} - aS_{1n})c_n = 0,$$
$$(A_{21} - aS_{21})c_1 + (A_{22} - aS_{22})c_2 + \ldots + (A_{2n} - aS_{2n})c_n = 0,$$
$$\vdots \qquad (6.64)$$
$$(A_{n1} - aS_{n1})c_1 + (A_{n2} - aS_{n2})c_2 + \ldots + (A_{nn} - aS_{nn})c_n = 0.$$

Cramer's rule tells us that all c_k's are zero unless the determinant of the coefficients vanishes; so we set

$$\begin{vmatrix} A_{11} - S_{11}a & A_{12} - S_{12}a & \cdots & A_{1n} - S_{1n}a \\ A_{21} - S_{21}a & A_{22} - S_{22}a & \cdots & A_{2n} - S_{2n}a \\ \cdot & \cdot & \cdot & \cdot \\ \cdot & \cdot & \cdot & \cdot \\ \cdot & \cdot & \cdot & \cdot \\ A_{n1} - S_{n1}a & A_{n2} - S_{n2}a & \cdots & A_{nn} - S_{nn}a \end{vmatrix} = 0. \qquad (6.65)$$

Equation (6.65) is called the *secular equation* for observable a. Since it is of the nth degree in a, it yields n roots, which need not be all distinct. Each root can be substituted back into Equations (6.64) and the normalization condition introduced to obtain the c_j's.

In the derivation, the n kets in (6.50) were taken to be fixed; only the coefficients in (6.49) were varied. The values found for these coefficients depend on the operator A and so on what is to be observed.

6.7
A Secular Equation for Energy

A common criterion for evaluating approximations is stability as measured by energy. Expression A is then taken to be the Hamiltonian operator and quantity a the system energy.

Now, the kinetic energy part of the Hamiltonian involves each particle in the system separately. The potential energy part, on the other hand, splits into one- and two-particle parts. Electric and magnetic interactions do not involve higher-order terms, so more-than-two particle operators are not needed.

In independent particle models, one considers the movement of each particle in the average field of all the rest. The two-particle interactions are thus approximated. For simplicity, one may not even require the fields for different particles to be consistent.

In any case, consider a system to be represented by a superposition of the kets

$$| 1 \rangle, | 2 \rangle, \ldots, | n \rangle, \tag{6.66}$$

as in Section 6.6.

Matrix element S_{jk} then measures how much ket $| j \rangle$ overlaps ket $| k \rangle$. So when $j \neq k$, it is called the *overlap integral*; when $j = k$, it is the *normalization integral*. Diagonal matrix element H_{jj} equals the energy if the system were in ket $| j \rangle$ and the kets were normalized to one. Off-diagonal matrix element H_{jk} is then a measure of the interaction between the jth and the kth kets.

When each ket is a separate atomic orbital and the Hamiltonian is an independent-electron operator, H_{jj} is called the jth *atomic parameter* or *Coulomb integral* and H_{jk} the jk-th *bond parameter* or *resonance integral*.

For any set (6.66) such that

$$S_{11} = S_{22} = \ldots = S_{nn} = 1, \tag{6.67}$$

the secular equation for energy E is

$$\begin{vmatrix} H_{11} - E & H_{12} - S_{12}E & \cdots & H_{1n} - S_{1n}E \\ H_{21} - S_{21}E & H_{22} - E & \cdots & H_{2n} - S_{2n}E \\ \cdot & \cdot & \cdot & \cdot \\ \cdot & \cdot & \cdot & \cdot \\ \cdot & \cdot & \cdot & \cdot \\ H_{n1} - S_{n1}E & H_{n2} - S_{n2}E & \cdots & H_{nn} - E \end{vmatrix} = 0. \tag{6.68}$$

As we noted before, this yields n roots, some of which may be degenerate. Each root may be substituted back into Equations (6.64) and the expansion coefficients c_j obtained.

In quantum chemistry, investigators have commonly employed schemes that consider one electron at a time. The procedures can be divided into three classes: Hückel-type methods, semiempirical self-consistent field models, and the ab-initio self-consistent field model.

6.8
Chemical Bonding and the Hückel Procedure

Qualitatively, chemical bonding may be understood as follows. Consider an atom A with valence orbital $|\,A\,\rangle$ and an atom B with valence orbital $|\,B\,\rangle$. When the atoms are far apart, an electron in $|\,A\,\rangle$ will stay there indefinitely, and an electron in $|\,B\,\rangle$ will stay in its orbital. But when the two atoms are brought close enough together so that there is some overlap of the orbitals, the electrons can move back and forth among the orbitals. At large distances, this movement will involve tunneling. When the electrons have opposing spins, each gains freedom and is thereby stabilized. When they have parallel spins, they become correspondingly restricted and destabilized.

When the atomic orbital on A is directed along the bond, towards atom B, and the atomic orbital on B is directed oppositely along the bond, towards atom A, the orbitals are called σ orbitals. But when the atomic orbitals on A and B are directed in parallel fashion perpendicular to the bond, the orbitals are called π orbitals.

With more atoms, one may have more than two π orbitals that interact. The occupying electrons may then gain more freedom and more stability. In quantum chemistry, *delocalization* refers, not to lack of position, but to lack of confinement to a single atom. When the delocalization extends over more than two atoms, *resonance* among the different classically bonded structures is said to exist.

In 1930, E. Hückel developed a quantum mechanical treatment for delocalized electrons in molecules. This involves the following specifications:

Hamiltonian: Independent electron, non-self-consistent field.

Basis: Minimal atomic orbital basis, ordinarily one valence orbital per atom, often taken to be mutually orthogonal.

Atomic Parameters: Standard values, possibly with corrections for inductive effects.

Bond Parameters: Standard values between directly linked atoms, zero otherwise.

Thus, in (6.68), the simple Hückel method involves

$$S_{jk} = \delta_{jk}, \tag{6.69}$$

and

$$H_{jj} = H_{AA} = \alpha_A, \tag{6.70}$$

$$H_{jk} = H_{AB} = \beta_{AB} \quad \text{for nearest neighbors,} \tag{6.71}$$

$$H_{jk} = 0 \quad \text{for non-nearest neighbors.} \tag{6.72}$$

However, the overlap integrals can be calculated for the given atomic orbitals. These can be substituted for (6.69) in the secular equation.

Example 6.3

Describe the most readily available atomic orbitals that can be used for π bonding in the benzene molecule.

The benzene molecule contains six carbon atoms held by σ and π bonds to a regular hexagonal ring. Attached to each carbon atom is a hydrogen atom in the plane of the ring as Figure 6.2 illustrates.

The most stable atomic orbitals employed for the π bonds are the real $2p$ orbitals of the carbon atoms with nodal planes in the plane of the ring, as Figure 6.3 shows. These constitute the minimal basis. Let us number the positions on the ring as in Figure 6.4 and label the pi atomic orbitals by their positions thus

$$|\,1\,\rangle, |\,2\,\rangle, |\,3\,\rangle, |\,4\,\rangle, |\,5\,\rangle, |\,6\,\rangle.$$

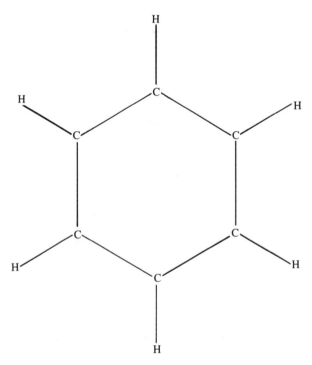

Figure 6.2. Arrangement of atoms and bonds in the benzene molecule.

A general superposition of these is

$$| g \rangle = \sum_{j=1}^{6} c_j | j \rangle.$$

Example 6.4

Develop the Hückel procedure for the superposition constructed in Example 6.3.

Since six kets are being superposed, the order of the determinant in (6.68) is six. Now, we consider the different atomic orbitals to be orthogonal and normalized. So

$$S_{jk} = 0 \qquad \text{when} \qquad j \neq k$$

and

$$S_{jk} = 1 \qquad \text{when} \qquad j = k.$$

Since the $2p\pi$ orbitals on the different carbon atoms are equivalent, the Coulomb integrals for the secular equation are all equal:

$$H_{ii} = \alpha.$$

Furthermore, the bond distances are all the same. So the relations between adjacent orbitals are the same and

$$H_{jk} = \beta \quad \text{when} \quad j \text{ and } k \text{ are nearest neighbors.}$$

We also take

$$H_{jk} = 0 \quad \text{when} \quad j \text{ and } k \text{ are not nearest neighbors.}$$

Substituting these parameters into the $n = 6$ form of (6.68) yields

$$\begin{vmatrix} \alpha - E & \beta & 0 & 0 & 0 & \beta \\ \beta & \alpha - E & \beta & 0 & 0 & 0 \\ 0 & \beta & \alpha - E & \beta & 0 & 0 \\ 0 & 0 & \beta & \alpha - E & \beta & 0 \\ 0 & 0 & 0 & \beta & \alpha - E & \beta \\ \beta & 0 & 0 & 0 & \beta & \alpha - E \end{vmatrix} = 0.$$

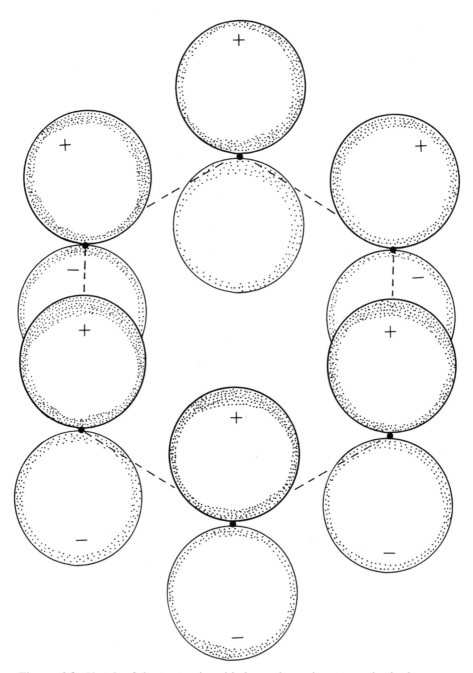

Figure 6.3. Sketch of the π atomic orbitals on the carbon atoms in the benzene molecule.

The six roots of this equation are

$$E = \alpha - 2\beta,$$
$$E = \alpha - \beta,$$
$$E = \alpha - \beta,$$
$$E = \alpha + \beta,$$
$$E = \alpha + \beta,$$
$$E = \alpha + 2\beta,$$

Matrix elements α and β are negative. So, putting the six available electrons into the three lowest molecular orbitals pairwise yields the Hückel energy

$$E = 2(\alpha + 2\beta) + 2(\alpha + \beta) + 2(\alpha + \beta) = 6\alpha + 8\beta.$$

Example 6.5

What is the energy of the six π electrons in the nonresonating benzene molecule of Figure 6.5?

In the given structure, a second bond is localized (a) between atoms 1 and 2, (b) between atoms 3 and 4, (c) between atoms 5 and 6. For a π electron in orbitals 1 and 2, we have the superposition

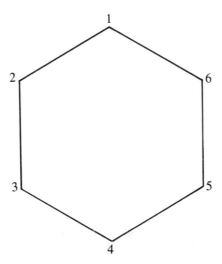

Figure 6.4. Numbering of positions on the benzene ring.

$$|g\rangle = c_1|1\rangle + c_2|2\rangle,$$

which leads to the secular equation

$$\begin{vmatrix} \alpha - E & \beta \\ \beta & \alpha - E \end{vmatrix} = 0$$

with the roots

$$E = \alpha - \beta,$$
$$E = \alpha + \beta.$$

The π electrons in orbitals 3 and 4, and those in orbitals 5 and 6, yield the same roots. Putting the six electrons pairwise into the 3 lowest orbitals gives us the energy

$$E = 2(\alpha + \beta) + 2(\alpha + \beta) + 2(\alpha + \beta) = 6\alpha + 6\beta.$$

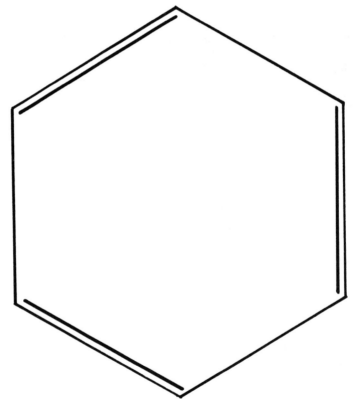

Figure 6.5. Bonds in the ring of nonresonating benzene.

Example 6.6

If the six π atomic orbitals in benzene did not combine coherently, what would the energy of the π electrons be?

The expectation value for the energy of an independent electron in the jth atomic orbital is

$$\langle H \rangle^j = \langle j \mid H \mid j \rangle = \alpha,$$

the Coulomb integral. By symmetry, the weight w_j of an electron in the jth π atomic orbital is 1.

But from (1.65) and (1.70), we have

$$E = \langle H \rangle = \sum_j w_j \langle H \rangle^j.$$

So here

$$E = \sum_{j=1}^{6} 1(\alpha) = 6\alpha,$$

and there is no bonding.

6.9
Parameters for the Hückel Method

Quantitites α_A and β_{AB} could be determined numerically, but because of the approximations in the method, the best values are obtained by fitting the calculated forms to experimental results on representative molecules.

Traditionally, beta has been obtained from thermochemical data. Indeed, it is found that hydrogenation of cyclohexene to cyclohexane liberates 28.6 kcal mole^{-1}. So if benzene consisted of alternating single and double bonds, its hydrogenation would presumedly liberate 85.8 kcal mole^{-1}. But only 49.8 kcal mole^{-1} is observed. Thus, benzene is 36 kcal mole^{-1} more stable than the single-bond double-bond structure.

Also, one may calculate the heat of combustion from bond energy data, including corrections for the different environments of bonds. From the single-bond double-bond structure, one finds 825.1 kcal mole^{-1}, but the experimental heat of combustion is 789.1 kcal mole^{-1}. These numbers yield 36 kcal mole^{-1} as the resonance stabilization.

From Examples 6.4 and 6.5, the Hückel theory stabilization equals 2β. For benzene, the empirical β is accordingly -18 kcal mole^{-1}. On averaging such values over common aromatic hydrocarbons, one finds that

$$\beta = -16 \text{ kcal mole}^{-1} \tag{6.73}$$

or

$$\beta = -0.69 \text{ eV} \tag{6.74}$$

However, these calculations do not allow for the stretching of the double bonds and the compression of the single bonds involved in going to the regular hexagonal rings. For benzene, Mulliken and Parr estimated this distortion energy to be 37 kcal mole^{-1}. The total resonance stabilization is then $36 + 37$ kcal mole^{-1} or 73 kcal mole^{-1}. This yields

$$\beta = -36.5 \text{ kcal mole}^{-1} \tag{6.75}$$

or

$$\beta = -1.58 \text{ eV}. \tag{6.76}$$

Now, one can determine both parameters from molecular electron affinities. The electron affinity EA is given by the negative of the energy of the lowest unfilled molecular orbital. If m_{n+1} is the root from the Hückel calculation, for this orbital, then we have

$$EA = -\alpha - m_{n+1}\beta. \tag{6.77}$$

One plots EA against $-m_{n+1}$ for representative compounds. The slope of the best straight line through the points yields β; the intercept at $-m_{n+1} = 0$ yields $-\alpha$. Wentworth, Kao, and Becker found that

$$\alpha = -1.42 \text{ eV} \tag{6.78}$$

and

$$\beta = -1.79 \text{ eV} \tag{6.79}$$

from data on benzenoid hydrocarbons.

In polyenes there are effects from electron correlations and the alternation of bond distances. As a consequence, there is less stabilization; the uncorrected thermochemical data yield

$$\beta = -6 \text{ kcal mole}^{-1}, \tag{6.80}$$

or

$$\beta = -0.26 \text{ eV}, \tag{6.81}$$

on the average, for these hydrocarbons.

6.10
Elements of an Extended Hückel Theory

One may extend the Hückel procedure to take into account more than one valence orbital on an atom. Thus, one may consider the σ bonds as well as the π bonds in a molecule. One can also consider a structure where this classification is not appropriate, as in various boron hydrides and complexes containing metal clusters.

Parameter $H_{jj} = \alpha_j$ varies not only with the atom but also with the valence state of the orbital. In the literature, this has been approximated by the Mulliken electronegativity

$$\chi_j = \tfrac{1}{2}(I_j + A_j). \tag{6.82}$$

Here I_j is the ionization potential of the valence electron from the orbital while A_j is the electron affinity of the orbital.

Shift of charge onto or away from a given atom on forming a molecule or complex alters its electronegativity and so the corresponding α. The alteration may be of the order of 2 eV per unit positive charge.

Parameter $H_{jk} = \beta_{jk}$ varies with the valence states of the two orbitals and with the relative arrangements of the orbitals. In 1952, Wolfsberg and Helmholz proposed a formula which we write in the form

$$\beta_{jk} = kS_{jk}\,\frac{\chi_j + \chi_k}{2} \quad \text{with} \quad k \simeq 1.5. \tag{6.83}$$

Here S_{jk} is the overlap between the atomic orbitals. The basis for (6.83) is the interpolation

$$\langle\, j \mid H \mid k \,\rangle \simeq \tfrac{1}{2}S_{jk}(\,\langle\, j \mid H \mid j \,\rangle + \langle\, k \mid H \mid k \,\rangle\,) \tag{6.84}$$

introduced by Mulliken.

However, the results obtained with (6.84) for a given system vary with the specific form of the basis orbitals. And, the formula holds much better for the potential energy part of H than for the kinetic energy part.

Accompanying use of (6.83), the overlaps should be systematicly introduced into the secular equation. Remember, the overlap integrals depend on the assumed bond distances and the orbital parameters, so they are to an extent empirical.

The main advantage of the extended Hückel procedure is that it provides a quick and qualitatively correct picture of molecules and complexes where classical valency theory does not apply; but in the process, parameters may be estimated incorrectly and essential terms may be neglected.

6.11
Possible Additional Improvements

In *semiempirical* methods, parameters for related structures are determined from experimental data, as Section 6.9 illustrates. In a thorough analysis the parameters vary, not only with the atomic orbitals involved and their relative orientations, but also with the internuclear distances and with the effective charges on the atoms. These properties change with the nature of the occupied molecular orbitals in an estimable manner.

However, the fields that the calculated orbitals present to each electron being considered may produce different values from those assumed. One should then alter the parameters in the amounts indicated and recalculate the orbitals. In a *self-consistent field* (SCF) model, the iterations are repeated until successive results check.

In addition, a molecular state function should be a superposition of anti-symmetrized products of its constituent one-electron orbitals. Each of the antisymmetrized combinations may be constructed as was done in Example 1.4 for two-constituent spin orbitals. The resulting determinant is called a *Slater determinant*.

In *ab initio* calculations on a given molecule, more basis functions are introduced. These may be Gaussian functions rather than Slater-type orbitals. Einstein relativistic effects are usually neglected and the Born-Oppenheimer approximation made. The Hamiltonian operator for the given structure is constructed. Matrix elements H_{jk} and S_{jk} are then calculated numerically and the resulting secular equation solved. The accuracy attainable depends on the number and kind of basis orbitals employed.

6.12
Natural Units

The conventional units employed in commerce and engineering have no fundamental significance. When they are used, physical constants and parameters come in odd sizes. Furthermore, these constants and parameters are often not known to the required number of significant figures. So mathematical manipulations and calculations are facilitated by replacing such units with ones fitting the problem at hand.

We have seen how symmetry considerations lead to commutation relations with the unit of energy multiplied by time chosen so that

$$\frac{h}{2\pi} = \hbar = 1. \tag{6.85}$$

Here, as usual, h is Planck's constant. Furthermore, one may take the unit of mass so that the pertinent particle's mass μ is unity:

$$\mu = 1. \tag{6.86}$$

For translation involving a fundamental period in space, one may set this distance equal to either 2π units or 1 unit:

$$a = 2\pi \text{ or } 1. \tag{6.87}$$

For rotation about a given axis, the reduced mass μ can be the mass unit. The unit of distance is then chosen so that the moment of inertia I equals 1 and

$$\sqrt{\frac{I}{\mu}} = 1. \tag{6.88}$$

For simple harmonic vibration, we employ the units in line (5.46).

For electronic states, the unit of charge e may be taken so

$$\frac{e^2}{4\pi\epsilon_0} = 1, \tag{6.89}$$

where ϵ_0 is the permittivity of free space. The unit of distance is often taken to be the Bohr radius of hydrogen:

$$\frac{4\pi\epsilon_0 \hbar^2}{m_e e^2} = 1. \tag{6.90}$$

In these units, the energy of a hydrogenlike atom, Equation (4.59), becomes

$$E = -\frac{Z^2}{2n^2} \tag{6.91}$$

with n the principal quantum number and Z the number of units of charge on the nucleus.

In Tables 6.1–6.4, the above choices have been collected and amplified.

Table 6.1
Natural Units for Translation

Property	Formula that Determines Unit
Planck's Constant	$h = 2\pi$
Mass	$\mu = 1$
Distance	$a = 2\pi$ or 1
Energy	$\dfrac{h^2}{\mu a^2} = 1$ or $4\pi^2$

Table 6.2
Natural Units for Rotation

Property	Formula that Determines Unit
Angular Momentum	$\hbar = 1$
Reduced Mass	$\mu = 1$
Distance	$\sqrt{\dfrac{I}{\mu}} = 1$
Energy	$\dfrac{\hbar^2}{I} = 1$

6.13
An Overview

A person may consider a given system together with a continuum of related systems introduced by varying parameters in its Hamiltonian H. Thus, one may construct the expression

$$H = H(\mathbf{X}, \mathbf{P}, \lambda) \qquad (6.92)$$

in which $\mathbf{X}$ is the operator for possible positions, $\mathbf{P}$ the operator for possible momenta, of the N particles in the system, while λ is a continuously variable real parameter. One then finds that

$$\left\langle \frac{\partial H}{\partial \lambda} \right\rangle = \frac{\partial E_j}{\partial \lambda}, \qquad (6.93)$$

where E_j is the energy of the jth state.

Table 6.3
Natural Units for Vibration

Property	Formula that Determines Unit
Reduced Mass	$\mu = 1$
Force Constant	$f = 1$
Distance	$\dfrac{\hbar^{1/2}}{(f\mu)^{1/4}} = 1$
Energy	$\hbar\omega = \hbar\sqrt{\dfrac{f}{\mu}} = 1$

Table 6.4
Natural Units for Electronic States in Atoms and Molecules

Property	Formula that Determines Unit	Name of Unit
Electron Mass	$m_e = 1$	—
Electron Charge	$\dfrac{e^2}{4\pi\epsilon_0} = 1$	—
Distance	$a_0 = \dfrac{4\pi\epsilon_0\hbar^2}{m_e e^2} = 1$	bohr
Energy	$\dfrac{e^2}{4\pi\epsilon_0 a_0} = 1$	hartree

With this Feynman-Hellmann equation, we derived the virial theorem for kinetic energy T,

$$2 \langle T \rangle = n \langle U \rangle, \tag{6.94}$$

which applies when potential U is a homogeneous function of the coordinates of degree n.

Alternatively, a person may consider a candidate ket $| f \rangle$ to vary in a way consistent with the boundary conditions for the given system. Then the expectation value for the observable of operator A would be

$$\langle A \rangle = \frac{\langle f | A | f \rangle}{\langle f | f \rangle}. \tag{6.95}$$

Wherever $\langle A \rangle$ is stationary,

$$\delta \langle A \rangle = 0, \tag{6.96}$$

with *no* additional constraints imposed, the eigenvalue equation

$$A \,|\, f \rangle = \langle A \rangle \,|\, f \rangle \tag{6.97}$$

would hold.

In practice, constraints are imposed by the form of representation for $|\, f \rangle$. Each stationary $\langle A \rangle$ is then the best approximation to the corresponding eigenvalue obtainable with the expression used.

Most convenient are linear variation functions

$$|\, g \rangle = \sum_{j=1}^{n} c_j \,|\, j \rangle = \sum_{k=1}^{n} c_k \,|\, k \rangle. \tag{6.98}$$

The known kets

$$|\, 1 \rangle, |\, 2 \rangle, \ldots, |\, n \rangle \tag{6.99}$$

are chosen so they describe possible independent contributions to the desired state kets. The scalar products

$$\langle j \,|\, A \,|\, k \rangle = A_{jk}, \tag{6.100}$$

$$\langle j \,|\, k \rangle = S_{jk} \tag{6.101}$$

are constructed.

Application of (6.96) then leads to the n simultaneous equations

$$\sum_{k=1}^{n} c_k (A_{lk} - a S_{lk}) = 0. \tag{6.102}$$

Cramer's rule tells us that these have nontrivial solutions only when

$$\begin{vmatrix} A_{11} - S_{11}a & A_{12} - S_{12}a & \cdots & A_{1n} - S_{1n}a \\ A_{21} - S_{21}a & A_{22} - S_{22}a & \cdots & A_{2n} - S_{2n}a \\ \cdot & \cdot & \cdot & \cdot \\ \cdot & \cdot & \cdot & \cdot \\ \cdot & \cdot & \cdot & \cdot \\ A_{n1} - S_{n1}a & A_{n2} - S_{n2}a & \cdots & A_{nn} - S_{nn}a \end{vmatrix} = 0. \tag{6.103}$$

The n roots of this secular equation are approximations to the eigenvalues of

$$A \,|\, j \rangle = a_j \,|\, j \rangle. \tag{6.104}$$

Any one of these roots may be substituted into the determinant. The jth equation of set (6.102) then results when the elements in the jth row are multiplied by the corresponding c_k's and added. From the set of equations, ratios of the coefficients can be calculated. Introducing a normalization condition and choosing a phase then leads to definite values.

Discussion Questions

6.1 How may parameters be used to describe conceivable alterations in a physical system?

6.2 When do the alterations lead to analytical changes in the eigenenergies?

6.3 How is the Feynman-Hellmann theorem established?

6.4 How can one calculate the expectation value for the kinetic energy?

6.5 How can the parameter λ be introduced so that it alters the potential function symmetricly? To what change in the kinetic energy is this alteration equivalent?

6.6 Show how $\langle T \rangle$ is related to $\langle U \rangle$ when the potential is a homogeneous function of degree n of the coordinates.

6.7 Why are representations of kets more complicated than the generating potentials?

6.8 Why can any trial ket $| g \rangle$ be expressed as a superposition of the eigenkets for the pertinent operator?

6.9 What does validity of the variation theorem depend on?

6.10 How can a person approximate a second derivative by differences?

6.11 Are the stationary values of the Ritz theorem minima?

6.12 Explain why the basis kets in the trial superposition

$$| g \rangle = \sum_{k=1}^{n} c_k | k \rangle$$

should be linearly independent. Why do they need not be mutually orthogonal?

6.13 What may the basis kets $| 1 \rangle, | 2 \rangle, \ldots, | n \rangle$ for the sum in Question 6.12 describe? Under what conditions would their choice not limit $| g \rangle$?

6.14 How may one differentiate with respect to a complex variable?

6.15 How did we introduce the stationarity condition

$$\delta a = 0$$

in this differentiation?

6.16 For a set of homogeneous simultaneous equations, as in (6.64), why do we set the determinant of the coefficients of the unknowns equal to zero?

6.17 Under what circumstances would the expansion coefficients c_1, $c_2, \ldots, c_n$ for a given state depend on the choice of operator A?

6.18 Why do we choose to let A in the secular equation be the Hamiltonian H?

6.19 Explain qualitatively how chemical bonding arises.

6.20 Define the atomic parameter, the bond parameter, delocalization, resonance.

6.21 Discuss the approximations employed in the Hückel method.

6.22 Why are α_A and β_{AB} determined empiricially?

6.23 How is the determinant in the secular equation for benzene multiplied out?

6.24 How may α and β be determined for benzenoid compounds?

6.25 Discuss the Wolfsberg-Helmholz formula.

6.26 Compare semiempirical procedures with ab initio procedures, noting the advantages and disadvantages of each.

6.27 What are appropriate natural units for translation, rotation, vibration, electronic states?

Problems

6.1 For the hydrogenlike atom, relate the expectation values for kinetic energy and for potential energy. How is the total energy related to the expectation value for the kinetic energy?

6.2 If a particle is confined between $x = 0$ and $x = 1$ in the state governed by

$$\psi = ax + bx^2 + cx^3,$$

how must a, b, and c be related? Why might a person set a equal to 1 when this ψ is used as a variation function?

6.3 Vary the coefficients in the function in Problem 6.2 to obtain approximations to the two lowest levels of the confined particle. Use the natural units in which a equals 1. Also, assume that V is 0 when $0 < x < 1$.

6.4 As an approximation to the ground state function for the harmonic oscillator, consider that

$$\psi = A(a^2 - x^2) \qquad \text{where} \qquad -a < x < a$$

and

$$\psi = 0 \qquad\qquad \text{elsewhere.}$$

Determine the best value for paramter a and calculate the corresponding energy.

6.5 Employ the variation theorem to find the value of parameter a in the function

$$\psi = A\,e^{-ar}$$

for the ground state of a hydrogenlike atom. Determine the corresponding energy. Employ natural units.

6.6 Calculate the fractional error in each variational energy obtained in Problems 6.3, 6.4, and 6.5.

6.7 Construct and solve the secular equation for the pi electrons in cyclobutadiene

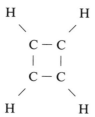

Then calculate the corresponding resonance energy.

6.8 Use the roots from Problem 6.7 to obtain the coefficients c_1, c_2, c_3, c_4 in the corresponding cyclobutadiene molecular orbitals.

6.9 Solve the secular equation for the pi electrons in normal butadiene

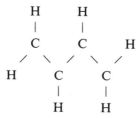

6.10 Multiply out the determinant in Example 6.4 and thus obtain the polynomial form of the secular equation for benzene. Show how the roots arise by factoring the polynomial. To simplify the procedure, you may set $\alpha - E$ equal to βx and cancel all β's. Manipulate the rows and columns to leave only one nonzero number in each of the first three columns (or rows) and then expand.

— — —

6.11 Consider an energy eigenstate of the quartic oscillator, for which

$$U = bx^4.$$

Relate the expectation value for its kinetic energy to that for its potential energy.

6.12 Employ the variation theorem to find the value of parameter b in the function

$$\psi = B\, e^{-br^2}$$

describing the ground state of the 3-dimensional oscillator for which

$$U = \tfrac{1}{2} fr^2.$$

Determine the corresponding energy. Use natural units.

6.13 As an approximation to the first-excited-state function for a harmonic oscillator, consider that

$$\psi = B \sin bx \qquad \text{where} \qquad -\frac{\pi}{b} < x < \frac{\pi}{b}$$

and

$$\psi = 0 \qquad\qquad\qquad \text{elsewhere.}$$

Determine the best value for b and calculate the corresponding energy.

6.14 Another approximation to the first-excited-state function for the harmonic oscillator, in natural units, is

$$\psi = B(b^2 x - x^3) \qquad \text{where} \qquad -b < x < b$$

and

$$\psi = 0 \qquad\qquad\qquad \text{elsewhere.}$$

Apply the variation theorem to obtain the best value for b. Then calculate the corresponding energy.

6.15 Calculate the fractional errors in the variational energies obtained in Problems 6.12, 6.13, and 6.14.

6.16 Consider a hydrogen atom in an electric field of uniform intensity **E** directed along the z axis. Break its Hamiltonian down into a term for the free atom and a term for the perturbing electric potential. Then determine the condition on parameter β in

$$\psi = (\cos \beta)\psi_{1s} + (\sin \beta)\psi_{2p_z}$$
$$= \left(\frac{\cos \beta}{a^{3/2}} 2\, e^{-\rho} + \frac{\sin \beta}{2^{3/2}a^{3/2}} \rho \cos \vartheta\, e^{-\rho/2} \right) \frac{1}{\sqrt{4\pi}}$$

that yields the best approximation to the energy of the system.

6.17 Use the result from Problem 6.16 to find an approximate formula for the energy of an unexcited hydrogen atom in the electric field.

6.18 Construct and solve the secular equation for the pi electrons in fulvene

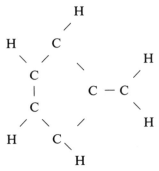

Then calculate the corresponding resonance energy.

   6.19 Calculate the resonance energies for the radicals

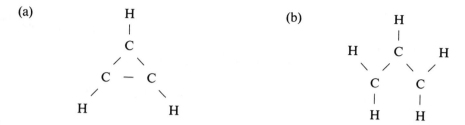

(a)   (b)

   6.20 Use the roots from Problem 6.19 to obtain the coefficients c_1, c_2, c_3 for the molecular orbitals of both radicals.

References

Books

Cohen-Tannoudji, C., Diu, B., and Laloe, F. (translated by Hemley, S. R., Ostrowsky, N., and Ostrowsky, D.): 1977. *Quantum Mechanics*, Wiley, New York, pp. 1148-1208. In this section of their treatise, the authors develop the conventional variational methods. Then they apply these to electrons in solids and in simple molecules.

Del Re, G., Berthier, G., and Serre, J.: 1980. *Electronic States of Molecules and Atom Clusters*, Springer-Verlag, Berlin, pp. 41-155. This detailed survey treats the various approximations that have been useful in studying the electronic states of complex molecules and atom clusters. Key parameters and results are tabulated.

I'Haya, Y.: 1964. "Recent Developments in the Generalized Hückel Method," in Löwdon, P.-O. (editor), *Advances in Quantum Chemistry*, vol. 1, Academic Press, New York, pp. 203-240. I'Haya's review is more limited than the preceding survey, covering only certain generalizations of the Hückel method for molecular orbitals.

Lowe, J. P.: 1978. *Quantum Chemistry*, Academic Press, New York, pp. 144-346. In this felicitous introductory text, Lowe presents the standard variational procedures and chemical applications. One may very well go to this reference before consulting the two preceding ones.

Moiseiwitsch, B. L.: 1961. "The Variational Method," in Bates, D. R. (editor), *Quantum Theory*, vol, I, Academic Press, New York, pp. 211-228. In this short chapter, Moiseivitsch develops formulas for lower bounds as well as for upper bounds on the energy of a system.

Rektorys, K.: 1980. *Variational Methods in Mathematics, Science and Engineering*, 2nd edn., Reidel, Dordrecht, pp. 441-501. In this section of his rigorous, comprehensive monograph, Rektorys treats the eigenvalue problem in a general, abstract manner.

Articles

Aebersold, D.: 1975, "Integral Equations in Quantum Chemistry", *J. Chem. Educ.* **52**, 434–436.

Baretty, R., and Garcia, C.: 1989, "Atomic Calculations with a One-Parameter, Single Integral Method," *J. Chem. Educ.* **66**, 45–46.

Crawford, F. S.: 1989, "Footnote to the Quantum Mechanical Virial Theorem," *Am. J. Phys.* **57**, 555–557.

Denham, S. A., and Harms, B. C.: 1982, "A Method for Numerical Determination of Eigenvalues", *Am. J. Phys.* **50**, 374–380.

Dewar, M. J. S., and Kelemen, J.: 1971, "LCAO MO Theory Illustrated by its Application to H_2", *J. Chem. Educ.* **48**, 494–501.

Dongpei, Z.: 1986, "Proof of the Quantum Virial Theorem", *Am. J. Phys.* **54**, 267–270.

Fernandez, F. M., and Castro, E. A.: 1984, "Approximate Energy Levels of Bound Quantum Systems", *Am. J. Phys.* **52**, 453–455.

Goodfriend, P. L.: 1987, "Diatomic Vibrations Revisited", *J. Chem. Educ.* **64**, 753–756.

Handy, C. R., and Bessis, D.: 1988, "Rapidly Converging Bounds for the Ground-State Energy of Hydrogenic Atoms in Superstrong Magnetic Fields," *Phys. Rev. Lett.* **60**, 253–256.

Keeports, D.: 1989, "Application of the Variational Method to the Particle-in-the-Box Problem," *J. Chem. Educ.* **66**, 314–318.

Keeports, D.: 1990, "Application of the Variational Method to One-Dimensional Infinite Wells with Internal cx^n Potentials," *Am. J. Phys.* **58**, 230–234.

Lee, J.: 1987, "The Upper and Lower Bounds of the Ground State Energies Using the Variational Method", *Am. J. Phys.* **55**, 1039–1040.

Li, W. -K.: 1988, "A Lesser Known One-Parameter Wave Function for the Helium Sequence and the Virial Theorem," *J. Chem. Educ.* **65**, 963–964.

Ogilvie, J. F.: 1990, "The Nature of the Chemical Bond – 1990: There are no such Things as Orbitals," *J. Chem. Educ.* **67**, 280–289.

Park, D.: 1973, "A Simple Variational Formula in Quantum Mechanics", *Am. J. Phys.* **41**, 1081–1083.

Ray, A., Mahata, K., and Ray, P. P.: 1988, "Moments of Probability Distribution, Wavefunctions, and their Derivatives at the Origin of N-Dimensional Central Potentials", *Am. J. Phys.* **56**, 462–464.

Robiette, A. G.: 1975, "The Variation Theorem Applied to H_2^+", *J. Chem. Educ.* **52**, 95–96.

Robiscoe, R. T.: 1975, "A Variational Hartree-Type Calculation", *Am. J. Phys.* **43**, 538–540.

Srivastava, M. K., and Bhaduri, R. K.: 1977, "Variational Method for Two-Electron Atoms", *Am. J. Phys.* **45**, 462–464.

Valk, H. S.: 1986, "Radial Expectation Values for Central Force Problems and the Feynman-Hellman Theorem", *Am. J. Phys.* **54**, 921–923.

7

STATIONARY-STATE
PERTURBATION THEORY

7.1
Reckoning with Complicating Influences

The free translator, the free rotator, the harmonic oscillator, and the hydrogenlike atom are simple systems for which standard exact eigenfunctions are known. Using such functions as guides, approximate representations of the states of more intricate systems can be formulated. For such systems, operators can be constructed and the expectation-value formula applied. One or more of the parameters may be varied and the best value or values determined.

A flexible, systematic method for improving the results exists. The part of the given operator for which the approximate solution is exact is identified. The difference between the whole operator and this part is considered to act as a perturber on the approximate solution; the effect produced is called a *perturbation*. The perturber is factored into a number λ that varies from 0 to some value that may be 1 and the remaining operator.

The eigenkets and eigenvalues are considered to be power series in this parameter λ. These series are formally introduced into the eigenvalue equation. Since the resulting equality must not determine λ, it is an identity in λ. Coefficients of each power of λ are equated and a sequence of relationships obtained. These are solved in order until the extra work involved in going further is not worth the gain in accuracy.

The procedure works whenever increasing the perturbation parameter from zero introduces changes from a determinable state smoothly and gradually.

Then the series obtained for the eigenvalues may converge without trouble. Or, this series may be asymptotic but still useable. On the other hand, the introduced changes may be abrupt or catastrophic and cause the series to be uselessly divergent.

7.2
Development from a Single Simplified State

The procedure will first be applied to an energy eigenstate that can be developed from a single, known, approximate eigenstate. Development from a cluster of approximate states, degenerate or nearly degenerate, will be considered later.

We recall from Equation (3.77), that the Hamiltonian H for a given system consists of the sum of a kinetic energy operator and a potential energy operator. When the complete operator acts on one of its eigenkets, $|i\rangle$, the result equals the energy E_i of the corresponding state times the ket:

$$H|i\rangle = E_i|i\rangle. \tag{7.1}$$

In general, this eigenvalue equation is difficult or impossible to solve directly.

But let us suppose that removing some part of the potential energy operator from H allows an exact solution to be obtained. If the residual approximate part is $H^{(0)}$ and the complicating part is expressed as $\lambda U^{(1)}$, where λ is a number, we have

$$H = H^{(0)} + \lambda U^{(1)}. \tag{7.2}$$

Now we restrict the discussion to a state ket $|i\rangle$ that can be generated from a single, readily calculable, ket $|i, 0\rangle$, for which we have

$$H^{(0)}|i, 0\rangle = E_i^{(0)}|i, 0\rangle. \tag{7.3}$$

We thus suppose that $|i, 0\rangle$ transforms smoothly into $|i\rangle$ as λ increases from 0 to its value in (7.2), which may conveniently be 1. Number λ is known as the *perturbation parameter*.

Without introducing further restrictions, we let the ket and the energy for the state be power series in this parameter:

$$|i\rangle = |i, 0\rangle + \lambda|i, 1\rangle + \lambda^2|i, 2\rangle + \ldots = \sum_{k=0}^{\infty} \lambda^k |i, k\rangle. \tag{7.4}$$

$$E_i = E_i^{(0)} + \lambda E_i^{(1)} + \lambda^2 E_i^{(2)} + \ldots = \sum_{j=0}^{\infty} \lambda^j E_i^{(j)}. \tag{7.5}$$

Note that series (7.4) and (7.5) may turn out to be divergent. However, they are still useful if they are of asymptotic form. But trouble may arise when $U^{(1)}$ introduces new singularities or alters existing singularities.

Substituting (7.2), (7.4), (7.5) into (7.1) yields

$$\sum_{j=0}^{\infty} \sum_{k=0}^{\infty} (H^{(0)} + \lambda U^{(1)} - \lambda^j E_i^{(j)})(\lambda^k \mid i, k \rangle) = 0 \tag{7.6}$$

Since this equation must not restrict λ, it is an identity in λ and the coefficient of each power of λ on the left must equal 0. Thus,

$$(H^{(0)} - E_i^{(0)}) \mid i, 0 \rangle = 0, \tag{7.7}$$

$$(H^{(0)} - E_i^{(0)}) \mid i, 1 \rangle + (U^{(1)} - E_i^{(1)}) \mid i, 0 \rangle = 0, \tag{7.8}$$

$$(H^{(0)} - E_i^{(0)}) \mid i, 2 \rangle + (U^{(1)} - E_i^{(1)}) \mid i, 1 \rangle - E_i^{(2)} \mid i, 0 \rangle = 0, \tag{7.9}$$

$$\cdot$$
$$\cdot$$
$$\cdot$$

$$(H^{(0)} - E_i^{(0)}) \mid i, n \rangle + (U^{(1)} - E_i^{(1)}) \mid i, n - 1 \rangle - E_i^{(2)} \mid i, n - 2 \rangle$$
$$- \ldots - E_i^{(n)} \mid i, 0 \rangle = 0. \tag{7.10}$$

$$\cdot$$
$$\cdot$$
$$\cdot$$

Equation (7.7) contains nothing new; it is (7.3) rearranged. But multiplying (7.8) by bra $\langle i, 0 \mid$ yields

$$\langle i, 0 \mid H^{(0)} - E_i^{(0)} \mid i, 1 \rangle + \langle i, 0 \mid U^{(1)} - E_i^{(1)} \mid i, 0 \rangle = 0. \tag{7.11}$$

Since $H^{(0)} - E_i^{(0)}$ is Hermitian and (7.7) holds, we have

$$\langle i, 0 \mid H^{(0)} - E_i^{(0)} \mid i, 1 \rangle = \langle i, 1 \mid H^{(0)} - E_i^{(0)} \mid i, 0 \rangle * = 0. \tag{7.12}$$

Thus, the first term in (7.11) vanishes and the equation rearranges to give

$$\langle i, 0 \mid U^{(1)} \mid i, 0 \rangle = \langle i, 0 \mid E_i^{(1)} \mid i, 0 \rangle = E_i^{(1)}, \tag{7.13}$$

as long as $\mid i, 0 \rangle$ is normalized to 1.

Substituting (7.13) into (7.5) yields

$$E_i = E_i^{(0)} + \lambda \langle i, 0 \mid U^{(1)} \mid i, 0 \rangle + \ldots \tag{7.14}$$

One can calculate higher terms similarly, using Equations (7.9)–(7.10). The power of λ in the last term that is retained is called the *order* of the approximation. Note that Equation (7.14) can be rewritten in the form

$$E_i = \langle i, 0 \mid H^{(0)} \mid i, 0 \rangle + \langle i, 0 \mid \lambda U^{(1)} \mid i, 0 \rangle + \ldots$$

$$= \langle i, 0 \mid H^{(0)} + \lambda U^{(1)} \mid i, 0 \rangle + \ldots$$

$$= \langle i, 0 \mid H \mid i, 0 \rangle + \ldots. \tag{7.15}$$

To first order, the change in the ith energy level caused by perturber $\lambda U^{(1)}$ equals the expectation value of the perturber in the corresponding unperturbed state; the energy itself equals the expectation value for H in the unperturbed state.

Example 7.1

A freely moving electron is confined in its ground state between walls 1.000×10^{-10} m apart. How much is its energy lowered by introduction of a well 18.0 eV deep and 1.000×10^{-11} m wide midway between the two walls, as Figure 7.1 shows? What is the net energy of the electron?

We consider this to be a 1-dimensional problem; thus, we neglect any motion in the y or z directions. We also consider that the potential U is piecewise constant, being zero between the walls except in the well, where it is -18.0 eV. The distance between the walls is designated a, while the width of the well is designated b, as Figure 7.1 shows.

Before the well was introduced, the momentum of the electron would have a constant magnitude, the de Broglie wavelength λ would be definite between the walls, and the state function ψ would be sinusoidal. At each wall, U would rise abruptly without limit and ψ would vanish.

In the ground state, the wavelength would be as long as possible; here we would have

$$(1/2)\lambda = a$$

and

$$k = \frac{2\pi}{\lambda} = \frac{\pi}{a}.$$

Equation (4.16) now yields the coordinate representation

$$\psi^{(0)} = \sqrt{\frac{2}{a}} \cos \frac{\pi x}{a}$$

of the ket. The corresponding energy is

$$E^{(0)} = \frac{p^2}{2\mu} = \frac{h^2}{2\mu\lambda^2} = \frac{h^2}{8\mu a^2}.$$

Expressing the mass of the electron and Planck's constant in appropriate units leads to

$$\frac{h^2}{2\mu} = 150.41 \text{ eV Å}^2.$$

Combining with the given interwall distance then gives us

$$E^{(0)} = \frac{150.41 \text{ eV Å}^2}{4(1.000 \text{ Å})^2} = 37.6 \text{ eV}.$$

The well is described by the perturbing potential

$$U^{(1)} = 0 \text{ where } \quad -\frac{a}{2} < x < -\frac{b}{2} \text{ and } \frac{b}{2} < x < \frac{a}{2}$$

and

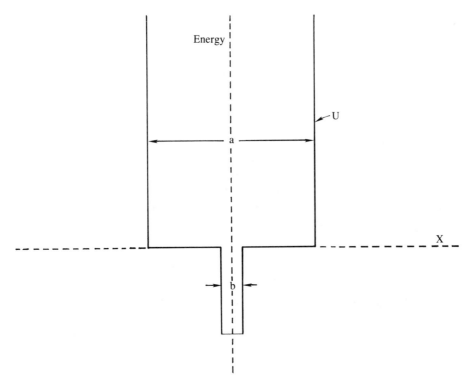

Figure 7.1. Small rectangular potential well midway between parallel confining walls.

$$U^{(1)} = -A \qquad \text{where} \qquad -\frac{b}{2} < x < \frac{b}{2}.$$

Substituting this into (7.13) yields

$$E^{(1)} = \int_{-b/2}^{b/2} \sqrt{\frac{2}{a}} \cos \frac{\pi x}{a} \left(-A \right) \sqrt{\frac{2}{a}} \cos \frac{\pi x}{a} \, dx$$

$$\simeq -\frac{2}{a} \int_{-b/2}^{b/2} A \, dx = -\frac{2A}{a} b.$$

The given parameters reduce this result to

$$E^{(1)} = -\frac{2(18.0 \text{ eV})}{10^{-10} \text{m}} 10^{-11} \text{m} = -3.6 \text{ eV},$$

whence the calculated energy is

$$E = (37.6 - 3.6) \text{ eV} = 34.0 \text{ eV}.$$

7.3
First-Order Description of the Ground State of a Helium-like Atom or Ion

Hydrogenlike structures have been described in section 4.8. Adding one electron to any one of these produces a heliumlike structure with an electron-electron interaction in addition to the electron-nucleus interactions. We will here consider the ground state for such a structure in the coordinate representation. Magnetic effects will be neglected. Also, any translational motion of the system as a whole will not be considered.

So, Cartesian axes are erected on the center of the nucleus, as Figure 7.2 shows. The corresponding spherical coordinates of the first electron are r_1, ϑ_1, φ_1, those of the second electron, r_2, ϑ_2, φ_2. The interelectron distance is labeled r_{12}.

In the natural units of Table 6.4, the Hamiltonian operator for the structure is

$$H = -\frac{1}{2} (\nabla_1^2 + \nabla_2^2) - \left(\frac{Z}{r_1} + \frac{Z}{r_2} \right) + \frac{1}{r_{12}}. \qquad (7.16)$$

Since the interelectronic potential is the source of difficulty in this problem, the

term that expresses it is considered to be the perturber. The simple approximate
Hamiltonian operator is

$$H^{(0)} = -\frac{1}{2}(\nabla_1^2 + \nabla_2^2) - \left(\frac{Z}{r_1} + \frac{Z}{r_2}\right), \qquad (7.17)$$

while the correction to it is

$$U^{(1)} = \frac{1}{r_{12}}. \qquad (7.18)$$

In the approximation that magnetic effects are negligible, the spin is independent of the orbital motion; then the state function representing the ket

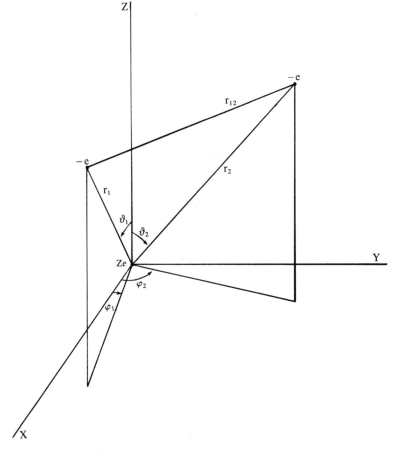

Figure 7.2. Coordinates for the heliumlike structure.

factors into an orbital function and a spin factor. Because electrons are fermions, this state function must be antisymmetric with respect to the two electrons. In the ground state and in some of the excited states, the antisymmetry is in the spin factor. Then the orbital factor is symmetric.

When such electrons occupy orthogonal orbitals $\psi_I(\mathbf{r}_j)$ and $\psi_{II}(\mathbf{r}_j)$, we have

$$\psi = \frac{1}{\sqrt{2}} [\psi_I(\mathbf{r}_1) \, \psi_{II}(\mathbf{r}_2) + \psi_{II}(\mathbf{r}_1)\psi_I(\mathbf{r}_2)]. \tag{7.19}$$

But when ψ_{II} does not differ from ψ_I, we have

$$\psi = \psi_I(\mathbf{r}_1) \, \psi_I(\mathbf{r}_2). \tag{7.20}$$

Since the most stable orbital is not degenerate, the ground state ψ fits (7.20).

Substituting (7.17) and (7.20) into (7.3) yields

$$\left[-\frac{1}{2} (\nabla_1^2 + \nabla_2^2) - \left(\frac{Z}{r_1} + \frac{Z}{r_2} \right) \right] \psi_I(\mathbf{r}_1) \, \psi_I(\mathbf{r}_2) = E^{(0)} \, \psi_I(\mathbf{r}_1) \, \psi_I(\mathbf{r}_2). \tag{7.21}$$

Since $\mathbf{r}_2$ is constant when ∇_1^2 acts, we have

$$\nabla_1^2 \, \psi_I(\mathbf{r}_1) \, \psi_I(\mathbf{r}_2) = \psi_I(\mathbf{r}_2)\nabla_1^2 \, \psi_I(\mathbf{r}_1). \tag{7.22}$$

Similarly,

$$\nabla_2^2 \, \psi_I(\mathbf{r}_1) \, \psi_I(\mathbf{r}_2) = \psi_I(\mathbf{r}_1)\nabla_2^2 \, \psi_I(\mathbf{r}_2). \tag{7.23}$$

Reducing (7.21) with (7.22) and (7.23), dividing by the product $\psi_I(\mathbf{r}_1) \, \psi_I(\mathbf{r}_2)$, and rearranging leads to

$$-\frac{\frac{1}{2}\nabla_1^2\psi_I(\mathbf{r}_1)}{\psi_I(\mathbf{r}_1)} - \frac{Z}{r_1} = \frac{\frac{1}{2}\nabla_2^2\psi_I(\mathbf{r}_2)}{\psi_I(\mathbf{r}_2)} + \frac{Z}{r_2} + E^{(0)}. \tag{7.24}$$

The left side of (7.24) may vary only when $\mathbf{r}_1$ varies; the right side may vary only when $\mathbf{r}_2$ varies. For the two sides to be equal, they must be constant. Letting the constant be $E_1^{(0)}$ and letting $E^{(0)} - E_1^{(0)}$ be $E_2^{(0)}$ produces

$$\left(-\frac{1}{2}\nabla_1^2 - \frac{Z}{r_1} \right)\psi_I(\mathbf{r}_1) = E_1^{(0)}\psi_I(\mathbf{r}_1), \tag{7.25}$$

$$\left(-\frac{1}{2}\nabla_2^2 - \frac{Z}{r_2} \right)\psi_I(\mathbf{r}_2) = E_2^{(0)}\psi_I(\mathbf{r}_2), \tag{7.26}$$

and

$$E^{(0)} = E_1^{(0)} + E_2^{(0)}. \tag{7.27}$$

Equations (7.25) and (7.26) govern hydrogenlike structures. If n is the principal quantum number for either, the energy is

$$E_j^{(0)} = -\frac{Z^2}{2n^2}. \tag{7.28}$$

In the lowest state, $n = 1$ and

$$E^{(0)} = -\frac{Z^2}{2} - \frac{Z^2}{2} = -Z^2. \tag{7.29}$$

From Tables 4.3 and 4.5, the eigenfunctions are

$$\psi_j(\mathbf{r}_j) = \frac{1}{\sqrt{4\pi}} Z^{3/2} 2e^{-Zr_j} = \left(\frac{Z^3}{\pi}\right)^{1/2} e^{-Zr_j}. \tag{7.30}$$

Multiplying these yields the zeroth-order state function

$$\psi_0 = \psi_j(\mathbf{r}_1)\psi_j(\mathbf{r}_2) = \frac{Z^3}{\pi} e^{-Z(r_1 + r_2)}. \tag{7.31}$$

With

$$U^{(1)} = \frac{1}{r_{12}}, \tag{7.32}$$

the first order correction (7.13) becomes

$$\begin{aligned}
E_0^{(1)} &= \int \int \frac{Z^3}{\pi} e^{-Z(r_1 + r_2)} \frac{1}{r_{12}} \frac{Z^3}{\pi} e^{-Z(r_1 + r_2)} \, d^3\mathbf{r}_1 \, d^3\mathbf{r}_2 \\
&= \frac{Z^6}{\pi^2} \int \int \frac{e^{-2Zr_1}e^{-2Zr_2}}{r_{12}} \, d^3\mathbf{r}_1 \, d^3\mathbf{r}_2. \tag{7.33}
\end{aligned}$$

The last integral may be interpreted as the repulsion energy of a charge cloud of density $\exp -2Zr_1$ on one of density $\exp -2Zr_2$. Since the first cloud is spherically symmetric, Gauss's law tells us that the electrostatic potential due to a spherical shell of the cloud of radius r_1 and thickness dr_1, in the appropriate units, is

$$4\pi r_1^2 e^{-2Zr_1} \, dr_1 \frac{1}{r_1} = 4\pi r_1 e^{-2Zr_1} \, dr_1 \tag{7.34}$$

when $r < r_1$ and

$$4\pi r_1^2 e^{-2Zr_1} \, dr_1 \frac{1}{r} \tag{7.35}$$

when $r > r_1$. The net electrostatic potential at radius r is

$$\phi(r) = \frac{4\pi}{r} \int_0^r e^{-2Zr_1} r_1^2 \, dr_1 + 4\pi \int_r^\infty e^{-2Zr_1} r_1 \, dr_1$$

$$= \frac{\pi}{Z^3} \frac{1}{r} [1 - e^{-2Zr}(1 + Zr)].$$ (7.36)

Integrating $\phi(r)$ over the density of the second cloud and multiplying by Z^6/π^2 yields

$$E_0^{(1)} = \frac{Z^6}{\pi^2} \frac{\pi}{Z^3} \int_0^\infty \frac{1}{r} (e^{-2Zr} - e^{-4Zr} - Zr \, e^{-4Zr}) \, 4\pi r^2 \, dr$$

$$= 4Z^3 \left(\frac{1}{4Z^2} - \frac{1}{16Z^2} - \frac{1}{32Z^2} \right) = \frac{5}{8} Z.$$ (7.37)

Substituting (7.29) and (7.37) into (7.5) yields the result

$$E_0 = - \left(Z^2 - \frac{5}{8} Z \right)$$ (7.38)

when the perturbation is exerting its full effect ($\lambda = 1$) and terms beyond the linear one are neglected.

7.4
The Average Effective Nuclear Charge in the Heliumlike Structure

In the independent-particle model for the heliumlike atom or ion, a person considers that each electron moves in the average field of the other one. But in the unperturbed state this field is spherically symmetric about the nucleus, so it has the effect of canceling part of the nuclear charge that the given electron experiences. The effective number of units on the nucleus would vary from Z when the electron is at $r = 0$ to $Z - 1$ when it is at $r = \infty$.

For simplicity, we here represent this variable quantity by a constant Z' to be determined by the variational method. The state of each electron is then approximated by function (7.30) with Z replaced by Z'. Function (7.31) becomes

$$\psi_0 = \frac{Z'^3}{\pi} e^{-Z'(r_1 + r_2)}.$$ (7.39)

This expression is the lowest eigenfunction for the Hamiltonian operator

$$H^{(0)} = -\tfrac{1}{2}(\nabla_1^2 + \nabla_2^2) - \left(\frac{Z'}{r_1} + \frac{Z'}{r_2}\right). \qquad (7.40)$$

The difference between (7.16) and (7.40) is the correction

$$U^{(1)} = \frac{1}{r_{12}} - (Z - Z')\left(\frac{1}{r_1} + \frac{1}{r_2}\right). \qquad (7.41)$$

Substituting this into perturbation formula (7.13) yields

$$E_0^{(1)} = \frac{Z'^6}{\pi^2} \int \int e^{-Z'(r_1 + r_2)} \frac{1}{r_{12}} e^{-Z'(r_1 + r_2)} \, d^3\mathbf{r}_1 \, d^3\mathbf{r}_2$$
$$- (Z - Z') \frac{Z'^6}{\pi^2} \int \int e^{-Z'(r_1 + r_2)} \left(\frac{1}{r_1} + \frac{1}{r_2}\right) e^{-Z'(r_1 + r_2)} \, d^3\mathbf{r}_1 \, d^3\mathbf{r}_2. \qquad (7.42)$$

The first term on the right amounts to (7.33) with Z' in place of Z; by (7.37) its value is $(5/8)Z'$. The second term rearranges to

$$- (Z - Z') \frac{Z'^6}{\pi^2} \int \frac{e^{-2Z'r_1}}{r_1} \, d^3\mathbf{r}_1 \int e^{-2Z'r_2} \, d^3\mathbf{r}_2$$
$$- (Z - Z') \frac{Z'^6}{\pi^2} \int e^{-2Z'r_1} \, d^3\mathbf{r}_1 \int \frac{e^{-2Z'r_2}}{r_2} \, d^3\mathbf{r}_2. \qquad (7.43)$$

But since

$$\int_0^\infty e^{-2Z'r} \, 4\pi r^2 \, dr = \frac{\pi}{Z'^3}, \qquad (7.44)$$

$$\int_0^\infty \frac{e^{-2Z'r}}{r} \, 4\pi r^2 \, dr = \frac{\pi}{Z'^2}, \qquad (7.45)$$

expression (7.43) reduces to

$$- 2(Z - Z') \frac{Z'^6}{\pi^2} \frac{\pi}{Z'^3} \frac{\pi}{Z'^2} = -2(Z - Z')Z'. \qquad (7.46)$$

Therefore, (7.42) reduces to

$$E_0^{(1)} = \frac{5}{8}Z' - 2(Z - Z')Z' \qquad (7.47)$$

Since (7.40) is obtained from (7.17) on replacing Z with Z', the zeroth-order energy now equals $-Z'^2$. Adding this to (7.47) yields

$$E_0 = Z'^2 + \frac{5}{8}Z' - 2ZZ'. \qquad (7.48)$$

According to the variation theorem, the best Z' is that which minimizes the expectation value for the energy. So we differentiate (7.48), set the derivative equal to zero,

$$\frac{dE_0}{dZ'} = 2Z' + \frac{5}{8} - 2Z = 0, \qquad (7.49)$$

solve for Z',

$$Z' = Z - \frac{5}{16} \qquad (7.50)$$

and substitute the result into (7.48):

$$E_0 = -Z^2 + \frac{5}{8} Z - \left(\frac{5}{16}\right)^2 = -\left(Z - \frac{5}{16}\right)^2. \qquad (7.51)$$

This formula does not allow for the deviations of the state function from form (7.39) caused by the interelectron repulsion. The error is greatest when Z is small.

The *binding energy* of an atom or ion is the energy needed to disperse the structure into its constituent entities, the nucleus and the individual electrons. Binding energies calculated from (7.38) and (7.51) for various heliumlike structures are compared with the observed binding energies in Table 7.1.

Table 7.1
Binding Energies of
Heliumlike Structures in Hartrees

Atom or Ion	Formula (7.38)	Formula (7.51)	Experiment
H^-	0.375	0.473	0.528
He	2.750	2.848	2.904
Li^+	7.125	7.223	7.280
Be^{2+}	13.500	13.598	13.656
B^{3+}	21.875	21.973	22.031
C^{4+}	32.250	32.348	32.406
N^{5+}	44.625	44.723	44.781
O^{6+}	59.000	59.098	59.157
F^{7+}	75.375	75.473	75.532
Ne^{8+}	93.750	93.848	93.907

Example 7.2

Calculate the first-order energies for the helium atom and express them in electron volts.

When all effects of interelectronic repulsion on the state function are ne-glected, formula (7.38) applies. With $Z = 2$, as we have in the helium atom, we obtain

$$E_0 = -\left(2^2 - \frac{10}{8}\right) = -2.750 \text{ hartrees.}$$

When the effect of interelectronic repulsion on the size of the state function is considered, as in (7.39), formula (7.51) applies and we get

$$E_0 = -\left(2 - \frac{5}{16}\right)^2 = -2.848 \text{ hartrees.}$$

Let us insert the known values of the fundamental constants into the ex-pression for the hartree

$$\frac{e^2}{4\pi\epsilon_0 a_0} = \frac{m_e e^4}{(4\pi\epsilon_0)^2 \hbar^2} = \frac{m_e c^4 \, 10^{-14} e^4}{\hbar^2}$$

$$= \frac{(9.1095 \times 10^{-31} \text{ kg})(2.9979 \times 10^8)^4 \, (10^{-14} \text{C}^{-4}\text{J}^2\text{m}^2)(1.6022 \times 10^{-19}\text{C})^4}{(1.0546 \times 10^{-34} \text{ J s})^2 \, [1.6022 \times 10^{-19} \text{ J(eV)}^{-1}]}$$

$$= 27.211 \text{ eV.}$$

Then (7.38) yields

$$E_0 = (-2.750 \text{ hartrees})(27.211 \text{ eV hartree}^{-1}) = -74.83 \text{ eV,}$$

while (7.51) yields

$$E_0 = (-2.848 \text{ hartrees})(27.211 \text{ eV hartree}^{-1}) = -77.50 \text{ eV.}$$

7.5
Second and Higher Order Corrections to the Energies

In perturbation theory all complicating parts of the Hamiltonian H for a given system may be gathered into expression $U^{(1)}$, with $\lambda = 1$. Then a com-plete set of orthonormalized eigenkets of the remaining $H^{(0)}$, consisting of

$$| 1, 0 \rangle, | 2, 0 \rangle, | 3, 0 \rangle, \dots, \tag{7.52}$$

is identified. Recall Equation (7.3). For a state generated from a single eigenket of this set, the energy to first order equals the expectation value for H which this eigenket yields. See (7.15).

However, adding $U^{(1)}$ to $H^{(0)}$, to get the H in (7.1), does alter the eigenkets. The effect is represented formally by (7.4). Now, any perturbation of a ket does introduce higher order changes in the corresponding energy.

Without loss of generality, we can require the kets on the right of (7.4) to be mutually orthogonal:

$$\langle i, \alpha \mid i, \beta \rangle = 0 \quad \text{when} \quad \alpha \neq \beta. \tag{7.53}$$

To get the second-order correction to E, multiply (7.9) by bra $\langle i, 0 \mid$,

$$\langle i, 0 \mid H^{(0)} - E_i^{(0)} \mid i, 2 \rangle + \langle i, 0 \mid U^{(1)} - E_i^{(1)} \mid i, 1 \rangle$$
$$- \langle i, 0 \mid E_i^{(2)} \mid i, 0 \rangle = 0. \tag{7.54}$$

As in (7.12), the first term vanishes. The last term on the left reduces to $-E_i^{(2)}$. Putting it on the other side and reducing the middle term, using (7.53), yields

$$E_i^{(2)} = \langle i, 0 \mid U^{(1)} \mid i, 1 \rangle. \tag{7.55}$$

Similarly, multiply (7.10) by bra $\langle i, 0 \mid$,

$$\langle i, 0 \mid H^{(0)} - E_i^{(0)} \mid i, n \rangle + \langle i, 0 \mid U^{(1)} - E_i^{(1)} \mid i, n - 1 \rangle$$
$$- \langle i, 0 \mid E_i^{(2)} \mid i, n - 2 \rangle - \ldots - \langle i, 0 \mid E_i^{(n)} \mid i, 0 \rangle = 0. \tag{7.56}$$

As before, the first term equals zero. The matrix elements for $E_i^{(1)}$, $E_i^{(2)}$, $\ldots$, $E_i^{(n-1)}$ vanish because of orthogonality condition (7.53). The last term on the left reduces to $-E_i^{(n)}$. Shifting it to the other side and reducing the remaining term yields

$$E_i^{(n)} = \langle i, 0 \mid U^{(1)} \mid i, n - 1 \rangle. \tag{7.57}$$

To apply formula (7.55) for $E_i^{(2)}$, one needs the first order correction $\mid i, 1 \rangle$. To apply (7.57) for $E_i^{(n)}$, one needs the $(n - 1)$th order correction $\mid i, n - 1 \rangle$. So we need to consider how these may be found or related to lower order corrections.

7.6
Corrections to the Eigenkets

For any system, a complete set of coordinates is involved in the kinetic energy part of H, and so in the simplified $H^{(0)}$. Adding the complicating part $U^{(1)}$ does not require the Hilbert space to be enlarged. Furthermore, the zeroth-order eigenkets (7.52) span this Hilbert space. As a consequence, each correction ket in (7.4) may be built up from these eigenkets.

When each $| i, n \rangle$ is chosen to be orthogonal to the unperturbed ket $| i, 0 \rangle$, we have

$$| i, n \rangle = \sum_{j \neq i} c_{ji}^{(n)} | j, 0 \rangle. \tag{7.58}$$

with no contribution from $| i, 0 \rangle$. For the first-order correction, (7.58) becomes

$$| i, 1 \rangle = \sum_{j \neq i} c_{ji}^{(1)} | j, 0 \rangle. \tag{7.59}$$

Equation (7.8) connects $| i, 1 \rangle$ to $| i, 0 \rangle$; so we substitute (7.59) into it to get

$$(H^{(0)} - E_i^{(0)}) \sum c_{ji}^{(1)} | j, 0 \rangle + (U^{(1)} - E_i^{(1)}) | i, 0 \rangle = 0 \tag{7.60}$$

whence

$$\sum c_{ji}^{(1)} (E_j^{(0)} - E_i^{(0)}) | j, 0 \rangle + (U^{(1)} - E_i^{(1)}) | i, 0 \rangle = 0. \tag{7.61}$$

Multiply (7.61) by $\langle k, 0 |$,

$$\sum c_{ji}^{(1)} (E_j^{(0)} - E_i^{(0)}) \langle k, 0 | j, 0 \rangle + \langle k, 0 | U^{(1)} - E_i^{(1)} | i, 0 \rangle = 0. \tag{7.62}$$

Since the zeroth-order eigenkets are orthonormalized, this reduces to

$$c_{ki}^{(1)} (E_k^{(0)} - E_i^{(0)}) + \langle k, 0 | U^{(1)} | i, 0 \rangle = 0, \quad k \neq i. \tag{7.63}$$

Let us solve for the coefficient

$$c_{ki}^{(1)} = \frac{\langle k, 0 | U^{(1)} | i, 0 \rangle}{E_i^{(0)} - E_k^{(0)}} \tag{7.64}$$

and substitute it into (7.59),

$$| i, 1 \rangle = \sum_{j \neq i} \frac{\langle j, 0 | U^{(1)} | i, 0 \rangle}{E_i^{(0)} - E_j^{(0)}} | j, 0 \rangle. \tag{7.65}$$

Then combine with (7.55) to obtain

$$E_i^{(2)} = \sum_{j \neq i} \frac{\langle i, 0 | U^{(1)} | j, 0 \rangle \langle j, 0 | U^{(1)} | i, 0 \rangle}{E_i^{(0)} - E_j^{(0)}}. \tag{7.66}$$

When $U^{(1)}$ is Hermitian, this rearranges to

$$E_i^{(2)} = \sum_{j \neq i} \frac{\langle i, 0 | U^{(1)} | j, 0 \rangle \langle i, 0 | U^{(1)} | j, 0 \rangle^*}{E_i^{(0)} - E_j^{(0)}}. \tag{7.67}$$

For the second-order correction, we substitute (7.58), with $n = 2$ and with $n = 1$, into (7.9) to construct

$$\sum_j c_{ji}^{(2)}(E_j^{(0)} - E_i^{(0)}) \mid j, 0 \rangle + \sum_j c_{ji}^{(1)}(U^{(1)} - E_i^{(1)}) \mid j, 0 \rangle$$

$$- E_i^{(2)} \mid i, 0 \rangle = 0. \qquad (7.68)$$

Then we multiply by bra $\langle k, 0 \mid$, reduce as before, and introduce (7.64), (7.13):

$$c_{ki}^{(2)}(E_k^{(0)} - E_i^{(0)}) + \sum_{j \neq i} \frac{\langle j, 0 \mid U^{(1)} \mid i, 0 \rangle}{E_i^{(0)} - E_j^{(0)}} \langle k, 0 \mid U^{(1)} \mid j, 0 \rangle$$

$$- \frac{\langle k, 0 \mid U^{(1)} \mid i, 0 \rangle}{E_i^{(0)} - E_k^{(0)}} \langle i, 0 \mid U^{(1)} \mid i, 0 \rangle = 0, \quad k \neq i. \qquad (7.69)$$

The formula that emerges for the kth second-order coefficient is

$$c_{ki}^{(2)} = \sum_{j \neq i} \frac{\langle k, 0 \mid U^{(1)} \mid j, 0 \rangle \langle j, 0 \mid U^{(1)} \mid i, 0 \rangle}{(E_i^{(0)} - E_j^{(0)})(E_i^{(0)} - E_k^{(0)})}$$

$$- \frac{\langle i, 0 \mid U^{(1)} \mid i, 0 \rangle \langle k, 0 \mid U^{(1)} \mid i, 0 \rangle}{(E_i^{(0)} - E_k^{(0)})^2}. \qquad (7.70)$$

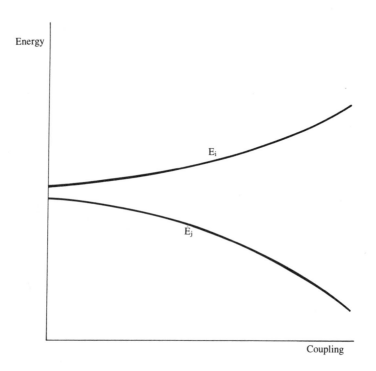

Figure 7.3. Effect of strength of coupling on neighboring levels.

Note that in each term of the series in (7.67), the numerator is real and positive. So when $E_j^{(0)}$ is below $E_i^{(0)}$, the jth term tends to move E_i away from E_j upward; when $E_j^{(0)}$ is above $E_i^{(0)}$, the jth term tends to move E_i away from E_j downward. For a given $E_i^{(0)} - E_j^{(0)}$, the net effect appears as Figure 7.3 indicates. The closer the jth state is to the ith state, and the stronger the coupling as measured by $| \langle i, 0 | U^{(1)} | j, 0 \rangle |$, the more the two levels tend to repel each other.

As $E_j^{(0)}$ approaches $E_i^{(0)}$, with nonzero coupling, the magnitude of $\langle j, 0 | U^{(1)} | i, 0 \rangle / (E_i^{(0)} - E_j^{(0)})$ increases, and by (7.64) the magnitude of $c_{ji}^{(1)}$ does also. At a certain point in the approach one or more correction ket $| i, n \rangle$ becomes unnormalizable and the perturbation process breaks down.

However, the process can be modified by considering the perturbation to evolve from determinable combinations of the states that are degenerate or nearly degenerate. See Section 7.8.

The basic relations of nondegenerate perturbation theory and the most important consequences have been collected in Table 7.2.

Table 7.2
Key Formulas for Perturbation from a Single Energy Eigenket

$$H | i \rangle = E_i | i \rangle \qquad H^{(0)} | i, 0 \rangle = E_i^{(0)} | i, 0 \rangle$$

$$H = H^{(0)} + \lambda U^{(1)}$$

$$| i \rangle = | i, 0 \rangle + \lambda | i, 1 \rangle + \lambda^2 | i, 2 \rangle + \dots$$

$$E_i = E_i^{(0)} + \lambda E_i^{(1)} + \lambda^2 E_i^{(2)} + \dots$$

$$\langle j, 0 | k, 0 \rangle = \delta_{jk}$$

$$\langle i, \alpha | i, \beta \rangle = 0 \quad \text{when} \quad \alpha \neq \beta$$

$$E_i^{(n)} = \langle i, 0 | U^{(1)} | i, n - 1 \rangle$$

$$| i, 1 \rangle = \sum_{j \neq i} \frac{\langle j, 0 | U^{(1)} | i, 0 \rangle}{E_i^{(0)} - E_j^{(0)}} | j, 0 \rangle$$

7.7
Polarizability of the Hydrogen Atom

An electric field uniform in a limited region alters the energy of a spherically symmetric atom located there by distorting its structure. Here we will consider such a field acting on a ground-state hydrogen atom.

In the natural units of Table 6.4, the Hamiltonian for an isolated hydrogen atom has the form

$$H^{(0)} = -\frac{1}{2}\nabla^2 - \frac{1}{r}. \tag{7.71}$$

A uniform electric field exerting force F on the orbital electron in the direction of the z axis introduces the perturbation

$$U^{(1)} = -Fz. \tag{7.72}$$

The ground state of the unperturbed atom is described by the eigenfunction

$$\psi^{(0)} = \left(\frac{1}{\pi}\right)^{1/2} e^{-r} \tag{7.73}$$

according to Tables 4.3 and 4.5. In the natural units, this state has the energy

$$E^{(0)} = -\frac{1}{2}. \tag{7.74}$$

Now from (7.13), the first-order correction to the energy is

$$E^{(1)} = \frac{1}{\pi} \int e^{-r}(-Fz)e^{-r}\, d^3\mathbf{r} = 0. \tag{7.75}$$

Equation (7.8) governs the first-order perturbed state function. With (7.71), (7.74), (7.72), and (7.75), it becomes

$$\left(-\frac{1}{2}\nabla^2 - \frac{1}{r} + \frac{1}{2}\right)\psi^{(1)} - Fz\psi^{(0)} = 0. \tag{7.76}$$

From the form of this equation, we are led to consider $\psi^{(1)}$ as a power series in r times $Fz\psi^{(0)}$. Introducing the product into (7.76) and determining the coefficients then leads to

$$\psi^{(1)} = Fz\left(\frac{r}{2} + 1\right)\psi^{(0)}. \tag{7.77}$$

Substituting (7.73), (7.72), and (7.77) into (7.55) gives us

$$E^{(2)} = \int \left(\frac{1}{\pi}\right)^{1/2} e^{-r}(-Fz)Fz \left(\frac{r}{2} + 1\right) \left(\frac{1}{\pi}\right)^{1/2} e^{-r}\, d^3\mathbf{r}$$

$$= \frac{F^2}{\pi} \int_0^\infty \left(\frac{r^5}{2} + r^4\right)e^{-2r}\, dr \int_0^\pi \cos^2\vartheta(-\sin\vartheta)d\vartheta \int_0^{2\pi} d\varphi. \tag{7.78}$$

Since

$$\int_0^{2\pi} d\varphi = 2\pi, \tag{7.79}$$

$$\int_0^\pi \cos^2 \vartheta (-\sin \vartheta) \, d\vartheta = \left. \frac{\cos^3 \vartheta}{3} \right|_0^\pi = -\frac{2}{3}, \qquad (7.80)$$

$$\int_0^\infty \frac{r^5}{2} e^{-2r} \, dr = 0 - \left. \frac{5 \cdot 4 \cdot 3 \cdot 2}{2 \cdot 2 \cdot 2 \cdot 2 \cdot 2 \cdot 2 \cdot 2} e^{-2r} \right|_0^\infty = \frac{15}{16}, \qquad (7.81)$$

$$\int_0^\infty r^4 e^{-2r} \, dr = 0 - \left. \frac{4 \cdot 3 \cdot 2}{2 \cdot 2 \cdot 2 \cdot 2 \cdot 2} e^{-2r} \right|_0^\infty = \frac{3}{4}, \qquad (7.82)$$

Equation (7.78) reduces to

$$E^{(2)} = \frac{F^2}{\pi} \frac{27}{16} \left(-\frac{2}{3} \right) (2\pi) = -\frac{9}{4} F^2. \qquad (7.83)$$

Introducing the full perturbation, by letting λ be 1, and neglecting higher terms in (7.5) yields

$$E = E^{(0)} - \frac{9}{4} F^2, \qquad (7.84)$$

where $E^{(0)}$ is the energy of the unperturbed hydrogen atom and F equals the force the applied electric field exerts on the orbital electron.

When the electric field to which an atom is exposed extends far enough so that the potential energy of an orbital electron moving with this field can drop below its state energy, the system becomes unstable. The system acquires a finite half life, being subject to spontaneous decay by tunneling. This phenomenon accompanies conventional observations of the Stark effect and leads to a broadening of the affected energy levels.

Example 7.3

Express the second-order perturbation energy for the electrically stressed hydrogen atom in general units.

From Table 6.4, the natural units for distance and energy are

$$a_0 \qquad \text{and} \qquad \frac{e^2}{4\pi\epsilon_0 a_0}.$$

Since potential energy decrease per unit distance is a force, the natural unit of force is

$$\frac{e^2}{4\pi\epsilon_0 a_0^2}.$$

So when the energy is E' and the force on the electron charge is F' in the chosen units, we have

$$E = \frac{E'}{e^2/4\pi\epsilon_0 a_0}, \qquad F = \frac{F'}{e^2/4\pi\epsilon_0 a_0^2};$$

Substituting into (7.83) and reducing now gives us

$$E'^{(2)} = -9\pi\epsilon_0 a_0^3 \left(\frac{F'}{e}\right)^2 = -9\pi\epsilon_0 a_0^3 \, \mathcal{E}'^2.$$

Here $\mathcal{E}'$ is the applied electric field intensity in the chosen units.

7.8
Development from a Cluster of States

In absence of particular interactions, or when the interactions are present at certain strengths, some orthogonal (independent) states of a system may appear at the same, or at nearly the same, energy. Introducing or altering the interactions may then lead to perturbations of coherent superpositions of these states. Indeed whenever any appreciable change is enough to impose a certain composition, the superposition must be considered, even in the first-order approximation.

Consider a system governed by the eigenvalue equation

$$H \,|\, j \rangle = E_j \,|\, j \rangle \tag{7.85}$$

which is difficult or impossible to solve. Also consider that removing the complicating potential $\lambda U^{(1)}$ from the Hamiltonian

$$H = H^{(0)} + \lambda U^{(1)} \tag{7.86}$$

leads to an equation

$$H^{(0)} \,|\, i, 0 \rangle = E_i^{(0)} \,|\, i, 0 \rangle \tag{7.87}$$

that is readily solvable.

Also, suppose that the state in which we are interested derives from a cluster of kets

$$|\, 1, 0 \rangle, |\, 2, 0 \rangle, \ldots, |\, i, 0 \rangle, \ldots, |\, n, 0 \rangle \tag{7.88}$$

at or near the same level in the unperturbed system. Recall formula (7.64) and its implications. Without loss of generality, these kets can be taken to be normalized and orthogonal:

$$\langle \alpha, 0 \mid \beta, 0 \rangle = \delta_{\alpha\beta}. \tag{7.89}$$

Instead of being related to a single $\mid i, 0 \rangle$, eigenket $\mid j \rangle$ may stem from a linear combination of kets (7.88). For, a small change in λ from 0 can impose such a superposition. Then each unperturbed ket must be described by a sum

$$\mid j, 0 \rangle = \sum_{\beta=1}^{n} c_\beta \mid \beta, 0 \rangle. \tag{7.90}$$

Equation (7.4) becomes

$$\mid j \rangle = \mid j, 0 \rangle + \lambda \mid j, 1 \rangle + \lambda^2 \mid j, 2 \rangle + \ldots = \sum_{k=0}^{\infty} \lambda^k \mid j, k \rangle, \tag{7.91}$$

while (7.5) is replaced by

$$E = E^{(0)} + \lambda E^{(1)} + \lambda^2 E^{(2)} + \ldots = \sum_{j=0}^{\infty} \lambda^j E^{(j)}. \tag{7.92}$$

Substituting (7.86), (7.91), (7.90), (7.92) into (7.85) yields

$$(H^{(0)} + \lambda U^{(1)})(\sum c_\beta \mid \beta, 0 \rangle + \lambda \mid j, 1 \rangle + \ldots)$$
$$= (E^{(0)} + \lambda E^{(1)} + \ldots)(\sum c_\beta \mid \beta, 0 \rangle + \lambda \mid j, 1 \rangle + \ldots)$$

$$= \sum_{\beta=1}^{n} [E_\beta{}^{(0)} + \lambda E^{(1)} + \ldots + (E^{(0)} - E_\beta{}^{(0)})]$$
$$\times (c_\beta \mid \beta, 0 \rangle + \lambda \mid j, 1 \rangle + \ldots). \tag{7.93}$$

Since this equation is true for arbitrary λ, it is an identity in λ. Equations (7.87) ensure that the λ^0 terms (the constant over λ terms) on both sides are equal if we choose

$$\sum_{\beta=1}^{n} (E^{(0)} - E_\beta{}^{(0)})c_\beta \mid \beta, 0 \rangle = 0. \tag{7.94}$$

Multiply both sides of (7.94) by $c_\alpha^* \langle \alpha, 0 \mid$ and sum over α to get

$$\sum_{\alpha=1}^{n} \sum_{\beta=1}^{n} (E^{(0)} - E_\beta{}^{(0)})c_\alpha^* c_\beta \langle \alpha, 0 \mid \beta, 0 \rangle = 0. \tag{7.95}$$

Orthonormality relationship (7.89) reduces this to

$$\sum_{\alpha=1}^{n} (E^{(0)} - E_\alpha{}^{(0)})c_\alpha^* c_\alpha = 0. \tag{7.96}$$

Since $c_\alpha^* c_\alpha$ is the statistical weight of $\mid \alpha, 0 \rangle$ in $\mid j, 0 \rangle$, we have

$$\sum_{\alpha=1}^{n} c_{\alpha}^{*} c_{\alpha} = 1 \tag{7.97}$$

and (7.96) yields

$$E^{(0)} = \sum_{\alpha=1}^{n} c_{\alpha}^{*} c_{\alpha} E_{\alpha}^{(0)} = \langle E_{j}^{(0)} \rangle. \tag{7.98}$$

Parameter $E^{(0)}$ equals the expectation value for $E_{j}^{(0)}$ in unperturbed state (7.90). When differences among the $E_{\alpha}^{(0)}$ that contribute significantly to $E^{(0)}$ are insignificant, formula (7.98) need not be used.

Setting the coefficient of λ on the left of (7.93) equal to the coefficient of λ on the right,

$$H^{(0)} | j, 1 \rangle + U^{(1)} \sum c_{\beta} | \beta, 0 \rangle = E^{(0)} | j, 1 \rangle + E^{(1)} \sum c_{\beta} | \beta, 0 \rangle, \tag{7.99}$$

and multiplying the result by $\langle \alpha, 0 |$ gives us

$$\langle \alpha, 0 | H^{(0)} | j, 1 \rangle + \sum \langle \alpha, 0 | U^{(1)} | \beta, 0 \rangle c_{\beta}$$
$$= E^{(0)} \langle \alpha, 0 | j, 1 \rangle + \sum E^{(1)} \langle \alpha, 0 | \beta, 0 \rangle c_{\beta} \tag{7.100}$$

or

$$\sum (U^{(1)}_{\alpha\beta} - E^{(1)}\delta_{\alpha\beta}) c_{\beta} = E^{(0)} \langle \alpha, 0 | j, 1 \rangle - \langle \alpha, 0 | H^{(0)} | j, 1 \rangle \tag{7.101}$$

where

$$U^{(1)}_{\alpha\beta} \equiv \langle \alpha, 0 | U^{(1)} | \beta, 0 \rangle. \tag{7.102}$$

Since $H^{(0)}$ is Hermitian, the last term in (7.101) is

$$\langle \alpha, 0 | H^{(0)} | j, 1 \rangle = \langle H^{(0)} | \alpha, 0 | | j, 1 \rangle = E_{\alpha}^{(0)} \langle \alpha, 0 | j, 1 \rangle. \tag{7.103}$$

As in (7.53), we take $| j, 1 \rangle$ to be orthogonal to $| j, 0 \rangle$. But for $| j, 1 \rangle$ to be orthogonal to $| j, 0 \rangle$, it must be orthogonal to each base ket contributing to $| j, 0 \rangle$. Therefore, we take

$$\langle \alpha, 0 | j, 1 \rangle = 0 \tag{7.104}$$

and reduce (7.101) to

$$\sum_{\beta=1}^{n} (U^{(1)}_{\alpha\beta} - E^{(1)}\delta_{\alpha\beta}) c_{\beta} = 0. \tag{7.105}$$

Formula (7.105) yields n simultaneous linear equations in the unkonwns c_{1}, $c_{2}, \ldots , c_{n}$. Since these equations are homogeneous, all the c_{α}'s are zero unless the determinant of the coefficients vanishes (recall Cramer's rule):

$$
\begin{vmatrix}
U^{(1)}{}_{11} - E^{(1)} & U^{(1)}{}_{12} & \cdots & U^{(1)}{}_{1n} \\
U^{(1)}{}_{21} & U^{(1)}{}_{22} - E^{(1)} & \cdots & U^{(1)}{}_{2n} \\
\vdots & \vdots & \ddots & \vdots \\
U^{(1)}{}_{n1} & U^{(1)}{}_{n2} & \cdots & U^{(1)}{}_{nn} - E^{(1)}
\end{vmatrix} = 0. \quad (7.106)
$$

Equation (7.106) fits into form (6.65); it is called the secular equation of degenerate perturbation theory.

Each root $E^{(1)}$ can be substituted back into (7.105) and the simultaneous equations solved for c_1/c_i, c_2/c_i, . . . , c_n/c_i. The normalization condition and an arbitrary phase can then be introduced and suitable individual coefficients constructed.

7.9
2sp Hybridization Caused by an Electric Field

Imposing an external electric field of limited extent on an excited hydrogen atom polarizes the atom by hybridizing the orbitals. When the principal quantum number n is 2, the most stable state involves one half the $2s$ orbital and one half the $2p$ orbital that lines up with the field. The least stable state involves the same orbitals in the orthogonal arrangement.

The Hamiltonian operator for an isolated hydrogen atom is

$$
H^{(0)} = -\frac{1}{2} \nabla^2 - \frac{1}{r} \quad (7.107)
$$

in the natural units of Table 6.4. Consider the hydrogen atom in a uniform electric field exerting force F on the orbital electron in the direction of the z axis. The perturbation of (7.107) is

$$
U^{(1)} = -Fz = -Fr \cos \vartheta. \quad (7.108)
$$

The kets that may contribute include the $2s$, $2p_z$, $2p_x$, and $2p_y$ ones. According to Tables 4.2 and 4.5, these are represented by the functions

$$
\psi_{2s} = \left(\frac{1}{32\pi}\right)^{1/2} (2 - r) \, e^{-r/2} = \psi_{1,0}, \quad (7.109)
$$

$$
\psi_{2p_z} = \left(\frac{1}{32\pi}\right)^{1/2} r \, e^{-r/2} \cos \vartheta = \psi_{2,0}, \quad (7.110)
$$

$$\psi_{2p_x} = \left(\frac{1}{32\pi}\right)^{1/2} r\, e^{-r/2} \sin\vartheta \cos\varphi = \psi_{3,0}, \qquad (7.111)$$

$$\psi_{2p_y} = \left(\frac{1}{32\pi}\right)^{1/2} r\, e^{-r/2} \sin\vartheta \sin\varphi = \psi_{4,0}, \qquad (7.112)$$

in the natural units. The energy of each of these states is

$$E^{(0)} = -\frac{1}{8}\ \text{hartree}. \qquad (7.113)$$

The matrix element between the 2s and $2p_z$ orbitals for the perturbation is

$$U^{(1)}{}_{12} = \int \left(\frac{1}{32\pi}\right)^{1/2}(2-r)e^{-r/2}(-Fr\cos\vartheta)\left(\frac{1}{32\pi}\right)^{1/2} r e^{-r/2}\cos\vartheta\, d^3\mathbf{r}$$

$$= \frac{F}{32\pi}\int_0^\infty (2r^4 - r^5)e^{-r}\, dr \int_0^\pi \cos^2\vartheta(-\sin\vartheta)d\vartheta \int_0^{2\pi} d\varphi. \qquad (7.114)$$

Equations (7.79) and (7.80) apply to the φ and ϑ integrals. We also have

$$\int_0^\infty r^5 e^{-r}\, dr = 0 - 5\cdot4\cdot3\cdot2\, e^{-r}\ \Big|_0^\infty = 5\cdot4\cdot3\cdot2, \qquad (7.115)$$

$$\int_0^\infty r^4 e^{-r}\, dr = 0 - 4\cdot3\cdot2\, e^{-r}\ \Big|_0^\infty = 4\cdot3\cdot2; \qquad (7.116)$$

consequently, (7.114) reduces to

$$U^{(1)}{}_{12} = \frac{F}{32\pi}(-3\cdot4\cdot3\cdot2)\left(-\frac{2}{3}\right)(2\pi) = 3F. \qquad (7.117)$$

The square of each orbital has even parity, while the perturbation $-Fz$ has odd parity. Consequently, the integrands of $U^{(1)}{}_{11}$, $U^{(1)}{}_{22}$, $U^{(1)}{}_{33}$, and $U^{(1)}{}_{44}$ all have odd parity; and the integral of each of these integrands over all space equals zero. The matrix elements among the 2s, $2p_x$, and $2p_y$ orbitals for the perturbation are all zero since the pertinent φ integrals equal zero:

$$\int_0^{2\pi} \cos\varphi\, d\varphi = \sin\varphi\ \Big|_0^{2\pi} = 0, \qquad (7.118)$$

$$\int_0^{2\pi} \sin\varphi\, d\varphi = -\cos\varphi\ \Big|_0^{2\pi} = 0, \qquad (7.119)$$

$$\int_0^{2\pi} \cos\varphi \sin\varphi\, d\varphi = -\frac{\cos^2\varphi}{2}\ \Big|_0^{2\pi} = 0, \qquad (7.120)$$

Substituting the perturbation matrix elements into (7.106) now yields

$$\begin{vmatrix} -E^{(1)} & 3F & 0 & 0 \\ 3F & -E^{(1)} & 0 & 0 \\ 0 & 0 & -E^{(1)} & 0 \\ 0 & 0 & 0 & -E^{(1)} \end{vmatrix} = 0 \qquad (7.121)$$

whence

$$(E^{(1)})^2 \, [(E^{(1)})^2 - 9F^2] = 0. \qquad (7.122)$$

The four roots of this secular equation are

$$E^{(1)} = 3F, \qquad (7.123)$$

$$E^{(1)} = 0, \qquad (7.124)$$

$$E^{(1)} = 0, \qquad (7.125)$$

$$E^{(1)} = -3F, \qquad (7.126)$$

Substituting (7.126) into (7.5), letting λ be 1, and neglecting higher terms leads to the result

$$E = -\frac{1}{8} - 3F \qquad (7.127)$$

for the most stable configuration when F is positive.

When the electric field acting on a bound electron extends far enough so that the electron's potential energy can drop below its state energy, the electron can tunnel out of the atom. Then there is a finite half life for the decay and the corresponding energy level is broadened.

Example 7.4

Express the first-order perturbation energy for the $n = 2$ electron in more general units.

Substitute the relationships constructed in Example 7.3 into (7.126) and reduce:

$$\frac{E'^{(1)}}{e^2/4\pi\epsilon_0 a_0} = - \frac{3F'}{e^2/4\pi\epsilon_0 a_0{}^2},$$

$$E'^{(1)} = -3ea_0\left(\frac{F'}{e}\right) = -3ea_0\,\mathcal{E}\,'.$$

As before, e is the magnitude of charge on the electron, a_0 is the Bohr radius of the hydrogen atom, and $\mathcal{E}\,'$ is the electric field intensity.

Example 7.5

Construct and interpret the zeroth-order state functions for a hydrogen atom perturbed from its first excited level by a uniform electric field pointing in the $+z$ direction.

For the first excited level, a complete set of basis functions consists of (7.109), (7.110), (7.111), and (7.112). These yield secular equation (7.121) with roots (7.123), (7.124), (7.125), (7.126).

Substituting roots (7.123) and (7.126) into (7.105) gives us

for $\alpha = 1$, $\mp 3Fc_1 + 3Fc_2 = 0.$

for $\alpha = 2$, $3Fc_1 \mp 3Fc_2 = 0.$

for $\alpha = 3$, $3Fc_3 = 0.$

for $\alpha = 4$, $3Fc_4 = 0.$

From the last two equations, we obtain

$$c_3 = c_4 = 0.$$

From the first two equations, we find that

$$c_2 = \pm c_1.$$

To normalize the zeroth-order state function ψ, we set

$$c_1^2 + c_2^2 = 1.$$

The phase for c_1 can be arbitrarily chosen. Let us take c_1 to be real and positive; then

$$c_1 = \frac{1}{\sqrt{2}}$$

and

$$\psi_{1,4} = \frac{1}{\sqrt{2}} \psi_{2s} \pm \frac{1}{\sqrt{2}} \psi_{2p_z}.$$

The positive sign corresponds to root $3F$. But for it, the orbital concentrates on the $-z$ side of the origin and the dipole moment points in the positive direction. This opposes the field when F is positive and the higher energy results. On the other hand, the negative sign corresponds to root $-3F$. For it, the orbital concentrates on the $+z$ side of the origin; the dipole moment points in the negative direction. When F is positive, this works with the field and the lower energy results.

Similarly, we substitute roots (7.124) and (7.125) into (7.105) to get

$$3Fc_2 = 0,$$
$$3Fc_1 = 0,$$
$$0c_3 = 0,$$
$$0c_4 = 0.$$

The secular equation now imposes no restrictions on c_3 or c_4. We choose $c_3 = 1$, $c_4 = 0$, whence

$$\psi_2 = \psi_{2p_x},$$

and $c_3 = 0$, $c_4 = 1$, whence

$$\psi_3 = \psi_{2p_y}.$$

With these two orbitals the center of electron charge is at the nucleus. Consequently, their energies are not affected by the field, as the roots indicate.

By the way, ψ_1 and ψ_4 are the functions one obtains if one constructs from the $n = 2$ functions two equivalent oppositely directed orbitals along the z axis.

7.10
The Non-Crossing Rule

The Hamiltonian operator for a system may depend on parameters that can be varied. In general, the energy levels will change with each of these. Since different levels will not vary at the same rate with respect to any one of the parameters, a person might expect that a large number of crossings would occur; but in many instances there will be coupling between the pertinent levels and the resulting repulsion will prevent crossing.

Indeed, in constructing the potential energy functions for the vibrational motions and the dissociations of a given molecule, one generally invokes the Born-Oppenheimer approximation. The electronic energies are then calculated for representative fixed relative positions of the nuclei. Coordinates specifying the positions would then be parameters. By interpolation and extrapolation, one obtains the complete functions. These will appear to cross at points, lines, surfaces, hypersurfaces, . . . , depending on the number of parameters.

Consider a system with parameters as described. Suppose that its energy has been determined at representative points in its parameter space. Now, various possible interpolations are being examined.

Consider that one of these leads to a crossing at a particular point. For this point the system exhibits its usual kinetic energy operator T and a particular

potential energy operator $U^{(0)}$. Construct the unperturbed Hamiltonian operator as

$$H^{(0)} = T + U^{(0)}.$$ (7.128)

Let the perturbation be the function one has to add to $U^{(0)}$ to get the actual
potential U; thus

$$U = U^{(0)} + \lambda U^{(1)}$$ (7.129)

and

$$H = H^{(0)} + \lambda U^{(1)}.$$ (7.130)

Now, under what conditions will the perturbation prevent the crossing from
occurring? To answer this question, we apply the theory from Section 7.8 with
$n = 2$. Secular equation (7.106) becomes

$$\begin{vmatrix} U^{(1)}{}_{11} - E^{(1)} & U^{(1)}{}_{12} \\ U^{(1)}{}_{21} & U^{(1)}{}_{22} - E^{(1)} \end{vmatrix} = 0.$$ (7.131)

For the two roots to be the same, and for crossing to occur, we must have
both

$$U^{(1)}{}_{11} = U^{(1)}{}_{22}$$ (7.132)

and

$$U^{(1)}{}_{12} = U^{(1)}{}_{21} = 0.$$ (7.133)

Presumedly, Equation (7.132) can be satisfied by choosing the right value for
the parameter that is being varied.

When the physical system generates no symmetry operation other than the
identity, the off-diagonal matrix elements will generally differ from zero:

$$U^{(1)}{}_{12} = U^{(1)}{}_{21} \neq 0.$$ (7.134)

We say that coupling between the states exists and this coupling acts to repel the
two states. Crossing does not then occur.

But when the system possesses elements of symmetry, symmetry operations
in addition to the identity exist and the state functions fall into two or more
symmetry species. In the language of group theory, they are bases for two or
more irreducible representations.

Since functions U and $U^{(0)}$ exhibit the symmetry of the system, perturbing
potential $U^{(1)}$ does also, in order that (7.129) be satisfied. It belongs to the
totally symmetric species.

So integrand

$$\psi_{1,0}{}^{*}U^{(1)}\psi_{2,0} \qquad (7.135)$$

exhibits the same symmetry as

$$\psi_{1,0}{}^{*}\psi_{2,0}. \qquad (7.136)$$

As a result, when $\psi_{1,0}$ and $\psi_{2,0}$ belong to the same symmetry species $U^{(1)}{}_{12} = U^{(1)}{}_{21}$ generally differs from zero. Then the two states repel each other and crossing cannot occur.

But when $\psi_{1,0}$ and $\psi_{2,0}$ belong to different symmetry species, the positive regions of the integral are canceled by equivalent negative regions and Equation (7.133) is satisfied. Then crossing can occur.

The results are summarized in the *non-crossing rule*: The energy functions for states belonging to the same symmetry species cannot cross in parameter space.

Example 7.6

If calculated energies fall as the small circles in Figure 7.4 indicate, can a person tell whether the levels cross?

If one believes that the levels do not cross, one would extend the solid lines as the dashes (– – – –) indicate. But if one chooses to make the curves as smooth as possible, one would extend the solid lines as the dots (· · · · ·) indicate.

From section 7.10, crossing cannot occur when the off-diagonal matrix element $U^{(1)}{}_{12}$, calculated for the point where $U^{(1)}{}_{11} = U^{(1)}{}_{22}$, differs from zero. This element measures the coupling between the two levels.

On the other hand, crossing occurs when this off-diagonal matrix element vanishes. The vanishing occurs when the levels belong to different symmetry species. However, this classification can be broken down by the presence of a neighboring or colliding system.

7.11
Perturbation at an Atom in a Hückel-Model Structure

The Hückel procedure considers the delocalization of certain valence electrons over chains and/or rings, cages, in a given molecule. The atoms over which the corresponding movement occurs form the skeleton for the system.

Since the groups attached to a given skeletal atom influence the electric potential about the atom, they affect its Coulomb integral or atomic parameter. So a change in a ligand alters this parameter. With perturbation methods, the effect on total energy and on electron distribution can be estimated.

Consider a molecule or radical for which a Hückel molecular orbital solution is known. Suppose that a substitution is made on the *k*th atom in the skeleton of the system.

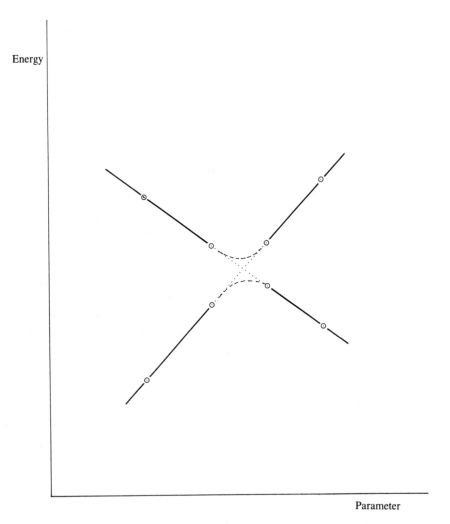

Figure 7.4. Alternative interpolations of discrete energy-level data.

Consider that before the substitution, the Coulomb integral for this atom is

$$H^{(0)}{}_{kk} = \alpha_k, \tag{7.137}$$

while the substitution introduces the change

$$H'_{kk} = \Delta\alpha_k. \tag{7.138}$$

Suppose that no other atomic parameters in the Hückel skeleton are altered. Also, suppose that the zeroth-order ith molecular ket is

$$| g_i \rangle = \sum_j c_{ji} | j \rangle. \tag{7.139}$$

From (7.13), the change in energy for an electron occupying this ket is

$$E_i^{(1)} = \sum_j \sum_l \langle c_{ji} | j || H' | c_{li} | l \rangle = \sum_j \sum_l c_{ji}{}^* c_{li} H'_{jl}. \tag{7.140}$$

Since H'_{jl} is zero except for $j = l = k$, where it equals $\Delta\alpha_k$, (7.140) reduces to

$$E_i^{(1)} = c_{ki}{}^* c_{ki} \Delta\alpha_k. \tag{7.141}$$

Summing over the occupied molecular orbitals gives the first-order correction to the energy

$$E^{(1)} = \left(\sum_i n_i c_{ki}{}^* c_{ki} \right) \Delta\alpha_k = q_k \Delta\alpha_k. \tag{7.142}$$

Here q_k is the unperturbed Hückel electron density at the kth atom while $\Delta\alpha$ is the change in H_{kk} at the kth atom.

From (7.65), the first-order correction to the ket is

$$| g'_i \rangle = \sum_{j \neq i} \frac{\sum_n \sum_m \langle c_{nj} | n || H' | c_{mi} | m \rangle}{E_i^{(0)} - E_j^{(0)}} \sum_l c_{lj} | l \rangle. \tag{7.143}$$

But H'_{nm} equals zero except for $n = m = k$, when it equals $\Delta\alpha_k$. So (7.143) reduces to

$$| g'_i \rangle = \sum_{j \neq i} \sum_l \frac{c_{kj}{}^* c_{ki} c_{lj} \Delta\alpha_k}{E_i^{(0)} - E_j^{(0)}} | l \rangle. \tag{7.144}$$

When the coefficients from (7.139) are real and the low lying, filled, molecular kets are doubly occupied, the change in the Hückel electron density at the lth atom, for *small* perturbations, is

$$\Delta q_l = 2 \sum_i [(c_{li} + c_{li}{}^{(1)})^2 - c_{li}^2] \approx 4 \sum_i c_{li} c_{li}{}^{(1)}. \tag{7.145}$$

Substituting the coefficient of $|\, l\, \rangle$ from (7.144) for $c_{li}^{(1)}$ into (7.145) now yields

$$\Delta q_l = 4\Delta\alpha_k \sum_i \sum_{j\neq i} \frac{c_{li}c_{kj}c_{ki}c_{lj}}{E_i^{(0)} - E_j^{(0)}}$$

$$= 4\Delta\alpha_k \sum_{i=1}^{m} \sum_{j=m+1}^{n} \frac{c_{li}c_{kj}c_{ki}c_{lj}}{E_i^{(0)} - E_j^{(0)}}. \qquad (7.146)$$

In the last step, terms with canceling mates have been dropped from the summations. The occupied kets have the indices $1, 2, \ldots, m$.

Our perturbation theory was based on a resolution of the Hamiltonian operator for the given system into a simplified, unperturbed, operator and the complicating part, the perturber. The basis kets were eigenkets of the simplified operator. The first-order correction to the energy was the expectation value for the perturbation in the appropriate unperturbed state. This is a diagonal matrix element. The contribution of the jth basis ket to the ith correction ket equals the corresponding ji-th matrix element divided by the energy offset. Thus, the key results involve only matrix elements and unperturbed energies. As long as these elements and energies are known, the theory can be applied. One need not know or define the unperturbed Hamiltonian operator.

Indeed, the Hückel molecular orbitals and orbital energies are not eigenfunctions and eigenvalues since they result from variational calculations with a very limited basis set. We could proceed, however, since we had the atomic parameters (Coulomb integrals) for the basis set.

7.12
Abstract

Through a parameter, a person can introduce complications into an eigenvalue equation continuously. For convenience the parameter may be introduced as a factor multiplying the complicating terms. Then each eigenket and eigenvalue would be developed as a power series in the parameter.

Suppose that solutions to the equation

$$H^{(0)} |\, i, 0\, \rangle = E_i^{(0)} |\, i, 0\, \rangle \qquad (7.147)$$

are known. Also, suppose that the Hamiltonian for the system under study can be written in the form

$$H = H^{(0)} + \lambda U^{(1)}, \qquad (7.148)$$

while the equation to be solved is

$$H \mid i \rangle = E_i \mid i \rangle. \tag{7.149}$$

Letting parameter λ vary from zero to its value for the given system (which may be 1) then transforms the original kets and eigenvalues. Wherever a solution develops from a single ket $\mid i, 0 \rangle$ in a smooth and continuous fashion, we may write

$$\mid i \rangle = \sum_{k=0}^{\infty} \lambda^k \mid i, k \rangle \tag{7.150}$$

and

$$E_i = \sum_{j=0}^{\infty} \lambda^j E_i^{(j)}. \tag{7.151}$$

Substituting (7.148), (7.150), and (7.151) into (7.149) gives us an identity in λ. Equating coefficients of each power of λ separately yields a sequence of equations. The first is (7.147). From the second, we obtain

$$E_i^{(1)} = \langle i, 0 \mid U^{(1)} \mid i, 0 \rangle. \tag{7.152}$$

From the third equation in the sequence, together with the orthogonality condition

$$\langle i, 0 \mid i, 1 \rangle = 0, \tag{7.153}$$

we find that

$$E_i^{(2)} = \langle i, 0 \mid U^{(1)} \mid i, 1 \rangle, \tag{7.154}$$

while from the second equation we get

$$\mid i, 1 \rangle = \sum_{j \neq i} \frac{\langle j, 0 \mid U^{(1)} \mid i, 0 \rangle}{E_i^{(0)} - E_j^{(0)}} \mid j, 0 \rangle. \tag{7.155}$$

Note that this $\mid i, 1 \rangle$ is not normalized to 1. But it is normalizable as long as no energy level $E_j^{(0)}$ is close to $E_i^{(0)}$. When one or more $E_j^{(0)}$ is near enough to $E_i^{(0)}$ to interfere with the normalizability, a person needs to consider the perturbed state as developing from a coherent superposition of these nearly degenerate and degenerate kets.

Indeed, consider a source cluster made up of the orthonormal kets

$$\mid 1, 0 \rangle, \mid 2, 0 \rangle, \ldots, \mid \beta, 0 \rangle, \ldots, \mid n, 0 \rangle. \tag{7.156}$$

Properly phased fractions of these add to give n orthonormal zeroth-order solutions of the form

$$| j, 0 \rangle = \sum_{\beta=1}^{n} c_\beta | \beta, 0 \rangle. \tag{7.157}$$

This combination would make up the λ^0 term in (7.150). The higher $| i, k \rangle$'s would involve the $| i, 0 \rangle$'s outside the cluster.

Substituting the resulting (7.150), together with the other expansions, into (7.149) gives us an identity in λ. The λ^0 terms yield

$$\sum_{\beta=1}^{n} (E^{(0)} - E_\beta{}^{(0)}) c_\beta | \beta, 0 \rangle = 0. \tag{7.158}$$

Multiplying both sides by $c_\alpha{}^* \langle \alpha, 0 |$, summing over α, and introducing the orthonormality condition leads to

$$\sum_{\alpha=1}^{n} (E^{(0)} - E_\alpha{}^{(0)}) c_\alpha{}^* c_\alpha = 0, \tag{7.159}$$

whence

$$E^{(0)} = \sum c_\alpha{}^* c_\alpha E_\alpha{}^{(0)} = \langle E_j{}^{(0)} \rangle. \tag{7.160}$$

The λ^1 terms in the identity, together with the orthonormality condition, yield

$$\sum_{\beta=1}^{n} (U^{(1)}{}_{\alpha\beta} - E^{(1)} \delta_{\alpha\beta}) c_\beta = 0 \tag{7.161}$$

with

$$U^{(1)}{}_{\alpha\beta} = \langle \alpha, 0 | U^{(1)} | \beta, 0 \rangle. \tag{7.162}$$

The n simultaneous equations have nontrivial solutions only when

$$\begin{vmatrix} U^{(1)}{}_{11} - E^{(1)} & U^{(1)}{}_{12} & \cdots & U^{(1)}{}_{1n} \\ U^{(1)}{}_{21} & U^{(1)}{}_{22} - E^{(1)} & \cdots & U^{(1)}{}_{2n} \\ \cdot & \cdot & \cdot & \cdot \\ \cdot & \cdot & \cdot & \cdot \\ \cdot & \cdot & \cdot & \cdot \\ U^{(1)}{}_{n1} & U^{(1)}{}_{n2} & \cdots & U^{(1)}{}_{nn} - E^{(1)} \end{vmatrix} = 0. \tag{7.163}$$

Each root of this secular equation offers a different first-order correction to energy $E^{(0)}$. It can be substituted into the determinant. Then the αth row would be multiplied by the corresponding c_β's and summed to get (7.161). From the set of equations and the normalization condition, values for the coefficients are obtained.

Note that the secular equation involves the kets in the cluster through the matrix elements $U^{(1)}{}_{\alpha\beta}$. Similar matrix elements are employed in calculating $E_i^{(1)}$ and $E_i^{(2)}$ in the development from $|\, i,\, 0\,\rangle$. In some problems only the matrix elements and the unperturbed energies are known. The perturbed energies can be calculated from these parameters.

Example 7.7

Compare the results of the two approaches for a system where only two levels are interacting appreciably.

Let the unperturbed kets be $|\, 1,\, 0\,\rangle$ and $|\, 2,\, 0\,\rangle$. For the development from $|\, 1,\, 0\,\rangle$, Equation (7.152) yields

$$E_1^{(1)} = U^{(1)}{}_{11},$$

while (7.154) and (7.155), or (7.67), have

$$E_1^{(1)} = \frac{U^{(1)}{}_{12} U^{(1)}{}_{21}}{E_1^{(0)} - E_2^{(0)}}.$$

The perturbation of the energy through second order is

$$\Delta E = U^{(1)}{}_{11} + \frac{U^{(1)}{}_{12} U^{(1)}{}_{21}}{E_1^{(0)} - E_2^{(0)}}.$$

For the development from both $|\, 1,\, 0\,\rangle$ and $|\, 2,\, 0\,\rangle$, Equation (7.163) yields

$$(U^{(1)}{}_{11} - E^{(1)})(U^{(1)}{}_{22} - E^{(1)}) = U^{(1)}{}_{12} U^{(1)}{}_{21},$$

whence

$$
\begin{aligned}
E^{(1)} &= U^{(1)}{}_{11} + \frac{U^{(1)}{}_{12} U^{(1)}{}_{21}}{E^{(1)} - U^{(1)}{}_{22}} \\
&= U^{(1)}{}_{11} + \frac{U^{(1)}{}_{12} U^{(1)}{}_{21}}{E^{(1)} - E_2^{(1)}}.
\end{aligned}
$$

Here $E^{(1)}$ is the perturbed energy of the first level.

In the second result, the difference between the unperturbed energies, $E_1^{(0)} - E_2^{(0)}$, has been replaced with $E^{(1)} - E_2^{(1)}$.

Discussion Questions

7.1 Why does one seek approximate solutions to physical problems? How can such solutions aid one in understanding physical systems?

7.2 Identify the kinetic energy and potential energy parts of the Hamiltonian operator. Why do complications appear only in the potential energy part when magnetic effects are neglected?

7.3 How may a perturbation parameter control the introduction of such a complication?

7.4 How are the perturbations in the eigenvalue and the eigenket introduced?

7.5 Why might a perturbation calculation fail?

7.6 In what approximation does the energy equal the expectation value for H in the unperturbed state?

7.7 Why does the position of a narrow rectangular potential well in a box affect the energy of a particle confined in the box?

7.8 Why is the interelectronic potential energy considered to be the perturbation in a heliumlike atom?

7.9 How does Gauss's law aid one in determining the first-order perturbation of the ground state energy in the heliumlike atom?

7.10 How does one electron shield the other from the nucleus in the heliumlike atom? To what extent can this shielding be represented by a fixed change in the nuclear charge?

7.11 Why should the state function depend on the interelectron distance in the heliumlike atom?

7.12 How are expectation values affected by a perturbation of a ket?

7.13 Why can we require the contributing kets $|\ i,\ k\ \rangle$ to the perturbed form to be mutually orthogonal?

7.14 Why can one express $|\ i,\ ,k\ \rangle$ as a linear combination of the unperturbed kets $|\ 1,\ 0\ \rangle,\ |\ 2,\ 0\ \rangle,\ |\ 3,\ 0\ \rangle, \ .\ .\ .\ ?$

7.15 When does a perturbing potential cause neighboring levels to repel each other?

7.16 What function measures the extent of this repulsion?

7.17 Why is the first-order perturbation of the energy of a $1s$ electron by a uniform electric field zero?

7.18 Why wasn't the first-order correction to the state function for the $1s$ electron in the uniform electric field expressed as a superposition of the zeroth-order state functions?

7.19 Under what conditions does an atom become unstable in an electric field?

7.20 Show that $\epsilon_0 a_0\ \mathcal{E}^2$ has the dimensions (units) of energy.

7.21 Why should a perturbed ket stem from a linear combination of unperturbed kets? Need the original kets be degenerate for such combining to occur?

7.22 When a perturber hybridizes states differing in a property, how is the zeroth-order value of the property obtained? How is the first-order correction to the property determined?

7.23 When one ket does not combine appreciably with the other kets in the zeroth-order approximation, how does the secular equation appear?

7.24 Why is the first-order perturbation of the energy of an excited hydrogen atom by an electric field different from zero?

7.25 How can superposition of part of a 2*p* orbital with part of a 2*s* one produce a dipole moment? How does this moment interact with an electric field?

7.26 What are the dimensions (units) of $-a_0 e \mathcal{E}'$?

7.27 What determines whether levels that approach each other, with a certain variation in conditions, and then recede, actually cross or not?

7.28 In an atom, the least stable electron is shielded from the nucleus by the other electrons by an amount that varies with its distance from the nucleus. Explain qualitatively how the states in which this electron appears as one goes through the periodic table differ in order from the states of the hydrogen atom.

7.29 If the potential field in which a particle moves is changed from the single-center Coulomb form to the form illustrated by Figure 7.5, how is the order of energy levels altered?

7.30 To apply perturbation theory, need one construct the unperturbed Hamiltonian $H^{(0)}$? Explain.

7.31 Why does substitution on an atom alter the atomic parameter for the atom?

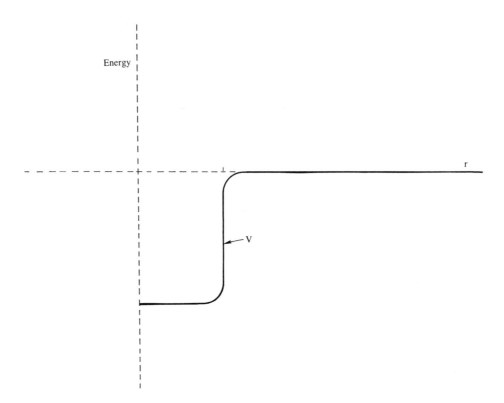

Figure 7.5. Approximate radial dependence of the potential energy for a nucleon in a nucleus.

Problems

7.1 Show how the energy obtained from first-order perturbation theory for the ground state of a system is related to the exact energy for this state.

7.2 A small uniform electric field is applied to an electron constrained to move along a straight-line segment of length a. As a result, the potential of the electron varies from 0 at one end to u at the other. Calculate the perturbation of the electron's energy by the field.

7.3 An electron is confined in its first excited state between walls 1.000×10^{-10} m apart. How much is its energy lowered by introduction of a well 18.0 eV deep and 1.000×10^{-11} m wide midway between the two walls, as in Figure 7.1?

7.4 If the well in Problem 7.3 were placed just inside one of the walls, instead of in the middle, how would it affect the ground-state energy?

7.5 The charge on the nucleus of an atom does not exist at a point but extends over a small volume. In a first approximation, one may replace this extension by a uniform distribution over a shell of proper radius. How much are the (a) $1s$ and (b) $2p$ levels of hydrogen raised if the nuclear charge is distributed uniformly over a small shell of radius 10^{-15} m?

7.6 Substitute a power series in r multiplied by $Fz\psi^{(0)}$ for $\psi^{(1)}$ into Equation (7.76), reduce, make an appropriate truncation, derive the remaining coefficients, to establish (7.77).

7.7 A one-dimensional periodic lattice is simulated by the potential

$$-U \cos \frac{2\pi}{a} x.$$

Let the lattice contain one mobile particle in distance a, a multiple of length $a/2$. Then the independent zeroth-order state functions may be chosen as

$$\sqrt{\frac{2}{a}} \cos kx \quad \text{and} \quad \sqrt{\frac{2}{a}} \sin kx,$$

and there are band gaps at

$$k = \frac{\pi}{a}, \frac{2\pi}{a}, \frac{3\pi}{a}, \ldots.$$

Use perturbation theory to find the first energy gap, at $k = \pi/a$.

7.8 An electron is confined within a square of edge 1.000×10^{-10} m with a potential trench 18.0 eV deep and 1.000×10^{-11} m wide running from the middle of one wall to the middle of the opposite wall. If the x and y axes run through the center parallel to the walls, how does the trench affect the $n_x = 1$, $n_y = 2$, and the $n_x = 2$, $n_y = 1$ levels?

Table 7.3
Molecular Orbital Coefficients for Fulvene

Root	c_1	c_2	c_3	c_4	c_5	c_6
−2.1149	0.4294	0.3851	0.3851	0.4294	0.5230	0.2473
−1.0000	0.0000	0.5000	0.5000	0.0000	−0.5000	−0.5000
−0.6180	0.6015	0.3718	−0.3718	−0.6015	0.0000	0.0000

7.9 A Hückel treatment of the pi electrons in fulvene yields the coefficients in Table 7.3 when all atomic parameters equal α and all bond parameters equal β. A better approximation involves, for the side-chain carbon, number 6, the form

$$\alpha_6 = \alpha + 0.5\beta.$$

What is the resulting first-order perturbation of the energy of the pi electrons?

--- --- ---

7.10 An electron is confined between two parallel walls distance a apart. If the potential energy is $-A$ when $-a/2 < x < 0$ and A when $0 < x < a/2$, how are the energy levels affected by the magnitude of A, in the first-order approximation?

7.11 A one-dimensional harmonic oscillator is perturbed by an electric field so that its potential is

$$U = \tfrac{1}{2}fx^2 - Fx.$$

Find the transformation of x that reduces the Schrödinger equation to the form for an unperturbed harmonic oscillator and determine the energy levels for the system.

7.12 Calculate the first-order correction to the ground-state energy of a hydrogen atom caused by the finite spatial extension of its nucleus. Assume the charge is uniformly distributed throughout a sphere of radius 10^{-15} m.

7.13 Obtain the ground state energy of a hydrogen atom perturbed by the potential g/r^2 hartree, accurate through first order in g.

7.14 For two particles bound by a potential varying only with the interparticle distance, variation of the distance in each eigenstate for energy is governed by the equation

$$HR = \left[-\frac{\hbar^2}{2\mu} \left(\frac{d^2}{dr^2} + \frac{2}{r}\frac{d}{dr} \right) + U(r) + \frac{\hbar^2 l(l+1)}{2\mu r^2} \right] R = ER.$$

Show how the last term in the brackets can be considered as a perturbation of the potential. Then derive an expression for the corresponding contribution to the energy E, with

$$\left\langle \frac{1}{\mu r^2} \right\rangle = \frac{1}{I}$$

in the unperturbed state.

7.15 Using first-order theory, determine how much the perturber ga^2x^4 shifts (a) the ground state and (b) the first excited state of a harmonic oscillator.

7.16 A one-dimensional harmonic oscillator is perturbed, in the natural units of Table 6.3, by the cubic potential

$$U^{(1)} = \sigma x^3.$$

Express operator $U^{(1)}$ in terms of the step operators in Table 5.2. Substitute into $\langle i, 0 \mid U^{(1)} \mid j, 0 \rangle$ and evaluate the nonzero matrix elements of $U^{(1)}$.

7.17 With the results from Problem 7.16, determine the energy of the oscillator through the second-order approximation.

7.18 A two-dimensional harmonic oscillator for which

$$U = \frac{1}{2}f(x^2 + y^2)$$

is in its first excited level, at which $v_x = 0$, $v_y = 1$, or $v_x = 1$, $v_y = 0$. How is this level split by the perturbation

$$U^{(1)} = gaxy$$

in the first-order approximation?

7.19 The effect of replacing one CH group in benzene by N, to form pyridine, is approximated by increasing the atomic parameter at that position from α to $\alpha + 0.5\beta$. Use first-order perturbation theory to calculate the effect on the π electron energies.

References

Books

Bohm, D.; 1951, *Quantum Theory*, Prentice-Hall, Englewood Cliffs, N.J., pp. 453–495. In Bohm, the order of topics differs from that in the present text; time-independent perturbations are considered after time-dependent ones. Nevertheless, the reference here may be useful because of the comments accompanying the mathematics.

Borowitz, S.: 1967, *Fundamentals of Quantum Mechanics,* Benjamin, Reading, Mass., pp. 317–353. Borowitz presents the theory simply, then illustrates its use with many examples. These include discussion of the hydrogen molecule ion and of the realistic hydrogen atom.

Cohen-Tannoudji, C., Diu, B., and Laloe, F. (translated by Hemley, S. R., Ostrowsky, N., and Ostrowsky, D.): 1977, *Quantum Mechanics*, Wiley, New York, pp. 1095–1147. This reference is noteworthly for the comments accompanying the development and for the examples considered. The latter include a harmonic oscillator subject to perturbing potentials, the interaction between magnetic dipoles, and the interaction between neutral atoms (van der Waals forces).

Dalgarno, A.: 1961, 'Stationary Perturbation Theory', in Bates, D. R. (editor), *Quantum Theory*, vol. I, Academic Press, New York, pp. 171–209. This short chapter is loaded with derivations and formulas. Dalgarno, for instance, shows how to calculate the $(2n + 1)$th order energy from the wave functions up to order n.

Herzberg, G.: 1950, *Molecular Spectra and Molecular Structure I. Spectra of Diatomic Molecules*, D. van Nostrand, Princeton, N.J., pp. 13–15, 280–298. Various interesting applications of the theory appear in this reference. Selection rules for perturbations are formulated.

Hirschfelder, J. O., Brown, W. B., and Epstein, S. T.: 1964, 'Recent Developments in Perturbation Theory', in Löwdin, P.-O. (editor), *Advances in Quantum Chemistry*, vol. 1, Academic Press, New York, pp. 255–374. This long chapter supplements that of Dalgarno. Much additional discussion, with derivations and formulas, is presented. Thus, Hirschfelder states that any theorem valid for the complete Schrödinger equation with any given λ has an analog for the solution of each perturbation equation separately.

Lowe, J. P.: 1978, *Quantum Chemistry*, Academic Press, New York, pp. 347–380. The account here is slanted towards chemical applications.

Migdal, A. B. (translated by Leggett, A. J.): 1977, *Qualitative Methods in Quantum Theory*, Benjamin, Reading, Mass., pp. 86–148. The word qualitative in the title may be misleading. It refers to the fact that Migdal emphasizes explanation rather than rigor. The section here is loaded with equations. Various kinds of perturbations are considered.

Articles

Adkins, G. S., and Hood, R. F.: 1989. "Use of the Schrödinger-Coulomb Green Function in the Evaluation of the Quadratic Stark Effect," *Eur. J. Phys.* **10**, 61–66.

Adler, C., and Rose, O.: 1979. "Generalized Bound-State Perturbation Theory," *Am. J. Phys.* **47**, 822–824.

Aguilera-Navarro, V. C., Iwamoto, H., Ley-Koo, E., and Zimerman, A. H.: 1981. "Quantum Bouncer in a Closed Court," *Am. J. Phys.* **49**, 648–651.

Baker, J. D., Freund, D. E., Hill, R. N., and Morgan III, J. D.: 1990. "Radius of Convergence and Analytic Behavior of the 1/Z Expansion," *Phys. Rev. A* **41**, 1247–1273.

Bes, D. R., Dussel, G. G., and Sofia, H. M.: 1977. "Normalization of States in Perturbation Theories," *Am. J. Phys.* **45**, 191–192.

Coronado,M., Dominguez, N., Flores, J., and Portilla, C. de la: 1982. "Effect of the Continuum States in Second-Order Perturbation Theory: An Example," *Am. J. Phys.* **50**, 27–29.

Coronado, M., Dominguez, N., Flores, J., and Portilla, C. de la: 1987. "Effect of the Boundary Conditions on the Accuracy of Perturbation Theory," *Am. J. Phys.* **55**, 924–929.

Dasgupta, B. B.: 1981. "Hydrogenlike Atom in the Potential $V = r^2(a + b \cos^2\Theta)$: Second-Order Perturbation Theory," *Am. J. Phys.* **49**, 764–767.

Doty, D. R., and Einstein, S. A.: 1974. "Perturbation Treatment for Two-Particle Symmetrized Systems with Interactions," *Am. J. Phys.* **42**, 985–991.

Fanelli, R., and Struzynski, R. E.: 1983. "Energy Eigenvalues of a Quantum Anharmonic Oscillator," *Am. J. Phys.* **51**, 561–564.

Ferrell, T. L.: 1980. "Diagrams for Quantum Oscillators," *Am. J. Phys.* **48**, 728–731.

Gottdiener, L.: 1978. "Derivation of the Perturbation Formulas for the Energy in Quantum Mechanics," *Am. J. Phys.* **46**, 893–895.

Grotch, H., and Kazes, E.: 1977. "Nonrelativistic Quantum Mechanics and the Anomalous Part of the Electron g Factor," *Am. J. Phys.* **45**, 618–623.

Jothan, R. W.: 1975. "Why Do Energy Levels Repel One Another?" *J. Chem. Educ.* **52**, 377–378.

Lain, L., and Torre, A.: 1987. "A Simple Derivation of the Series in Perturbation Theory," *Eur. J. Phys.* **8**, 178–181.

Muynck, W. M. de: 1977. "On the Unambiguity of Stationary Perturbation Theory," *Am. J. Phys.* **45**, 1191–1193.

Naqvi, K. R.: 1987. "Derivation of Energy Expressions in the Rayleigh-Schrödinger Perturbation Theory," *Am. J. Phys.* **55**, 269–271.

Purcell, K.M., and Henneberger, W. C.: 1978. "Aharonov-Bohm Effect in Perturbation Theory," *Am. J. Phys.* **46**, 1255–1256.

Tang, A. Z., Lieber, M., and Chan, F. T.: 1985. "Simple Example in Second-Order Perturbation Theory," *Am. J. Phys.* **53**, 595–596.

Urumov, V., and Ivanovski, G. J.: 1983. "The Role of Continuum States in Perturbation Theory," *Am. J. Phys.* **51**, 950–952.

8

VARYING-STATE PERTURBATION THEORY

8.1
Kinds of Processes

Coherent systems not only exist in states but also undergo changes. The possibilities include (a) steady-state transformations, (b) induced shifts, and (c) spontaneous processes.

A steady-state transformation may be described by a wave function. A simple example is provided by a homogeneous beam; a more complicated example, by steady scattering from such a beam.

Field-caused absorption and emission, together with reaction processes, are induced. With these, the pertinent influence may be brought to bear at a particular time and the effect calculated. Thus, the process is made to result from a change in the Hamiltonian. The change keeps the time from being separable from the coordinates in the relevant Schrödinger equation and in the state function.

Simple emission and decay processes are the common spontaneous ones. Each may be considered as the limit of an induced process, where the inducing field has been reduced to zero. As a consequence the process may be introduced mathematically at a specified time.

In calculations a person may consider the varying part of the Hamiltonian to be the perturber. As long as the varying influence does not alter the space of the kets, it acts to cause transitions from each initially occupied state to other states of the unperturbed system. The actual system is represented by a continually varying superposition of such states.

One finds that effective procedures can be developed (a) when the varying part of the Hamiltonian acts over a short enough time interval, or (b) when this part is relatively small, or (c) when the perturber acts at a sufficiently slow rate.

8.2
General Dependence on Time

While stationary states are governed by the time-independent Schrödinger equation, processes involve the time-dependent one.

If $| G \rangle$ is the temporally varying ket describing a given system, then from (2.137), we have

$$H | G \rangle = -\frac{\hbar}{i} \frac{\partial}{\partial t} | G \rangle. \qquad (8.1)$$

In the coordinate-time representation, (8.1) becomes

$$H \Psi (\mathbf{r}, t) = -\frac{\hbar}{i} \frac{\partial}{\partial t} \Psi (\mathbf{r}, t) \qquad (8.2)$$

whence

$$d \Psi = -\frac{i}{\hbar} dt \, H \, \Psi. \qquad (8.3)$$

For an infinitesimal step in time, from $t = 0$, we have

$$\Psi (\mathbf{r}, dt) = (1 - \frac{i}{\hbar} dt \, H) \, \Psi (\mathbf{r}, 0). \qquad (8.4)$$

If H is independent of time, over a finite step Δt, we see that

$$\Psi_1 (\mathbf{r}, t) = \lim_{\Delta t \to 0} \left(1 - \frac{i}{\hbar} \Delta t \, H \right)^{1/\Delta t} \Psi (\mathbf{r}, 0)$$

$$= \exp \left(-\frac{it}{\hbar} H \right) \Psi (\mathbf{r}, 0). \qquad (8.5)$$

In the last step, the definition of the exponential function has been employed.

When the Hamiltonian for the given system of particles varies, the variation appears in the potential part. If we let the varying part be ΔH, we have

$$H(t) = H(0) + \Delta H(t) \qquad (8.6)$$

and (8.5) is replaced by

$$\Psi_2(\mathbf{r}, t) = \exp\left(-\frac{itH}{\hbar} - \frac{i}{\hbar} \int_0^t \Delta H\, dt \right) \Psi\,(\mathbf{r}, 0)$$

$$= \exp\left(-\frac{i}{\hbar} \int_0^t \Delta H\, dt \right) \Psi_1\,(\mathbf{r}, t). \qquad (8.7)$$

Since H is an operator, Equations (8.5) and (8.7) have limited utility.

Example 8.1

How is a state function altered by a shift in the potential U introduced over a time interval Δt?

The Hamiltonian for the given system is the time-independent expression

$$H_1 \qquad \text{when} \qquad t < 0,$$

the varying expression

$$H_1 + \Delta U \qquad \text{during the interval} \qquad 0 < t < \Delta t,$$

and the time-independent expression

$$H_2 \qquad \text{when} \qquad t > \Delta t.$$

Over the interval that the Hamiltonian is changing, the system is governed by the time-dependent Schrödinger equation. For $t > \Delta t$, the integrated form (8.7) becomes

$$\Psi = \exp\left(-\frac{i}{\hbar} \int_0^{\Delta t} \Delta U\, dt \right) \Psi_1.$$

8.3
The Sudden Approximation

Any change in the potential function for a system alters the Hamiltonian operator and its eigenkets. But if the change occurs within a small enough interval of time, the state ket is not altered. The prevailing ket can be projected onto either the initial eigenkets or the final eigenkets. This ket may be taken to be one of the initial eigenkets and its destiny determined.

Let us consider a system for which the potential U varies as described in Example 8.1. As long as the function ΔU is limited and the time interval Δt is very small, the exponential factor in the last equation does not differ appreciably from unity. The transformed state function Ψ and the corresponding ket then do not deviate appreciably from the original function and ket.

Now, the Schrödinger equation for the system has the form

$$H_1 \mid i, 1 \rangle = E_i \mid i, 1 \rangle \tag{8.8}$$

when $t < 0$ and the form

$$H_2 \mid j, 2 \rangle = E_j \mid j, 2 \rangle \tag{8.9}$$

when $t > \Delta t$. Both sets of eigenkets are base kets for the same Hilbert space and span it. So the prevailing ket can be expressed as a linear function of either set of eigenkets:

$$\mid g \rangle = \Sigma \, c_{1i} \mid i, 1 \rangle = \Sigma \, c_{2j} \mid j, 2 \rangle. \tag{8.10}$$

When the system is initially in a definite eigenstate, described by $\mid i, 1 \rangle$, Equation (8.10) yields

$$\mid i, 1 \rangle = \Sigma \, c_j \mid j, 2 \rangle. \tag{8.11}$$

With the eigenkets mutually orthogonal and normalized to 1, multiplying both sides of (8.11) by bra $\langle k, 2 \mid$ and reducing leads to

$$\langle k, 2 \mid i, 1 \rangle = c_k, \tag{8.12}$$

whence

$$c_k^* \, c_k = \mid \langle k, 2 \mid i, 1 \rangle \mid^2. \tag{8.13}$$

Following Axiom (VI) and Equation (1.61), we interpret (8.13) as the probability that a system initially in state $\mid i, 1 \rangle$ ends up in the state $\mid k, 2 \rangle$.

Example 8.2

A tritium atom in its ground state disintegrates into a helium ion, a beta particle, and an antineutrino,

$$^3H \rightarrow \, ^3He^+ + \beta^- + \bar{\nu},$$

with release of 18,000 eV energy. What is the probability that the electron in the helium ion is left in its $2s$ state?

The energy released is large enough so that the time interval during which the beta particle is near the orbital electron is generally small. On the other hand, the energy released is small enough so that the recoil of the nucleus with respect to the orbital electron can be neglected.

From Tables 4.3, 4.5, and 6.4, the initial $1s$ state function for ^{3}H in natural units is

$$\psi_I = \left(\frac{1}{\pi}\right)^{\frac{1}{2}} e^{-r},$$

while the final $2s$ state function for ^{3}He$^+$ is

$$\psi_{II} = \left(\frac{1}{\pi}\right)^{\frac{1}{2}} (1 - r) \, e^{-r}.$$

Substituting these into (8.12) yields

$$c_2 = \frac{1}{\pi} \int_0^\infty (1 - r) \, e^{-2r} \, 4\pi r^2 \, dr = 0 + \frac{e^{-2r}}{2} \bigg|_0^\infty = -\frac{1}{2}.$$

The corresponding weight is

$$w = c_2^* \, c_2 = \left(-\frac{1}{2}\right)^2 = 0.25.$$

8.4
Rates at which Expansion Coefficients Vary

The ket describing an arbitrary state of a coherent system can be resolved into a sum of fractions of standard stationary eigenkets, phased appropriately. Changes in the physical state involve alterations in these fractions, in the contributions of the eigenkets, and in their relative phasings. Such changes may be induced by influences that depend on time.

Let us consider a coherent system subject to varying interactions, which imply a varying Hamiltonian H. The resulting processes are governed by Equations (8.1) and (8.2). Let us split H into the part $H^{(0)}$ that does not vary with time and the part $\lambda H^{(1)}$ that does so vary:

$$H = H^{(0)} + \lambda \, H^{(1)}. \tag{8.14}$$

Here λ is a perturbation parameter.

The time-independent part of the Hamiltonian leads to the Schrödinger equation

$$H^{(0)} \mid j \rangle = E_j \mid j \rangle \qquad (8.15)$$

satisfied by a set of orthonormal, time-independent eigenkets $\mid 1 \rangle$, $\mid 2 \rangle, \ldots$

From (8.2), the complete wave function for a state of definite energy E_j consists of the spatial function multiplied by the temporal factor

$$T = e^{-i\omega_j t} \qquad (8.16)$$

in which

$$\omega_j = \frac{E_j}{\hbar} . \qquad (8.17)$$

Since multiplying (8.15) by (8.16) gives rise to

$$H^{(0)} \mid j \rangle e^{-i\omega_j t} = E_j \mid j \rangle e^{-i\omega_j t}, \qquad (8.18)$$

the time-dependent ket

$$\mid J \rangle = \mid j \rangle e^{-i\omega_j t} \qquad (8.19)$$

is an eigenket of $H^{(0)}$.

The $\mid J \rangle$'s span the vector space for the varying system as long as the perturbation does not add or remove volumes or dimensions to the space. Then any possible $\mid G \rangle$ is the superposition

$$\mid G \rangle = \Sigma \, c_k \mid k \rangle \, e^{-i\omega_k t}, \qquad (8.20)$$

whence

$$\frac{\partial}{\partial t} \mid G \rangle = \Sigma \left(\frac{dc_k}{dt} \mid k \rangle \, e^{-i\omega_k t} - i \frac{E_k}{\hbar} c_k \mid k \rangle \, e^{-i\omega_k t} \right). \qquad (8.21)$$

Substituting (8.14), (8.20), (8.21) into (8.1) and reducing the result with (8.18) gives us

$$\Sigma \left(c_k E_k \mid k \rangle \, e^{-i\omega_k t} + c_k \lambda H^{(1)} \mid k \rangle \, e^{-i\omega_k t} \right)$$

$$= \Sigma \left(-\frac{\hbar}{i} \frac{dc_k}{dt} \mid k \rangle \, e^{-i\omega_k t} + E_k c_k \mid k \rangle \, e^{-i\omega_k t} \right) \qquad (8.22)$$

or

$$\Sigma \, \lambda c_k H^{(1)} \mid k \rangle \, e^{-i\omega_k t} = \Sigma \, i\hbar \frac{dc_k}{dt} \mid k \rangle \, e^{-i\omega_k t}. \qquad (8.23)$$

Let us multiply both sides of (8.23) by bra $\langle j \mid$, introduce the orthonormality of the vectors,

$$\Sigma \lambda c_k \langle j \mid H^{(1)} \mid k \rangle e^{-i\omega_k t} = i\hbar \frac{dc_j}{dt} e^{-i\omega_j t}, \qquad (8.24)$$

and rearrange to form

$$\frac{dc_j}{dt} = \Sigma_k \frac{\lambda}{i\hbar} c_k H^{(1)}{}_{jk} e^{i(\omega_j - \omega_k)t}, \qquad (8.25)$$

in which

$$H^{(1)}{}_{jk} = \langle j \mid H^{(1)} \mid k \rangle. \qquad (8.26)$$

Expression (8.26) is the *jk*th matrix element for perturbation $H^{(1)}$.

One has an Equation (8.25) for each state that contributes appreciably to the evolving composite state being considered. The resulting set of equations must be satisfied simultaneously.

8.5
Managing Secular (Aperiodic) Perturbing Expressions

When the perturber $H^{(1)}$ becomes predominately positive, or negative, and appreciable, the cumulative effects of the *j*th term on the right side of (8.25),

$$\frac{\lambda}{i\hbar} c_j H^{(1)}{}_{jj}, \qquad (8.27)$$

grow and become considerably larger than the effects of any other term. Its effects then swamp the effects of the other terms and the system of equations becomes difficult to solve.

To alleviate the situation, one may transfer the offensive term to the left side and introduce a periodic exponential factor:

$$\left(\frac{dc_j}{dt} - c_j \frac{\lambda}{i\hbar} H^{(1)}{}_{jj} \right) \exp \left(\frac{i\lambda}{\hbar} \int_0^t H^{(1)}{}_{jj} \, dt \right)$$

$$= \Sigma_{k' \neq j} \frac{\lambda}{i\hbar} c_{k'} H^{(1)}{}_{jk'} e^{i(\omega_j - \omega_{k'})t} \exp \left(\frac{i\lambda}{\hbar} \int_0^t H^{(1)}{}_{jj} \, dt \right)$$

$$= \Sigma_{k' \neq j} \frac{\lambda}{i\hbar} c_{k'} \exp \left(\frac{i\lambda}{\hbar} \int_0^t H^{(1)}{}_{k'k'} \, dt \right) H^{(1)}{}_{jk'} \exp \left(\frac{i}{\hbar} \int_0^t \gamma_{jk'} \, dt \right). \qquad (8.28)$$

The prime on index k indicates that the term for $k = j$ has been removed from the sum. The integrand in the last integral represents a difference:

$$\gamma_{jk'} = \hbar\omega_j + \lambda H^{(1)}_{jj} - \hbar\omega_{k'} - \lambda H^{(1)}_{k'k'}. \tag{8.29}$$

Indeed if we let

$$\hbar\omega_k + \lambda H^{(1)}_{kk} = E'_k, \tag{8.30}$$

then (8.29) becomes

$$\gamma_{jk'} = E'_j - E'_{k'}. \tag{8.31}$$

Expression (8.30) is the energy of the kth state to first order if no other state contributes to it, following formula (7.14). So $\gamma_{jk'}$ is said to be the difference between the perturbed energies of the jth state and the k'th state.

Since the left side of (8.28) is the time derivative of

$$b_j = c_j \exp\left(\frac{i\lambda}{\hbar} \int_0^t H^{(1)}_{jj}\, dt\right), \tag{8.32}$$

Equation (8.28) reduce to

$$\frac{db_j}{dt} = \sum_{k' \neq j} \frac{\lambda}{i\hbar}\, b_{k'}\, H^{(1)}_{jk'} \exp\left(\frac{i}{\hbar} \int_0^t \gamma_{jk'}\, dt\right). \tag{8.33}$$

If the perturbing potential is weak enough, or appropriately distributed, so that

$$\left| \int_0^t (\lambda H^{(1)}_{jj} - \lambda H^{(1)}_{k'k'})\, dt \right| \ll \left| \int_0^t (\hbar\omega_j - \hbar\omega_{k'})\, dt \right|, \tag{8.34}$$

we have

$$\frac{1}{\hbar} \int_0^t \gamma_{jk'}\, dt \simeq \frac{1}{\hbar} \int_0^t (\hbar\omega_j - \hbar\omega_{k'})\, dt = (\omega_j - \omega_{k'})t. \tag{8.35}$$

Then we let

$$\omega_j - \omega_{k'} = \omega_{jk'}. \tag{8.36}$$

Alternatively, if the perturbing potential is simply constant while the time increases from 0 to t, then

$$H^{(1)}_{jj} = \text{constant}, \tag{8.37}$$

perturbed energies E'_j and $E'_{k'}$ are constant, and

$$\frac{1}{\hbar} \int_0^t \gamma_{jk'}\, dt = \frac{1}{\hbar} \int_0^t (E'_j - E'_{k'})\, dt = \frac{E'_j - E'_{k'}}{\hbar}t$$

$$= (\omega'_j - \omega'_{k'})\, t = \omega_{jk'} t, \tag{8.38}$$

if we let

$$\omega'_j - \omega'_{k'} = \omega_{jk'}. \tag{8.39}$$

Under either set of circumstances, Equation (8.33) can be rewritten in the form

$$\frac{db_j}{dt} = \sum_{k' \neq j} \frac{\lambda}{i\hbar}\, b_k H^{(1)}_{jk'}\, e^{i\omega_{jk'}t}, \tag{8.40}$$

which is similar to (8.25). Letting j for the significant states be $1, 2, \ldots, n$ then yields the matrix equation

$$\frac{d}{dt}\begin{pmatrix} b_1 \\ b_2 \\ \cdot \\ \cdot \\ \cdot \\ b_n \end{pmatrix} = \frac{\lambda}{i\hbar} \begin{pmatrix} 0 & H^{(1)}_{12}\, e^{i\omega 12 t} & \cdots \\ H^{(1)}_{21}\, e^{i\omega 21 t} & 0 & \cdots \\ \cdot & \cdot & \cdot \\ \cdot & \cdot & \cdot \\ \cdot & \cdot & \cdot \\ H^{(1)}_{n1} e^{i\omega n1 t} & \cdots & 0 \end{pmatrix} \begin{pmatrix} b_1 \\ b_2 \\ \cdot \\ \cdot \\ \cdot \\ b_n \end{pmatrix}. \tag{8.41}$$

If each side of (8.32) is multiplied by its complex conjugate, there ensues the relationship

$$b_j^* b_j = c_j^* c_j. \tag{8.42}$$

With Axiom (VI) and Equation (1.61), the probability that the jth state is occupied at any given time is

$$w_j = b_j^* b_j. \tag{8.43}$$

Example 8.3

A system is initially in a state that is orthogonal to, and energetically degenerate with, one other state. At time $t = 0$, a perturber symmetric with respect to the two states is introduced. How does the system evolve as long as the intrusion of other states is negligible?

Let the kets for the two states be

$$| \, 1 \, \rangle \, e^{-i\omega t} \quad \text{and} \quad | \, 2 \, \rangle \, e^{-i\omega t},$$

while the perturber after $t = 0$ is

$$H^{(1)}.$$

Because the perturber acts symmetrically on the two states, we have

$$H^{(1)}{}_{11} = H^{(1)}{}_{22}$$

and condition (8.35) applies. Because the states are equivalent in energy, we also have

$$\omega_{12} = \omega_{21} = 0.$$

Consequently, the system is governed by (8.41) in the form

$$\frac{db_1}{dt} = \frac{1}{i\hbar} H^{(1)}{}_{12} b_2,$$

$$\frac{db_2}{dt} = \frac{1}{i\hbar} H^{(1)}{}_{21} b_1.$$

Because the states are affected similarly by the perturber, we also have

$$H^{(1)}{}_{21} = H^{(1)}{}_{12}.$$

Let us first suppose that $H^{(1)}$ is constant over the period of integration. Differentiation of the first differential equation and substitution of the next two equations into it then yields

$$\frac{d^2 b_1}{dt^2} = \frac{1}{i\hbar} H^{(1)}{}_{12} \frac{1}{i\hbar} H^{(1)}{}_{12} \, b_1 = -\left(\frac{H^{(1)}{}_{12}}{\hbar} \right)^2 b_1.$$

The solution to this equation with $b_1 = 1$ and $b_2 = 0$ at $t = 0$ is

$$b_1 = \cos \frac{H^{(1)}{}_{12}}{\hbar} t.$$

Substituting this result into the first differential equation leads to

$$b_2 = -i \sin \frac{H^{(1)}{}_{12}}{\hbar} t.$$

With (8.43), we obtain

$$w_1 = b_1 {}^* b_1 = \cos^2 \frac{H^{(1)}{}_{12}}{\hbar} t$$

and

$$w_2 = b_2{}^* b_2 = \sin^2 \frac{H^{(1)}{}_{12}}{\hbar} \, t.$$

Under the given conditions, the system moves periodically from the first state to the second state and back again.

Secondly, let us consider that $H^{(1)}$ varies with time. Combining the first two differential equations with the subsequent equation yields

$$\frac{db_1}{db_2} = \frac{b_2}{b_1},$$

whence

$$b_1{}^2 - b_2{}^2 = 1,$$

if the constant of integration is chosen so that $b_2 = 0$ when $b_1 = 1$. A transformation that embodies this result is

$$b_1 = \cos \theta, \qquad b_2 = -i \sin \theta.$$

Applying this transformation to the first differential equation yields

$$- \sin \theta \, \frac{d\theta}{dt} = \frac{1}{i\hbar} \, H^{(1)}{}_{12} \, (-i \sin \theta)$$

whence

$$d\theta = \frac{H^{(1)}{}_{12}}{\hbar} \, dt$$

and

$$\theta = \frac{1}{\hbar} \int_0^t H^{(1)}{}_{12} \, dt.$$

Substituting the result back into the transformation equations gives us

$$b_1 = \cos \left(\frac{1}{\hbar} \int_0^t H^{(1)}{}_{12} \, dt \right)$$

and

$$b_2 = -i \sin \left(\frac{1}{\hbar} \int_0^t H^{(1)}{}_{12} \, dt \right).$$

With (8.43), the probability that the system is in the first state at time t is

$$w_1 = b_1 * b_1 = \cos^2 \left(\frac{1}{\hbar} \int_0^t H^{(1)}{}_{12} \, dt \right),$$

while the probability that it is in the second state at the same time t is

$$w_2 = b_2 * b_2 = \sin^2 \left(\frac{1}{\hbar} \int_0^t H^{(1)}{}_{12} \, dt \right).$$

The effect of the perturbation is to cause the system to move periodically from the first state to the second state and back again.

8.6
Perturbation Terms for the Transformed Coefficients

As long as introduction of the varying part of the Hamiltonian does not cause an abrupt or catastrophic change in the pertinent ket, a person may consider the transformed coefficients to be power series in the perturbation parameter. Substitution into the set of equations just derived then yields a sequence of equations that can be solved in principle.

Followng the procedure in Section 7.2, we consider that each transformed coefficient b_j is an analytic function of λ:

$$b_j = b_j^{(0)} + \lambda b_j^{(1)} + \lambda^2 b_j^{(2)} + \ldots = \sum_{l=0} \lambda^l b_j^{(l)}. \tag{8.44}$$

Use of Equation (8.44) implies that superposition (8.20) is valid and that coefficient c_j varies smoothly as λ increases from zero. Thus, it implies that introducing $H^{(1)}$ does not expand the Hilbert space, allowing kets with contributions orthogonal to all the original eigenkets.

Series (8.44) converts Equation (8.40) to

$$\sum_{l=0}^{\infty} \lambda^l \dot{b}_j^{(l)} = \sum_{\substack{k'=0 \\ k' \neq j}}^{\infty} \sum_{l=0}^{\infty} \frac{1}{i\hbar} H^{(1)}{}_{jk'} \, e^{i\omega jk't} \lambda^{l+1} b_{k'}^{(l)}, \tag{8.45}$$

if a dot over a letter represents total differentiation with respect to time t. Since (8.45) is to be true for an arbitrary λ, it is an identity in λ and the coefficients of each power of λ must be equal on both sides. Thus, we obtain

$$\dot{b}_j^{(0)} = 0, \tag{8.46}$$

$$\dot{b}_j^{(1)} = \sum_{k' \neq j} \frac{1}{i\hbar} H^{(1)}{}_{jk'} \, b_{k'}^{(0)} \, e^{i\omega jk't}, \tag{8.47}$$

$$\dot{b}_j^{(2)} = \sum_{k' \neq j} \frac{1}{i\hbar} H^{(1)}_{jk'} \, b_{k'}^{(1)} \, e^{i\omega jk't}, \tag{8.48}$$

.

.

.

When perturbation parameter λ is zero, Equation (8.44) reduces to

$$b_j(t) = b_j^{(0)}(t). \tag{8.49}$$

But condition (8.46) tells us that $b^{(0)}(t)$ is constant. So we have

$$b_j^{(0)}(t) = b_j(0) \tag{8.50}$$

and (8.47) becomes

$$\dot{b}_j^{(1)} = \sum_{k' \neq j} \frac{1}{i\hbar} H^{(1)}_{jk'} \, b_{k'}(0) \, e^{i\omega jk't}, \tag{8.51}$$

At the time it is introduced the perturbation has not had time to alter the state of the system. Consequently, we set

$$b_j(0) = b_j^{(0)}(0) \tag{8.52}$$

regardless of λ. On increasing λ from zero in (8.44), we now find that

$$b_j^{(1)}(0) = b_j^{(2)}(0) = \ldots = 0. \tag{8.53}$$

Thus, the initial value of each higher-order coefficient is also zero.

In Table 8.1, the basic relations of our varying-state perturbation theory have been collected.

Table 8.1
Key Formulas of
Time-Dependent Perturbation Theory

$$H \,|\, G \rangle = -\frac{\hbar}{i} \frac{\partial}{\partial t} \,|\, G \rangle \qquad\qquad H^{(0)} \,|\, j \rangle = E_j \,|\, j \rangle$$

$$H = H^{(0)} + \lambda H^{(1)}$$

$$|\, G \rangle = \sum c_k \,|\, k \rangle \, e^{-i\omega_k t}$$

$$b_j = c_j \exp\left(\frac{i\lambda}{\hbar} \int_0^t H^{(1)}_{jj} \, dt \right)$$

$$\omega_{jk'} = \omega'_j - \omega'_{k'}$$

$$w_j = b_j^* b_j$$

$$\dot{b}_j^{(1)} = \sum_{k' \neq j} \frac{1}{i\hbar} H^{(1)}_{jk'} \, b_{k'}(0) \, e^{i\omega jk't}$$

8.7
Steady Perturbers

A simple perturber is one that depends only on the coordinates from the time of its application on. Its value for a certain configuration of the particles in the system appears as Figure 8.1 shows. Equations (8.51) and (8.48) are then readily integrable.

Let the perturber for $\lambda = 1$ have the form

$$H^{(1)} = \eta(t)V^{(1)}(\text{particle coordinates}) \tag{8.54}$$

where

$$\eta(t) = 0 \quad \text{when time } t \text{ is negative} \tag{8.55}$$

and

$$\eta(t) = 1 \quad \text{when time } t \text{ is positive} \tag{8.56}$$

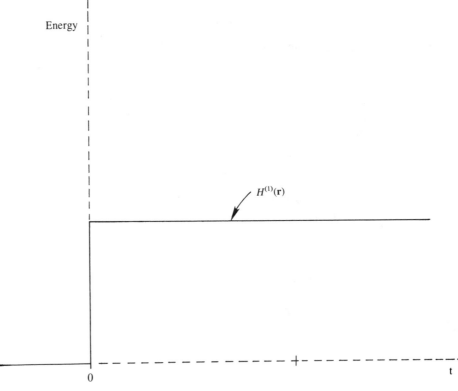

Figure 8.1. A step perturber for a given configuration of the particles in a system.

Then

$$H^{(1)}_{jk} = \langle j \mid H^{(1)} \mid k \rangle = \eta(t) \langle j \mid V^{(1)} \mid k \rangle = \eta(t)(\text{constant}). \quad (8.57)$$

Because

$$\int_0^t e^{i\omega_{jk'}t} \, dt = \frac{e^{i\omega_{jk'}t} - 1}{i\omega_{jk'}}, \quad (8.58)$$

the integral of (8.51) now becomes

$$b_j^{(1)} = \sum_{k' \neq j} -\frac{1}{\hbar} H^{(1)}_{jk'} \, b_k(0) \frac{e^{i\omega_{jk'}t} - 1}{\omega_{jk'}}. \quad (8.59)$$

Substituting (8.59) into (8.48) yields the equation

$$\dot{b}_j^{(2)} = \sum_{k' \neq j} \sum_{l' \neq k'} -\frac{1}{i\hbar^2} H^{(1)}_{jk'} \, H^{(1)}_{k'l'} \, b_l(0) \frac{e^{i\omega_{k'l'}t} - 1}{\omega_{k'l'}} e^{i\omega_{jk'}t}, \quad (8.60)$$

which integrates to

$$b_j^{(2)} = \sum_{k' \neq j} \sum_{l' \neq k'} \frac{1}{\hbar^2} \frac{H^{(1)}_{jk'} \, H^{(1)}_{k'l'} \, b_l(0)}{\omega_{k'l'}} \left(\frac{e^{i\omega_{jl'}t} - 1}{\omega_{jl'}} - \frac{e^{i\omega_{jk'}t} - 1}{\omega_{jk'}} \right). \quad (8.61)$$

Note that in the summations, k' and l' skip over numbers j and k', respectively.

From (8.43), the probability that the nth state is occupied at any given time equals the square of the absolute value of the coefficients:

$$w_n = b_n{}^* b_n. \quad (8.62)$$

If only the mth state is occupied initially, then

$$b_m(0) = 1, \quad (8.63)$$

$$b_n(0) = 0 \quad \text{when} \quad n \neq m, \quad (8.64)$$

and to first order, we have

$$b_n = \lambda b_n^{(1)}. \quad (8.65)$$

When the perturbation parameter λ equals 1, this becomes

$$b_n = b_n^{(1)}. \quad (8.66)$$

So when only the mth state is occupied initially, Equation (8.59) reduces to

$$
\begin{aligned}
b_n^{(1)} &= -\frac{1}{\hbar} H^{(1)}_{nm} \frac{e^{\frac{1}{2}i\omega_{nm}t}(e^{\frac{1}{2}i\omega_{nm}t} - e^{-\frac{1}{2}i\omega_{nm}t})}{\omega_{nm}} \\
&= -\frac{2i}{\hbar} H^{(1)}_{nm} \frac{e^{\frac{1}{2}i\omega_{nm}t}}{\omega_{nm}} \sin \tfrac{1}{2}\omega_{nm}t.
\end{aligned} \quad (8.67)
$$

The probability that a transition from the mth state to the nth state occurs in time t is

$$P_{mn} = b_n^* b_n = \frac{4 \, | \, H^{(1)}{}_{nm} \, |^2}{\hbar^2} \left(\frac{\sin \frac{1}{2} \omega_{nm} t}{\omega_{nm}} \right)^2, \qquad (8.68)$$

when terms higher in order the first are neglected. Introducing the pertinent half-angle formula and the definition of $H^{(1)}{}_{nm}$ leads to the result

$$P_{mn} = \frac{2}{\hbar^2} | \, \langle \, n \, | \, H^{(1)} \, | \, m \, \rangle \, |^2 \, \frac{1 - \cos \omega_{nm} t}{\omega_{nm}{}^2} . \qquad (8.69)$$

Generally, $H^{(1)}$ is Hermitian. Furthermore, interchanging n and m merely reverses the sign of ω_{nm}. Consequently, we have the symmetry

$$P_{mn} = P_{nm}. \qquad (8.70)$$

The probability for transition from the mth state to the nth state equals the probability for the reverse transition.

In the first-order approximation, transition probability P_{mn} is a sinusoidal function of time t with the period $| \, 2\pi/\omega_{mn} \, |$. The amplitude varies directly with the square of the matrix element $H^{(1)}{}_{nm}$ and inversely with the square of the difference between the perturbed energies

$$\hbar \omega_{nm} = E'_n - E'_m. \qquad (8.71)$$

See Figure 8.2.

The energy of the nth composite state generally differs from that of the mth state; the elementary transition then violates the conservation-of-energy law. But from Figure 8.2, the period over which it is violated equals $| \, 2\pi/\omega_{mn} \, |$, a limited interval. Furthermore, when the transition takes place to a band of states, participants in an ensemble suffer both upward and downward transitions. On the average, the losses cancel the gains and the total energy appears to be conserved.

Example 8.4

How does the transition probability to a state having nearly the same energy as the initial state increase over the short term?

As long as

$$| \, \omega_{nm} t \, | \ll 1,$$

we have

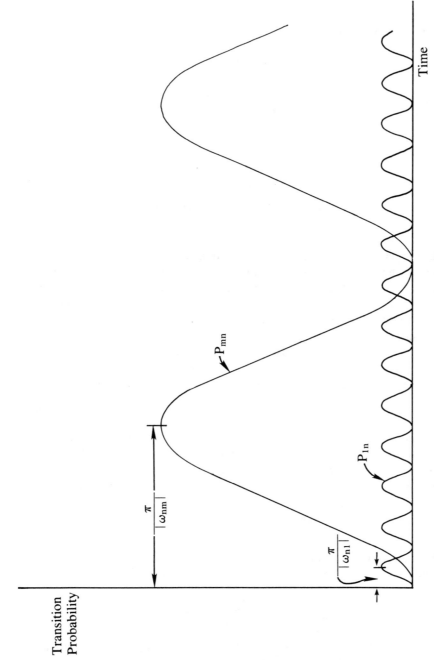

Figure 8.2. Temporal variation of the weight of the *n*th state when the source state is very close to, and when it is farther from, this *n*th state.

$$\sin \tfrac{1}{2}\omega_{nm}t \simeq \tfrac{1}{2}\omega_{nm}t.$$

Then (8.68) reduces to

$$P_{mn} = \frac{4 \mid \langle n \mid H^{(1)} \mid m \rangle \mid^2 (\tfrac{1}{4}\omega_{nm}^2 t^2)}{\hbar^2 \omega_{nm}^2}$$

$$= \frac{1}{\hbar^2} \mid \langle n \mid H^{(1)} \mid m \rangle \mid^2 t^2.$$

8.8
Transitions To or From a Band of States

How an ensemble of systems evolves into or out of a set of states depends on the varying probabilities of the transitions into or out of the set. When the set forms a nearly homogeneous band, the net rate of transition per system varies not only with the typical perturbation matrix element squared but also with the density of states in the band.

Consider systems that evolve from or to a definite state labeled m as the result of a step perturbation. Also, suppose that various n states with similar matrix elements $H^{(1)}_{nm}$ form a band having a *density* with respect to energy of $\rho(E_n)$. Thus, the number of such states in the energy range dE_n, per system, is

$$\rho(E_n) \, dE_n. \tag{8.72}$$

Interchanging m and n does not alter the value of expression (8.69). Consequently (8.69) gives the probability of transition either from the mth state to the nth state or to the mth state from the nth state. Replacing the nth state by the postulated band of states and integrating (8.69) over this band yields

$$P = \frac{2}{\hbar^2} \int |H^{(1)}_{nm}|^2 \, \frac{1 - \cos \omega_{nm}t}{\omega_{nm}^2} \, \rho(E_n) \, dE_n \tag{8.73}$$

for the total transition probability per system. The time rate of change in this probability is

$$\frac{dP}{dt} = \frac{2}{\hbar^2} \int |H^{(1)}_{nm}|^2 \, \frac{\sin \omega_{nm}t}{\omega_{nm}} \, \rho(E_n) \, dE_n. \tag{8.74}$$

Differentiating the relationship between energy and angular frequency, and recognizing that the mth frequency is constant, yields the substitution

$$dE_n = \hbar\, d\omega'_n = \hbar\, d(\omega'_n - \omega'_m) = \hbar\, d\omega_{nm}. \qquad (8.75)$$

When transitions occur to or from an essentially uniform band, with $H^{(1)}_{nm}$ and $\rho(E_n)$ approximately constant, formula (8.74) reduces to

$$\frac{dP}{dt} = \frac{2}{\hbar} |H^{(1)}_{nm}|^2 \, \rho(E_n) \int_{-\infty}^{\infty} \frac{\sin \omega_{nm} t}{\omega_{nm}} \, d\omega_{nm}. \qquad (8.76)$$

From Example 8.5, the last integral equals π. Substituting this into (8.76) yields the simple rate expression

$$\frac{dP}{dt} = \frac{2\pi}{\hbar} |H^{(1)}_{nm}|^2 \, \rho(E_n) \qquad (8.77)$$

Remember that P equals the probability of transition from the mth state to the band of states labeled by index n, or vice versa, in time t. So dP/dt equals the time rate for this transition, per unchanged submicroscopic system. Because of the fundmental role it plays in applications, formula (8.77) was called the *golden rule* of time-dependent perturbation theory by Enrico Fermi.

In Table 8.2, the basic relations for step-perturbation theory have been collected.

Table 8.2
Formulas for Transitions
Induced by a Steady Perturber

$$H^{(1)} = \eta(t) V^{(1)}(\mathbf{r}) \qquad\qquad \eta(t) = \text{unit step function}$$

$$b_j^{(1)} = \sum_{k' \neq j} -\frac{1}{\hbar} H^{(1)}_{jk'} \, b_{k'}(0) \, \frac{e^{i\omega_{jk'}t} - 1}{\omega_{jk'}}$$

When $b_m(0) = 1$ and $b_n(0) = 0$ for $n \neq m$,

$$P_{mn} = \frac{2}{\hbar^2} |\langle n | H^{(1)} | m \rangle|^2 \, \frac{1 - \cos \omega_{nm} t}{\omega_{nm}^2}$$

$$P_{mn} = P_{nm}$$

$$\rho(E_n) = \text{density over energy of states labeled by } n$$

$$P = \int P_{mn} \rho(E_n) \, dE_n$$

$$\frac{dP}{dt} = \frac{2\pi}{\hbar} |H^{(1)}_{nm}|^2 \, \rho(E_n)$$

Example 8.5

Evaluate the expression

$$\int_{-\infty}^{\infty} \frac{\sin x}{x}\, dx.$$

This integral forms an identifiable part of a line integral over a closed path in the complex plane. Cauchy's integral theorem enables us to relate the pertinent pieces.

We recall that a general complex number z can be represented as

$$z = x + iy = r\, e^{i\varphi}.$$

Thus, it can be plotted in the xy plane. Since

$$\frac{e^{ix}}{x} = \frac{\cos x + i \sin x}{x},$$

the quantity sought is the imaginary part of

$$\int_{-\infty}^{\infty} \frac{e^{iz}}{z}\, dz$$

evaluated along the x axis.

The integrand e^{iz}/z has one singularity, a pole at $z = 0$. To allow for it, we set

$$\int_{-\infty}^{\infty} \frac{e^{iz}}{z}\, dz = \lim_{\substack{r \to 0 \\ R \to \infty}} \left[\int_{-R}^{-r} \frac{e^{ix}}{x}\, dx + \int_{r}^{R} \frac{e^{ix}}{x}\, dx \right]$$

and investigate the expression in brackets.

The two integrals, before the limits are taken, are evaluated along the indicated parts of the x axis. To the two straight segments, we add semicircle C_1 passing from $-r$ to r and semicircle C_2 passing from R to $-R$ as Figure 8.3 shows.

Since the complete circuit encloses no singularities, Cauchy's theorem tells us that

$$\int_{-R}^{-r} \frac{e^{ix}}{x}\, dx + \int_{C_1} \frac{e^{iz}}{z}\, dz + \int_{r}^{R} \frac{e^{ix}}{x}\, dx + \int_{C_2} \frac{e^{iz}}{z}\, dz = \oint_{C} \frac{e^{iz}}{z}\, dz = 0.$$

Making r small enough causes

$$e^{iz} \simeq 1$$

and the integral around the small semicircle to become

$$\int_{C_1} \frac{e^{iz}\,dz}{z} \simeq \int_{\pi}^{0} \frac{1 \cdot ir\,e^{i\varphi}\,d\varphi}{r\,e^{i\varphi}} = i\varphi \Big|_{\pi}^{0} = -i\pi.$$

When R is made large enough, the integral around the large semicircle reduces to zero:

$$\int_{C_2} \frac{e^{iz}\,dz}{z} = \int_{0}^{\pi} e^{iR\,\exp\,i\varphi}\,i\,d\varphi = \int_{0}^{\pi} i\,e^{iR\,\cos\,\varphi}\,e^{-R\,\sin\,\varphi}\,d\varphi \simeq 0.$$

A rearrangement of the equation obtained from applying Cauchy's theorem is

$$\int_{-R}^{-r} \frac{e^{ix}}{x}\,dx + \int_{r}^{R} \frac{e^{ix}}{x}\,dx = -\int_{C_1} \frac{e^{iz}}{z}\,dz - \int_{C_2} \frac{e^{iz}}{z}\,dz.$$

Substituting in the limiting values on each side leads to

$$\int_{-\infty}^{\infty} \frac{e^{iz}}{z}\,dz = i\pi.$$

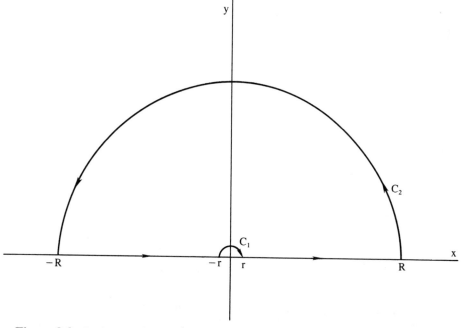

Figure 8.3. An integration path along the x axis that bypasses a singularity at the origin, together with a large semicircular completing path.

Since the desired integral is the imaginary part of this integral, we obtain

$$\int_{-\infty}^{\infty} \frac{\sin x}{x}\, dx = \pi.$$

Example 8.6

What is the density over momentum for freely translating particles?

We suppose that the particles are moving randomly with one per volume L^3, on the average. Also, we assume that conditions in a cubic region with edge length L are repeated periodically in adjacent blocks without end. Then each block edge is an integral number of de Broglie wavelengths.

Let us place the x axis along one edge. For this edge, we then have

$$L = n_x \lambda_x = n_x \frac{2\pi}{k_x} \qquad \text{with} \qquad n_x \text{ any integer,}$$

whence

$$\frac{k_x}{2\pi} = \frac{n_x}{L}.$$

Furthermore, de Broglie's relationship is

$$p_x = h\frac{k_x}{2\pi}.$$

Combining these equations yields

$$p_x = n_x \frac{h}{L} \qquad \text{with} \qquad n_x = \dots, -2, -1, 0, 1, 2, \dots .$$

Similarly,

$$p_y = n_y \frac{h}{L} \qquad \text{with} \qquad n_y = \dots, -2, -1, 0, 1, 2, \dots ,$$

$$p_z = n_z \frac{h}{L} \qquad \text{with} \qquad n_z = \dots, -2, -1, 0, 1, 2, \dots .$$

In a three-dimensional plot of momenta, each cubic section through which n_x increases by 1, n_y increases by 1, and n_z increases by 1 is assigned to an independent state. The size of the region allotted to a single state is thus

$$\frac{h}{L}\frac{h}{L}\frac{h}{L} = \frac{h^3}{L^3} .$$

The density over momentum equals the number of states per unit volume in the momentum plot. But this equals the reciprocal of the above expression:

$$\rho(\mathbf{p}) = \frac{L^3}{h^3} .$$

8.9
The Born Approximation for Scattering Processes

The scattering of particles from a uniform beam by small heterogeneities, such as atomic nuclei, is a steady-state process. Nevertheless, the deviations from homogeneity in the path can be considered as perturbers introduced at the zero of time.

Consider a macroscopically homogeneous beam of particles striking a macroscopically homogeneous target. Suppose that the submicroscopic nonhomogeneities in the target scatter the particles without transfer of energy. Let the potential representing a typical submicroscopic nonhomogeneity be $V(\mathbf{r})$. Let us introduce this potential as in (8.54). Furthermore, let us place the origin for $\mathbf{r}$ at the center of this nonhomogeneity.

If in the incident beam the volume per particle is L^3 and the wavevector is $\mathbf{k}_0$, the corresponding normalized state function is

$$\psi_m = \frac{1}{L^{3/2}} e^{i\mathbf{k}_0 \cdot \mathbf{r}} \tag{8.78}$$

as long as spin is neglected. Similarly, the normalized state function for particles scattered with wavevector $\mathbf{k}$, at the same density, is

$$\psi_n = \frac{1}{L^{3/2}} e^{i\mathbf{k} \cdot \mathbf{r}}, \tag{8.79}$$

again neglecting spin.

Since we are assuming that the particles do not lose or gain kinetic energy, the length of $\mathbf{k}$ equals the length of $\mathbf{k}_0$, as Figure 8.4 shows. As a result, the distance between the end points of the vectors is

$$K = 2k_0 \sin \tfrac{1}{2}\vartheta. \tag{8.80}$$

From Example 8.6, the density over momentum of the scattered particles is

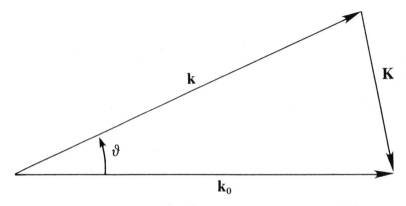

Figure 8.4. Wavevector $\mathbf{k}_0$ for an incident particle and wavevector $\mathbf{k}$ for a scattered particle drawn from the same initial point, with their difference $\mathbf{K}$.

$$\rho(\mathbf{p}) = \frac{L^3}{h^3} \, . \tag{8.81}$$

But since the particle momentum is

$$p = \frac{h}{2\pi} k \tag{8.82}$$

and the differential kinetic energy is

$$dE = d\left(\frac{p^2}{2\mu}\right) = \frac{\hbar^2 k}{\mu} dk, \tag{8.83}$$

we have

$$\rho(\mathbf{p}) \, d^3\mathbf{p} = \frac{L^3}{h^3} p^2 \sin\vartheta \, dp \, d\vartheta \, d\varphi = \frac{L^3}{(2\pi)^3} k^2 \sin\vartheta \, dk \, d\vartheta \, d\varphi$$

$$= \frac{\mu L^3}{(2\pi)^3 \hbar^2} k \sin\vartheta \, d\vartheta \, d\varphi \, dE_n = \rho(E_n) \, dE_n. \tag{8.84}$$

The density over energy is thus given by

$$\rho(E_n) = \frac{\mu L^3}{(2\pi)^3 \hbar^2} k \, d\Omega \tag{8.85}$$

where the element of solid angle is

$$d\Omega = \sin\vartheta \, d\vartheta \, d\varphi. \tag{8.86}$$

Substituting (8.78) and (8.79) into (8.57), with the assumed potential $V(\mathbf{r})$, yields the perturbation matrix element

$$H^{(1)}{}_{nm} = \frac{1}{L^3} \int V(\mathbf{r})\, e^{\,i(\mathbf{k}_0 - \mathbf{k})\cdot\mathbf{r}}\, d^3\mathbf{r} \tag{8.87}$$

when $t > 0$. The number of particles crossing unit area of the incident beam in unit time equals the distance a typical particle travels in unit time divided by the volume per particle:

$$I = \frac{v}{L^3} = \frac{\hbar k}{\mu L^3}. \tag{8.88}$$

The *differential cross section* $d\sigma$ is the fraction of the initial beam intensity scattered into element $d\Omega$ about the scattering center, by definition. The rate of scattering is given by (8.77), in the first-order approximation. Combining this with (8.87), (8.85), and (8.88) yields

$$d\sigma = \frac{\dfrac{2\pi}{\hbar}\dfrac{1}{L^6}\left| \int V(\mathbf{r})\, e^{\,i\mathbf{K}\cdot\mathbf{r}} d^3\mathbf{r} \right|^2 \dfrac{\mu L^3}{(2\pi)^3\hbar^2}\, k\, d\Omega}{\dfrac{\hbar k}{\mu L^3}}$$

$$= \left(\frac{\mu}{2\pi\hbar^2}\right)^2 \left| \int V(\mathbf{r})\, e^{\,i\mathbf{K}\cdot\mathbf{r}} d^3\mathbf{r} \right|^2 d\Omega. \tag{8.89}$$

Here μ is the mass of a typical particle in the beam, $\mathbf{K}$ the negative wavevector change $\mathbf{k}_0 - \mathbf{k}$, $V(\mathbf{r})$ the potential of the particle about a typical scattering center.

This calculation in which the potential of interaction between the incident particle and the struck nucleus is regarded as a perturbation carried only through first order is known as a *Born approximation*. It is accurate when the potential is relatively small, that is, when the incident velocity is large enough. Consequently, it supplements the method of partial waves, which is useful at low velocities, where only a few phase shifts are significant.

The magnitude of *scattering amplitude f* is defined by

$$d\sigma = |f|^2\, d\Omega. \tag{8.90}$$

Comparing (8.89) and (8.90), and choosing the phase so that a positive small potential reduces ψ appropriately around the origin, leads to

$$f = -\frac{\mu}{2\pi\hbar^2} \int V(\mathbf{r})\, e^{\,i\mathbf{K}\cdot\mathbf{r}}\, d^3\mathbf{r}. \tag{8.91}$$

For scattering in a given direction, vector $\mathbf{K}$ is fixed. Let us point the z axis from the center in this direction, so

$$e^{\,i\mathbf{K}\cdot\mathbf{r}} = e^{\,iKr\cos\vartheta}. \tag{8.92}$$

When the perturbation potential is spherically symmetric, the integration can be readily carried out over the solid angle about the center. For then, (8.91), with (8.92), becomes

$$f = -\frac{\mu}{2\pi\hbar^2} \int V(r)\, e^{\,iKr\,\cos\,\vartheta}\, r^2 \sin\vartheta\; d\vartheta\; d\varphi\; dr. \qquad (8.93)$$

Furthermore,

$$\int_0^{2\pi} d\varphi = 2\pi, \qquad (8.94)$$

$$\int_0^{\pi} e^{\,iKr\,\cos\,\vartheta} \sin\vartheta\; d\vartheta = -\frac{e^{\,iKr\,\cos\,\vartheta}}{iKr} \Bigg|_0^{\pi} = \frac{2\sin Kr}{Kr}. \qquad (8.95)$$

Combining these yields

$$f = -\frac{2\mu}{\hbar^2} \int V(r)\, \frac{\sin Kr}{Kr}\, r^2\, dr. \qquad (8.96)$$

Example 8.7

Calculate amplitude f for the scattering potential

$$V(r) = V_0 \frac{e^{-\alpha r}}{\alpha r}$$

in which V_0 and α are constants.

Substitute the given potential into (8.96):

$$f = -\frac{2\mu V_0}{\hbar^2 \alpha} \frac{1}{K} \int_0^\infty \sin Kr\, e^{-\alpha r}\, dr.$$

Carry out two integrations by parts on the integral,

$$\int_0^\infty \sin Kr\, e^{-\alpha r}\, dr = -\frac{K}{\alpha^2} \cos Kr\, e^{-\alpha r} \Bigg|_0^\infty - \frac{K^2}{\alpha^2} \int_0^\infty \sin Kr\, e^{-\alpha r}\, dr,$$

and rearrange:

$$\frac{1}{K} \int_0^\infty \sin Kr\, e^{-\alpha r}\, dr = \frac{1/\alpha^2}{(K^2/\alpha^2) + 1} = \frac{1}{K^2 + \alpha^2}.$$

With (8.80), the expression for the scattering amplitude reduces to

$$f = -\frac{2\mu V_0}{\hbar^2 \alpha} \frac{1}{K^2 + \alpha^2} = -\frac{2\mu V_0}{\hbar^2 \alpha} \frac{1}{4k_0^2 \sin^2 \tfrac{1}{2}\vartheta + \alpha^2}.$$

Example 8.8

Calculate the differential cross section for the scattering of point charge q_2 by point charge q_1.

The pertinent Coulomb field is obtained from the potential in Example 8.7 when

$$\frac{V_0}{\alpha} = \frac{q_1 q_2}{4\pi\epsilon_0}$$

and

$$\alpha \to 0.$$

But then, the square of the final expression in the example becomes

$$\frac{d\sigma}{d\Omega} = |f|^2 = \left(\frac{q_1 q_2}{4\pi\epsilon_0}\right)^2 \frac{4\mu^2}{\hbar^4 k_0^4} \frac{1}{(2 \sin \tfrac{1}{2}\vartheta)^4}.$$

Since

$$\frac{\hbar^2 k_0^2}{2\mu} = \frac{p^2}{2\mu} = E,$$

the kinetic energy of an incident particle, we also have

$$\frac{d\sigma}{d\Omega} = \left(\frac{q_1 q_2}{4\pi\epsilon_0 E}\right)^2 \frac{1}{(2 \sin \tfrac{1}{2}\vartheta)^4}.$$

This result agrees with the classical expression for Rutherford scattering.

Example 8.9

If the probability for a certain transition to occur to a particle per unit time is κ and if κ is fixed, what is the rate law for the transition?

Suppose that a time $t = 0$, there are N particles in the initial state in the given region. Then the number of particles undergoing the pertinent jump in time dt, on the average, is

$$\kappa N \, dt.$$

But if we treat the material as a continuum, disregarding its particle nature, the number transiting in time dt can be represented as

$$-dN.$$

Equating these expressions gives us

$$dN = -\kappa N \, dt$$

or

$$\frac{dN}{dt} = -\kappa N$$

for the rate law.

In some discussions, the rate per particle is expressed as the reciprocal of a time τ:

$$\kappa = \frac{1}{\tau}.$$

One can show that τ is the mean life for the initial state when one can neglect the reverse transition, as we have done here.

8.10
Beta Decay

The theory we have developed considers the evolution of an assemblage of submicroscopic systems from a given initial state to a band of states. Each elementary process may involve a change in the number of particles. Thus, a fermion may break down into a fermion and a boson. Or, instead of the boson a particle-antiparticle fermion pair may be produced.

A common process is

$$A \rightarrow C + d + \bar{b} \tag{8.97}$$

where A and C are relatively massive fermions, d is a light fermion, and $\bar{b}$ a light antifermion. The fermion corresponding to the antifermion, moving in such a way as to cancel the antifermion, can be added to both sides of the equation to yield

$$A + b \rightarrow C + d. \tag{8.98}$$

For example, a neutron in a nucleus can disintegrate to a proton in the nucleus, a beta particle, and an antineutrino:

$$n \rightarrow p + e^- + \bar{\nu}_e. \tag{8.99}$$

Adding a neutrino with the proper momentum to each side would convert (8.99) to form (8.98).

Let us suppose that the potential causing decay (8.99) is introduced at the zero of time. The subsequent decay rate is then given by (8.77) in the approximation that the first-order theory is adequate. We expect it to be because the interaction involved is known to be weak.

In the rest frame of the initial fermion, the momentum of the product fermion $\mathbf{P}$, the momentum of the product electron $\mathbf{p}$, and the momentum of the product antineutrino $\mathbf{q}$, add to zero:

$$\mathbf{P} + \mathbf{p} + \mathbf{q} = 0. \tag{8.100}$$

In this frame, the kinetic energies of the product fermion T, of the electron E, and of the antineutrino $E_{\bar{\nu}}$ add to give the energy of the final state E_0:

$$T + E + E_{\bar{\nu}} = E_0. \tag{8.101}$$

Since the energy need not be conserved in any individual decay, E_0 is not a fixed number. Indeed for (8.77), we need the density of states over this number.

From Example 8.6, the density in momentum space for a given particle is

$$\frac{L^3}{h^3} \tag{8.102}$$

if L^3 is the physical volume per particle and if spin is neglected. Consequently, the number of states for the beta particle appearing in solid angle $d\Omega$ about the source fermion in momentum interval dp is

$$\frac{L^3}{h^3} d^3\mathbf{p} = \frac{L^3}{h^3} p^2 \, d\Omega \, dp. \tag{8.103}$$

Integrating over the whole solid angle yields

$$\frac{4\pi p^2 L^3}{h^3} dp \tag{8.104}$$

Similarly,

$$\frac{4\pi q^2 L^3}{h^3} dq \qquad (8.105)$$

equals the number of states for the antineutrino in momentum interval dq, again neglecting spin.

We consider $\mathbf{p}$ and $\mathbf{q}$ to be independent, with (8.100) determining $\mathbf{P}$. Thus, the state of the proton depends on that of the electron and the antineutrino; there is no additional contribution to the number of states. From (8.104) and (8.105), the number of final states is

$$d^2N = \frac{4\pi p^2 L^3}{h^3} dp \, \frac{4\pi q^2 L^3}{h^3} dq = \frac{16\pi^2 L^6}{h^6} p^2 q^2 \, dp \, dq \qquad (8.106)$$

in momentum area $dp\, dq$.

For use in (8.77), this expression has to be integrated over all states consistent with the total energy E_0. But in (8.101), term T is negligible. Therefore,

$$E_{\bar\nu} = E_0 - E. \qquad (8.107)$$

Since the antineutrino travels at the speed of light, we have

$$E_{\bar\nu} = (mc)c = qc \qquad (8.108)$$

and

$$q^2 = \frac{(E_0 - E)^2}{c^2}, \qquad (8.109)$$

$$dq = \frac{dE_{\bar\nu}}{c} = \frac{dE_0}{c}. \qquad (8.110)$$

The last equality applies for each given p.

Substituting (8.109) and (8.110) into (8.106) yields

$$\frac{d^2N}{dE_0} = \frac{16\pi^2 L^6}{h^6 c^3} p^2 (E_0 - E)^2 \, dp, \qquad (8.111)$$

whence the density over energy for (8.77) is

$$\frac{dN}{dE_0} = \frac{16\pi^2 L^6}{h^6 c^3} \int p^2 (E_0 - E)^2 \, dp. \qquad (8.112)$$

Process (8.99) presumably involves a nucleon going from the neutron state to the proton state because of a weak-interaction potential $H^{(1)}$ between the initially virtual proton and electron. The energy loss is materialized as the e^- and $\bar\nu_e$ particles. However, the exact nature of the interaction potential is not known. For simplicity, we merely assume that $H^{(1)}$ is inversely proportional to the normalization volume for the beta particle. Then (8.26) becomes

$$H^{(1)}{}_{jk} = \frac{G}{L^3} \int \psi_p{}^* \psi_n \, d^3\mathbf{r} \equiv \frac{G}{L^3} M, \qquad (8.113)$$

in which ψ_p and ψ_n are the state functions for the final proton and the initial neutron. Parameter G is called the *coupling constant* for the process.

Substituting (8.111) and (8.113) into (8.77) yields

$$d\kappa = \frac{4\pi^2}{h} \frac{G^2}{L^6} \mid M \mid^2 \frac{16\pi^2 L^6}{h^6 c^3} p^2 (E_0 - E)^2 \, dp$$

$$= \mid M \mid^2 G^2 \frac{64\pi^4}{h^7 c^3} p^2 (E_0 - E)^2 \, dp \qquad (8.114)$$

for the differential transition rate. Integrating (8.114) or employing (8.112) instead of (8.111) gives us

$$\kappa = \frac{1}{\tau} = \mid M \mid^2 G^2 \frac{64\pi^4}{h^7 c^3} \int_0^{p_{\max}} p^2 (E_0 - E)^2 \, dp \qquad (8.115)$$

for the net transition rate.

Formula (8.115) implies that the radioactive law in Example 8.9 holds. Constant τ is the mean life for the decay process. Matrix element M is simply

$$M = \int \psi_p{}^* \psi_n \, d^3\mathbf{r} \qquad (8.116)$$

the overlap of the final spatial state function with the initial state function of the nucleon in the pertinent nucleus, h is Planck's constant, c the speed of light, p the momentum of the beta particle, and $E_0 - E$ the energy lost in forming the antineutrino.

Rate law (8.114) does not allow for electric interaction between the beta particle and the nucleus. However, the beta particle is retarded (or accelerated) by an amount dependent on the sign of its charge, its final momentum p, and the number of charges on the daughter nucleus Z. Consequently, expression (8.114) has to be multiplied by a dimensionless function

$$F(Z, p) \qquad (8.117)$$

which we will not determine. Integrating the result over all allowed momenta yields

$$\kappa = \frac{1}{\tau} = \mid M \mid^2 G^2 \frac{64\pi^4}{h^7 c^3} \int_0^{p_{\max}} F(Z, p) p^2 (E_0 - E)^2 \, dp. \qquad (8.118)$$

To reduce this integral to dimensionless form, one introduces the electron mass m and writes

$$p = p^1 mc, \tag{8.119}$$

$$E = E^1 mc^2, \tag{8.120}$$

$$F(Z, p) = F^1(Z, p^1). \tag{8.121}$$

Then (8.118) becomes

$$\kappa = \frac{1}{\tau} = |M|^2 G^2 \frac{64\pi^4 m^5 c^4}{h^7} f \tag{8.122}$$

where

$$f = \int_0^{p^1 \max} F^1(Z, p^1)(p^1)^2(E^1{}_0 - E^1)^2 \, dp^1. \tag{8.123}$$

Solving (8.122) for $f\tau$ yields

$$f\tau = \frac{h^7}{64 G^2 \pi^4 m^5 c^4} \; \frac{1}{|M|^2} . \tag{8.124}$$

Since the half life for the process is

$$t_{\frac{1}{2}} = (\ln 2)\tau, \tag{8.125}$$

we also have

$$ft_{\frac{1}{2}} = \frac{h^7 \ln 2}{64 G^2 \pi^4 m^5 c^4} \; \frac{1}{|M|^2} . \tag{8.126}$$

In allowed transitions, where

$$|M|^2 \simeq 1, \tag{8.127}$$

it is found that

$$\log ft_{\frac{1}{2}} \simeq 3\text{-}4, \tag{8.128}$$

when $t_{\frac{1}{2}}$ is in seconds. Each degree of forbiddenness tends to increase $\log ft_{\frac{1}{2}}$ by about $4\frac{1}{2}$.

Example 8.10

Estimate f for a high-energy beta decay.

At high energies the Coulomb factor F does not differ much from 1 and the energy follows the Einstein law

$$E \simeq pc.$$

So (8.123) becomes

$$f \simeq \int_0^{p_{max}} \frac{p^2}{m^2c^2} \frac{(E_0 - E)^2}{m^2c^4} \frac{dp}{mc}$$

$$= \frac{1}{(mc^2)^5} \int_0^{E_0} E^2(E_0 - E)^2 \, dE$$

$$= \frac{1}{(mc^2)^5} \left(\frac{1}{3} E_0^2 E^3 - \frac{1}{2} E_0 E^4 + \frac{1}{5} E^5 \right) \Bigg|_0^{E_0}$$

$$= \frac{E_0^5}{30(mc^2)^5} \, .$$

The rest energy of an electron is 0.511 MeV. So when E_0 is in MeV, we have

$$f \simeq \frac{E_0^5}{1.045} \, .$$

Example 8.11

If the $ft_{1/2}$ for the allowed beta decay of ^{14}O is 3.10×10^3s, what does coupling constant G equal?

The nucleus of ^{14}O consists of a relatively stable ^{12}C core plus two protons, either of which may decay. So the $ft_{1/2}$ for the system is one half what it would be for a single proton. And since the decay is allowed, we consider that $| M |^2$ is approximately 1.

Rearranging (8.126) and putting in the pertinent values leads to

$$G^2 = \frac{\hbar^7 \ln 2}{64\pi^4 m^5 c^4 ft_{1/2}} = \frac{2\pi^3 \hbar^7 \ln 2}{m^5 c^4 ft_{1/2}}$$

$$= \frac{2(3.1416)^3(1.0546 \times 10^{-34} \text{ J s})^7(0.69315)}{(9.1095 \times 10^{-31} \text{ kg})^5(2.9979 \times 10^8 \text{ m s}^{-1})^4(6.2 \times 10^3 \text{ s})}$$

$$= 1.98 \times 10^{-124} \text{ J}^2 \text{ m}^6,$$

whence

$$G = 1.4 \times 10^{-62} \text{ J m}^3.$$

8.11
Basic Ideas

A change in the environment to which a system is subjected alters the Hamiltonian governing the system.

If the change is sudden, there is not time for its ket to be altered appreciably. But in general, the set of eigenkets would be affected. A pure ket for the initial Hamiltonian would in general be a composite ket for the final Hamiltonian.

When the changes are relatively small, or at a sufficiently slow rate, the state may be expressed in terms of eigenkets of the unaffected system. The introduced change may be considered a perturbation.

Let the time-independent part of the Hamiltonian be $H^{(0)}$. The eigenvalue equation for the unaffected system is then

$$H^{(0)} \mid j \rangle \, e^{-i\omega_j t} = E_j \mid j \rangle \, e^{-i\omega_j t}, \tag{8.129}$$

where the exponential factor describes the temporal dependence of the eigenket.

We write the acting Hamiltonian in the form

$$H = H^{(0)} + \lambda H^{(1)}, \tag{8.130}$$

with $\lambda H^{(1)}$ the part varying with time, while λ is a perturbation parameter. We presume that the perturbation does not alter the Hilbert space for the system. Then the perturbed state may be represented by the superposition

$$\mid G \rangle = \Sigma \, c_k \mid k \rangle \, e^{-i\omega_k t}. \tag{8.131}$$

On combining these equations and reducing, we obtain the simultaneous equations

$$\dot{c}_j = \sum_k \frac{\lambda}{i\hbar} \, c_k \, H^{(1)}{}_{jk} \, e^{i(\omega_j - \omega_k)t}, \tag{8.132}$$

in which

$$H^{(1)}{}_{jk} = \langle \, j \mid H^{(1)} \mid k \, \rangle. \tag{8.133}$$

To improve the mathematical behavior of the set of equations, transfer the jth term in each to the left side and multiply both sides by a periodic exponential factor. Construct b_j as the product of c_j and this factor:

$$b_j = c_j \exp \left(\frac{i\lambda}{\hbar} \int_0^t H^{(1)}{}_{jj} \, dt \right). \tag{8.134}$$

Equation (8.132) now reduces to

$$\dot{b}_j = \sum_{k' \neq j} \frac{\lambda}{i\hbar} b_{k'} H^{(1)}{}_{jk'} e^{i\omega_{jk'}t}. \tag{8.135}$$

When the perturbing potential is small, we set

$$\omega_{jk'} = \omega_j - \omega_{k'}. \tag{8.136}$$

On the other hand, when the perturbing potential is constant, we have

$$\omega_{jk'} = \omega'_j - \omega'_{k'}. \tag{8.137}$$

For a perturbing potential depending only on the particle coordinates, but applied at time $t = 0$,

$$H^{(1)}{}_{jk} = \eta(t)(\text{constant}), \tag{8.138}$$

where $\eta(t)$ is the unit step function. Substituting into (8.135) and integrating to time t yields

$$b_j^{(1)} = \sum_{k' \neq j} -\frac{1}{\hbar} H^{(1)}{}_{jk'} b_{k'}(0) \frac{e^{i\omega_{jk'}t} - 1}{\omega_{jk'}}. \tag{8.139}$$

When only the mth state is occupied initially, we get

$$
\begin{aligned}
b_n^{(1)} &= -\frac{1}{\hbar} H^{(1)}{}_{nm} \frac{e^{\frac{1}{2}i\omega_{nm}t}(e^{\frac{1}{2}i\omega_{nm}t} - e^{-\frac{1}{2}i\omega_{nm}t})}{\omega_{nm}} \\
&= -\frac{2i}{\hbar} H^{(1)}{}_{nm} \frac{e^{\frac{1}{2}i\omega_{nm}t}}{\omega_{nm}} \sin \tfrac{1}{2}\omega_{nm}t.
\end{aligned} \tag{8.140}
$$

The probability that transition occurs in time t, from m to n, is

$$
\begin{aligned}
P_{mn} &= b_n{}^* b_n \simeq b_n^{(1)*} b_n^{(1)} \\
&= \frac{4 \,|\, H^{(1)}{}_{nm} \,|^2}{\hbar^2} \left(\frac{\sin \frac{1}{2}\omega_{nm}t}{\omega_{nm}} \right)^2 \\
&= \frac{2}{\hbar^2} |H^{(1)}{}_{nm}|^2 \frac{1 - \cos \omega_{nm}t}{\omega_{nm}{}^2}.
\end{aligned} \tag{8.141}
$$

Interchanging m and n does not alter this expression.

In practice, a person often considers transitions to or from a band of states. Consider the density of these states over energy to be $\rho(E_n)$. Replacing the nth state by this band of states and integrating (8.141) gives us

$$P = \frac{2}{\hbar^2} \int |H^{(1)}{}_{nm}|^2 \frac{1 - \cos \omega_{nm}t}{\omega_{nm}{}^2} \rho(E_n) \, dE_n \tag{8.142}$$

for the transition probability per system.

In the approximation that the band is uniform, $\rho(E_n)$ and $| H^{(1)}{}_{nm} |$ are constant. Differentiating (8.142) with respect to time and integrating then yields

$$\frac{dP}{dt} = \frac{2\pi}{\hbar} | H^{(1)}{}_{nm} |^2 \rho(E_n). \qquad (8.143)$$

Discussion Questions

8.1 What kinds of processes occur?

8.2 In which of these is the Hamiltonian independent of time?

8.3 How might a perturber alter the Hilbert space for a given system?

8.4 How may the same equation describe conditions before and after a jump transition?

8.5 How can the time-dependent Schrödinger equation describe transitions as continuous processes?

8.6 When is the temporal dependence not separable from the spatial dependence in Ψ?

8.7 When is the ket for a system unchanged by a time-dependent perturber?

8.8 How can a ket remain unchanged even though the system changes from a single eigenstate to a composite state?

8.9 When is the ket for a system a varying superposition of the eigenkets for a time-independent (invariant) Schrödinger equation?

8.10 How can the time-dependent Schrödinger equation, a partial differential equation, be replaced by a set of simultaneous ordinary differential equations?

8.11 What term causes difficulty in solving these equations? What transformation of the expansion coefficients serves to eliminate this term?

8.12 Why is

$$E'_j = \hbar\omega_j + \lambda H^{(1)}{}_{jj}$$

called the perturbed energy of the *j*th state? How can this energy vary with time?

8.13 When does

$$\exp\left[\frac{i}{\hbar} \int_0^t (E'_j - E'_k)\, dt \right]$$

reduce to the form

$$e^{i\omega_{jk}t}?$$

8.14 At what frequency does a system oscillate between two orthogonal states at the same energy, under the influence of a symmetric perturber, if no other states intrude?

8.15 When can the transformed expansion coefficients be represented by a perturbation expansion? What sequence of differential equations then results?

8.16 How do we consider a steady perturber to act? What behavior does such a perturber cause (a) over a short run, (b) over the long run?

8.17 Explain why the probability for transition from the mth state to the nth state equals the probability of transition from the nth state to the mth state.

8.18 How is it that the conservation-of-energy law is violated in individual transitions? In what way is energy conserved in an ensemble of transitions?

8.19 How can a given steady perturber cause transitions (a) from a certain state to various states in a band, (b) from various states in a band to a certain state, (c) from various states in one band to various states in another band?

8.20 When does

$$\frac{\sin \omega_{nm}t}{\omega_{nm}} \qquad \text{equal} \qquad \pi\delta(\omega_{nm})?$$

Why does

$$\int_{-\infty}^{\infty} \frac{\sin \omega_{nm}t}{\omega_{nm}} d\omega_{nm} \qquad \text{equal} \qquad \pi$$

even when

$$\frac{\sin \omega_{nm}t}{\omega_{nm}} \qquad \text{is not} \qquad \pi\delta(\omega_{nm})?$$

8.21 Why is

$$P = \int P_{mn}\rho(E_n)\, dE_n?$$

8.22 Interpret the golden rule of perturbation theory.

8.23 For the elastic scattering from a homogeneous beam, identify the definite state and the band of states.

8.24 How does one calculate the density over energy of the final states in the elastic scattering experiment?

8.25 How can the intensity in a homogeneous beam be described and expressed?

8.26 Explain the formula that describes the differential cross section.

8.27 What is the Born approximation? What cross-section expression does it yield?

8.28 What differential cross section does the scattering potential

$$V = V_0 \frac{e^{-\alpha r}}{\alpha r}$$

produce?

8.29 How can the cross section for Rutherford scattering be calculated using perturbation theory?

8.30 What is beta decay? How is weak interaction involved in such decay?

8.31 How is the density of final states found for beta decay?

8.32 How are observed beta-decay spectra explained?

8.33 How does perturbation theory uphold the accepted radioactive decay law?

8.34 How does $ft_{1/2}$ vary with the matrix element for the overlap between final and initial nuclear state functions?

Problems

8.1 When a tritium atom in its $1s$ state disintegrates into a helium ion, a beta particle, and an antineutrino, what is the probability that the He$^+$ appears in its $1s$ state?

8.2 A system initially in a single state is subjected to a perturbation from the zero of time on. This perturbation tends to cause the system to change to one other state. What is the probability that the system is in this second state at time t when the probability is still small?

8.3 A system is modeled by a particle of mass μ and charge q moving in the parabolic potential

$$U = \tfrac{1}{2}fx^2 = \tfrac{1}{2}\mu\omega^2 x^2.$$

It is perturbed by the electric field

$$\frac{A}{\sqrt{\pi}\,\tau}\,e^{-(t/\tau)^2}$$

directed along the axis of motion. Determine the probability that the electric field causes transition from the ground state to the first excited state of the oscillator.

8.4 A perturbation potential acts only within the short interval $t_1 < t < t_2$ on a system initially in the mth state. If Equation (8.40) is valid for the nth state, how is coefficient b_n related to the rate of change of $H^{(1)}{}_{nm}$?

8.5 In a scattering experiment, the incident beam is represented by

$$\Psi_i = A\,e^{\,ikz}\,e^{-i\omega t}$$

and the part scattered into angle $d\Omega$ by

$$\Psi_s = A\,\frac{f}{r}\,e^{\,ikr}\,e^{-i\omega t}.$$

Show that

$$d\sigma = |f|^2\,d\Omega.$$

8.6 What is the differential cross section for scattering of particles by the field

$$V = \frac{A}{r^2}$$

in the Born approximation?

8.7 Show that the total cross section obtained from the differential cross section for Coulomb scattering diverges.

8.8 For each of the following positron emitters, estimate the $ft_{1/2}$ product.

Nuclide	^{11}C	^{15}O	^{19}Ne	^{23}Mg	^{27}Si	^{31}S	^{35}Ar
E_0, MeV	0.98	1.70	2.24	3.10	3.85	4.39	4.93
$t_{1/2}$, s	1218	124	17.5	12.1	4.2	2.7	1.83

Are the matrix elements for these transitions similar?

– – –

8.9 A ^{6}He atom in its ground state disintegrates into a ^{6}Li ion, a beta particle, and an antineutrino. What is the probability that the Li$^+$ is produced in the $(1s)^2$ state?

8.10 A system initially in one state is subjected to a perturbation that tends to shift it into a second state, with

$$H^{(1)}{}_{12} = A \text{ sech } Bt.$$

If condition (8.34) applies and $\omega_{21} = \omega_2 - \omega_1$ is small, what is the probability that the system eventually reaches the second state?

8.11 A harmonic oscillator of angular frequency ω is modeled by a particle of mass μ and charge q moving along the x axis. Calculate the probability that the electric field

$$\frac{A}{\sqrt{\pi}\,\tau}\, e^{-(t/\tau)^2}$$

directed along the x axis raises the vibrational quantum number v from 1 to 2.

8.12 A hydrogen atom is perturbed by the electric field

$$\frac{A}{e\pi}\frac{\tau}{\tau^2+t^2},$$

directed along the z axis. Determine the probability that this field causes transition from the 1s to the pertinent 2p state.

8.13 Employ the Born approximation in calculating the differential and total scattering cross sections for the potential

$$V(r) = V_0 e^{-\alpha^2 r^2}.$$

8.14 What are the differential and total cross sections for scattering of particles by the field

$$V(r) = V_0 e^{-\alpha r}$$

in the Born approximation?

8.15 What is the differential cross section of a spherical well for which

$$V(r) = -V_0 \qquad \text{when} \qquad 0 \le r \le a$$

and

$$V(r) = 0 \qquad \text{when} \qquad r > a?$$

References

Books

Bates, D. R.: 1961, "Transitions," in Bates, D. R. (editor), *Quantum Theory*, vol. I, Academic Press, New York, pp. 251–297. Standard topics are covered in a terse but useful manner. Besides the various kinds of time-dependent perturbations, detailed balancing is discussed.

Bohm, D.: 1951, *Quantum Theory*, Prentice-Hall, New York, pp. 407–417, 448–453, 533–541. Bohm develops and discusses the common time-dependent perturbation methods.

Cohen-Tannoudji, C., Diu, B., and Laloe, F. (translated by Hemley, S. R., Ostrowsky, N., and Ostrowsky, D.): 1977, *Quantum Mechanics*, Wiley, New York, pp. 1285–1302. The basic theory for time-dependent problems is here treated, together with very useful comments.

Articles

Aub, M. R., and Niederreiter, H.: 1983, "A Criterion for Stationary States in Quantum Mechanics," *Am. J. Phys.* **51**, 818–819.

Berman, P. R.: 1974, "Two-Level Approximation in Atomic Systems," *Am. J. Phys.* **42**, 992–997.

Castano, F., Lain, L., Sanchez Rayo, M. N., and Torre, A.: 1983, "Does Quantum Mechanics Apply to One or Many Particles?," *J. Chem. Educ.* **60**, 377–378.

Forst, W.: 1989, "Collisional Relaxation via Eigenfunction-Eigenvalue Expansion," *J. Chem. Educ.* **66**, 142–146.

Macomber, J. D.: 1977, "Quantum Transitions without Quantum Jumps," *Am. J. Phys.* **45**, 522–532.

Moretti, P.: 1988, "The Calculation of Second-Order Terms in Time-Dependent Perturbation Theory," *Am. J. Phys.* **56**, 172–174.

Moretti, P.: 1989, "Complex Analysis and Quantum Mechanics: A Perturbative Approach for the Evolution Operator," *Am. J. Phys.* **57**, 77–78.

Morris, J. R.: 1988, "Boundary-Induced Perturbations," *Am. J. Phys.* **56**, 49–51.

Robinson, E. J.: 1984, "Nonobservability of Early-Time Departures from Fermi's 'Golden Rule,'" *Phys. Rev. Lett.* **52**, 2309–2312.

Venanzi, T. J.: 1983, "Time-Dependent Shifts in the Schrödinger Equation," *Am. J. Phys.* **51**, 624–627.

9

ABSORPTION, EMISSION, AND SPONTANEOUS DECAY

9.1
Interaction with the Electromagnetic Field

*I*n applying quantum mechanics, a person first considers systems isolated from their surroundings. The appropriate Hamiltonian operator may be constructed and the eigenvalue equation solved. One finds that each isolated system exists in various states. Some of the states are discrete; others form a continuum.

However, knowledge of the possible states for a given system comes from its interaction with the outside. Such intervention involves additional terms in the Hamiltonian operator. When the interaction is not too strong, these terms can be considered as perturbers. For convenience they may be introduced at a given point in time. They then induce transitions among the isolated-system states.

In this chapter we will consider the interaction between the charged particles of a given system and an inposed electromagnetic wave. The interaction induces changes in the system and in the wave. We will also consider the situation when no real wave is present at time $t = 0$. The potentiality of creating a wave in the field has then to be considered.

Indeed, this potentiality allows us to honor the principle of detailed balance.

In each elementary process the system goes stepwise from one state to another. The electromagnetic field apparently receives or donates one or more photons. Thus, the states of both the system and the wave are quantized. This quantization will be taken into account in calculating the pertinent density of states over energy.

But in constructing the perturber for the Hamiltonian operator, we will consider the wave to be Maxwellian, that is, classical. The variations in such a wave can be analyzed into orthogonal polarized monochromatic components.

Each such component is found to propagate independently through free space. Furthermore, each component interacts independently with the submicroscopic system. As a consequence the resolution into polarized single frequency components is physically significant.

9.2
Hamiltonian Operator for a Charged Particle

A charged particle interacts with both the scalar potential and the vector potential of the Maxwellian field, so both potentials need to appear in the Hamiltonian operator. Nearby electric charges act through the scalar potential and the vector potential, nearby spins through the vector potential. The far radiation field is transverse and varying. In the classical approximation, it is governed by a varying transverse $\mathbf{A}$.

When a given particle has mass μ and charge q, the Lorentz force law states that

$$\mu \frac{d\mathbf{v}}{dt} = q(\mathbf{E} + \mathbf{v} \times \mathbf{B}) \tag{9.1}$$

with

$$\mathbf{E} = -\nabla\phi - \frac{\partial \mathbf{A}}{\partial t}, \tag{9.2}$$

$$\mathbf{B} = \nabla \times \mathbf{A}. \tag{9.3}$$

In quantum mechanics, these formulas have only statistical validity.

From (3.77), the Hamiltonian function has the form

$$H = \frac{1}{2\mu} (\mathbf{P} - q\mathbf{A}) \cdot (\mathbf{P} - q\mathbf{A}) + q\phi$$

$$= \frac{1}{2\mu} (\mathbf{P} \cdot \mathbf{P} - q\mathbf{P} \cdot \mathbf{A} - q\mathbf{A} \cdot \mathbf{P} + q^2\mathbf{A} \cdot \mathbf{A}) + q\phi. \tag{9.4}$$

From (3.79), the operator for momentum $\mathbf{P}$ is

$$\mathbf{P} = \frac{\hbar}{i} \nabla. \tag{9.5}$$

So

$$\mathbf{P} \cdot \mathbf{P} = -\hbar^2 \nabla^2. \tag{9.6}$$

Being a function of coordinates and time only, the vector $\mathbf{A}$ acting on the charged particle serves as is.

Substituting into (9.4) yields

$$H = \frac{1}{2\mu} [-\hbar^2 \nabla^2 + i\hbar q(\nabla \cdot \mathbf{A} + \mathbf{A} \cdot \nabla) + q^2 A^2] + q\phi. \tag{9.7}$$

But we have

$$\nabla \cdot \mathbf{A}\Psi = \nabla \cdot \underline{\mathbf{A}}\Psi + \nabla \cdot \mathbf{A}\underline{\Psi} = \Psi\nabla \cdot \mathbf{A} + \mathbf{A} \cdot \nabla\Psi, \tag{9.8}$$

where each underline indicates the factor on which the differentiator in the del acts. Consequently, (9.7) reduces to

$$H = \frac{1}{2\mu} [-\hbar^2 \nabla^2 + i\hbar q(\nabla \cdot \mathbf{A}) + 2i\hbar q\mathbf{A} \cdot \nabla + q^2 A^2] + q\phi. \tag{9.9}$$

In the unperturbed part of the Hamiltonian, $H^{(0)}$, we will include the effects of the other charges and spins nearby, those in the molecule, ion, or nucleus under consideration. The perturber $H^{(1)}$ will then consist of the radiation field's contribution to the second, third, and fourth terms on the right of (9.9).

9.3
Matrix Element for Perturbation by Monochromatic Interaction

Following the classical description, we consider the far radiation field to be governed by a varying transverse vector potential. Because such a field obeys the superposition principle, one may analyze it into sinusoidal polarized components; each of these acts on the charged particle. In perturbation theory, the effect is mediated principally through one matrix element and its complex conjugate, which we will now develop.

A typical component of the vector potential, moving along the z axis in the positive or negative direction, with its polarization in the direction of the x axis, is

$$\mathbf{A} = A_x^0 \cos (kz \mp \omega t + a) \mathbf{x}. \tag{9.10}$$

Here A_x^0 is the amplitude of the component, k its wavevector, ω its angular frequency, and a the phase angle when z and t are zero.

If c is the speed at which a given phase of the wave moves, the phase initially at the origin appears at $z = \pm ct$ at time t. The argument of the cosine must still be a, the first two terms $k(\pm ct) \mp \omega t$ add to zero, and

$$k = \frac{\omega}{c}. \tag{9.11}$$

Let us introduce (9.11) into (9.10) with the wave traveling in the positive direction and the zero of time chosen so that a is zero:

$$\mathbf{A} = A_x^0 \cos \left(\frac{\omega}{c} z - \omega t \right) \hat{\mathbf{x}} = A_x^0 \cos \left[\omega \left(t - \frac{z}{c} \right) \right] \hat{\mathbf{x}}. \qquad (9.12)$$

Since vector $\mathbf{A}$ does not vary in the direction in which it points, the direction of $\mathbf{x}$, we have

$$\nabla \cdot \mathbf{A} = 0. \qquad (9.13)$$

Wherever the greatest amplitude of this component is not too large, we also have

$$q^2 A^2 \simeq 0. \qquad (9.14)$$

Then (9.9) yields the perturbation operator

$$H^{(1)} = \frac{i\hbar q}{\mu} \mathbf{A} \cdot \nabla$$

$$= \frac{i\hbar q A_x^0}{\mu} \cos \left[\omega \left(t - \frac{z}{c} \right) \right] \frac{\partial}{\partial x}$$

$$= \frac{i\hbar q A_x^0}{2\mu} \left\{ \exp \left[i\omega \left(t - \frac{z}{c} \right) \right] + \exp \left[-i\omega \left(t - \frac{z}{c} \right) \right] \right\} \frac{\partial}{\partial x}. \quad (9.15)$$

The right side of (9.15) has the form

$$H^{(1)} = F\, e^{i\omega t} + G\, e^{-i\omega t} \qquad (9.16)$$

with

$$F = \frac{i\hbar q A_x^0}{2\mu}\, e^{-i\omega z/c}\, \frac{\partial}{\partial x}, \qquad (9.17)$$

$$G = \frac{i\hbar q A_x^0}{2\mu}\, e^{i\omega z/c}\, \frac{\partial}{\partial x}. \qquad (9.18)$$

Operators (9.17) and (9.18) are related by the equation

$$G_{jk} = F_{kj}^*. \qquad (9.19)$$

With (9.17), we construct the matrix element

$$F_{nm} = \frac{i\hbar q A_x^0}{2\mu} \left\langle n \,\middle|\, e^{-i\omega z/c}\, \frac{\partial}{\partial x} \,\middle|\, m \right\rangle$$

$$\simeq \frac{i\hbar q A_x^0}{2\mu}\, e^{-i\omega z/c} \left\langle n \,\middle|\, \frac{\partial}{\partial x} \,\middle|\, m \right\rangle. \qquad (9.20)$$

We treat (9.18) similarly. For the approximations in the last step to be valid, the wavelength of the radiation must be considerably larger than the effective range of the eigenkets and $\langle n \,|\, \partial/\partial x \,|\, m \rangle$ should differ from zero.

9.4
The Density in Energy
of Electromagnetic States in a Band

In our development we will also need an expression for $\rho(E_n)$, the density over energy E_n of the pertinent composite states. Quantity E_n consists of the energy for system ket $|n\rangle$ plus the photon energy $\hbar\omega$ associated with a constituent state of the radiation field.

A typical polarized monochromatic component of the radiation field is governed by (9.12). The corresponding electric and magnetic intensities are derived using (9.2) and (9.3). Thus,

$$\mathbf{E} = -\frac{\partial \mathbf{A}}{\partial t} = A_x{}^0\omega \sin\left[\omega\left(t - \frac{z}{c}\right)\right]\hat{\mathbf{x}} \tag{9.21}$$

and

$$\mathbf{H} = c^2\epsilon_0\mathbf{B} = c^2\epsilon_0 \, \nabla \times \mathbf{A}$$

$$= c^2\epsilon_0 \begin{vmatrix} \hat{\mathbf{x}} & \hat{\mathbf{y}} & \hat{\mathbf{z}} \\ \dfrac{\partial}{\partial x} & \dfrac{\partial}{\partial y} & \dfrac{\partial}{\partial z} \\ A_x{}^0 \cos\left[\omega\left(t - \dfrac{z}{c}\right)\right] & 0 & 0 \end{vmatrix}$$

$$= c\epsilon_0 A_x{}^0\omega \sin\left[\omega\left(t - \frac{z}{c}\right)\right]\hat{\mathbf{y}}. \tag{9.22}$$

A conventional treatment of the rate at which energy is stored in a typical small region of a field tells us that the Poynting vector

$$\mathbf{P} = \mathbf{E} \times \mathbf{H} \tag{9.23}$$

gives the rate at which energy travels by a given point per unit cross section. Substituting (9.21), (9.22) into (9.23) leads to

$$\mathbf{P} = c(A_x{}^0\omega)^2\epsilon_0 \sin^2\left[\omega\left(t - \frac{z}{c}\right)\right]\hat{\mathbf{z}}. \tag{9.24}$$

Since the average value of the sine squared over an integral number of cycles is ½, the mean rate of energy flow per unit cross section is

$$\bar{P} = \frac{c}{2}(A_x{}^0\omega)^2\,\epsilon_0. \tag{9.25}$$

Since the component wave moves distance c in unit time, the corresponding energy density is

$$\tfrac{1}{2}(A_x{}^0\omega)^2\,\epsilon_0. \tag{9.26}$$

Associated with the composite wave are photons of differing frequencies. For these, consider a unit volume centered on the given submicroscopic system. Also, let

$$\rho(E)\,\mathrm{d}E \tag{9.27}$$

be the number of states in the photon-energy interval

$$\mathrm{d}E = h\,\mathrm{d}\nu \tag{9.28}$$

in the unit physical volume. Then if quantity

$$u_x(\nu)\mathrm{d}\nu \; = \; u_x(\nu)\,\frac{\mathrm{d}E}{h} \tag{9.29}$$

is the radiation-energy density in physical space over the frequency interval $\mathrm{d}\nu$, and over the corresponding energy interval $\mathrm{d}E$, we have

$$\frac{u_x(\nu)}{h}\,\mathrm{d}E \; = \; \tfrac{1}{2}(A_x{}^0\omega)^2\,\epsilon_0\rho(E)\,\mathrm{d}E, \tag{9.30}$$

whence

$$\rho(E) \; = \; \frac{2\,u_x(\nu)}{(A_x{}^0\omega)^2\,\epsilon_0 h}\;. \tag{9.31}$$

9.5
Oscillatory Perturbers

The next most simple perturber beyond the steady one of Section 8.7 is a perturber that varies sinusoidally at a single frequency after it has been turned on.

As in Chapter 8, we consider a system governed by the Schrödinger equation

$$H^{(0)}\,|\,j\rangle\,e^{-i\omega_j t} \; = \; E_j\,|\,j\rangle\,e^{-i\omega_j t} \tag{9.32}$$

in absence of the perturbation. The complete system is governed by the Hamiltonian

$$H = H^{(0)} + \lambda H^{(1)} \tag{9.33}$$

in which $H^{(0)}$ does not vary with time, λ is the parameter that controls the amount of perturbation, and $H^{(1)}$ varies with time. The evolution of the complete system follows Equation (8.1):

$$H\,|\,I\rangle \; = \; -\,\frac{\hbar}{i}\,\frac{\partial}{\partial t}\,|\,I\rangle. \tag{9.34}$$

We let

$$H^{(1)} = 0 \qquad \text{as long as time } t \text{ is negative} \qquad (9.35)$$

and

$$H^{(1)} = F\, e^{i\omega t} + G\, e^{-i\omega t} \qquad \text{when time } t \text{ is positive,} \qquad (9.36)$$

as Section 9.3 details.

Coefficients F and G are operators that do not vary with time. But for the perturber to be Hermitian,

$$H^{(1)}{}_{jk} \equiv \langle j | H^{(1)} | k \rangle = \langle k | H^{(1)} | j \rangle *$$
$$\equiv H^{(1)}{}_{kj}*, \qquad (9.37)$$

the matrix elements of these coefficients must be related in the manner

$$G_{jk} = F_{kj}*, \qquad (9.38)$$

as we saw in (9.19). Then matrix element (8.57) becomes

$$H^{(1)}{}_{jk} = \eta(t)\, (F_{jk} e^{i\omega t} + F_{kj}* e^{-i\omega t}) \qquad (9.39)$$

with $\eta(t)$ the unit step function, as before.

As long as the perturbation does not alter the space of the kets, solutions of Equation (9.32) superpose to yield the appropriate variable ket $| I \rangle$ for (9.34). The coefficient of each $| j \rangle$ in the superposition is determined by Equation (9.34) together with the initial conditions. Transformation (8.32) replaces each coefficient c_j with b_j and simplifies the governing simultaneous equations. However, this transformation does not alter the expression for the weight of a state; indeed,

$$w_j = c_j* c_j = b_j* b_j. \qquad (9.40)$$

Because of the cyclic nature of the perturber, approximation (8.34) may be introduced; then (8.35) holds,

$$\omega_{jk'} = \omega_j - \omega_{k'}, \qquad (9.41)$$

and Equations (8.41) apply. If increasing λ from zero to its final value does not introduce abrupt changes, we may also consider that

$$b_j = b_j^{(0)} + \lambda b_j^{(1)} + \lambda^2 b_j^{(2)} + \ldots \qquad (9.42)$$

as in (8.44). The simultaneous differential equations then yield (8.51).

Substituting (9.39) into (8.51) leads to the equation

$$\dot{b}_j^{(1)} = \sum_{k' \neq j} \frac{\eta(t)}{i\hbar}\, b_{k'}(0)\, (F_{jk'} e^{i(\omega_{jk'} + \omega)t} + F_{k'j}* e^{i(\omega_{jk'} - \omega)t}) \qquad (9.43)$$

which integrates, with (8.58), to

$$b_j^{(1)} = \sum_{k' \neq j} - b_{k'}(0) \left(F_{jk'} \frac{e^{i(\omega_{jk'} + \omega)t} - 1}{\hbar(\omega_{jk'} + \omega)} + F_{k'j}^* \frac{e^{i(\omega_{jk'} - \omega)t} - 1}{\hbar(\omega_{jk'} - \omega)} \right). \quad (9.44)$$

When angular frequency ω is not too small, the term in which the magnitude of the denominator is smaller predominates.

If we also assume that only the mth state is occupied initially and that the perturbation parameter equals 1, then (9.44) yields

$$b_n \simeq b_n^{(1)} = -\frac{1}{\hbar} F_{mn}^* \frac{e^{\frac{1}{2}i(\omega_{nm} - \omega)t} \left(e^{\frac{1}{2}i(\omega_{nm} - \omega)t} - e^{-\frac{1}{2}i(\omega_{nm} - \omega)t} \right)}{\omega_{nm} - \omega}$$

$$= -\frac{2i}{\hbar} F_{mn}^* \frac{e^{\frac{1}{2}i(\omega_{nm} - \omega)t}}{\omega_{nm} - \omega} \sin \frac{1}{2}(\omega_{nm} - \omega)t \quad (9.45)$$

when ω_{nm} is positive and

$$b_n \simeq -\frac{2i}{\hbar} F_{nm} \frac{e^{\frac{1}{2}i(\omega_{nm} + \omega)t}}{\omega_{nm} + \omega} \sin \frac{1}{2}(\omega_{nm} + \omega)t \quad (9.46)$$

when ω_{nm} is negative.

The probability that transition from the mth to the nth state occurs in time t is given by

$$P_{mn} = b_n^* b_n = \frac{4|F_{mn}|^2}{\hbar^2} \left[\frac{\sin \frac{1}{2}(\omega_{nm} - \omega)t}{\omega_{nm} - \omega} \right]^2$$

$$= 2|F_{mn}|^2 \frac{1 - \cos(\omega_{nm} - \omega)t}{\hbar^2(\omega_{nm} - \omega)^2} \quad (9.47)$$

when

$$\omega_{nm} - \omega \ll \omega_{nm} + \omega \quad (\omega_{nm} \text{ pos.}) \quad (9.48)$$

and

$$P_{mn} = \frac{4|F_{nm}|^2}{\hbar^2} \left[\frac{\sin \frac{1}{2}(\omega_{nm} + \omega)t}{\omega_{nm} + \omega} \right]^2$$

$$= 2|F_{nm}|^2 \frac{1 - \cos(\omega_{nm} + \omega)t}{\hbar^2(\omega_{nm} + \omega)^2} \quad (9.49)$$

when

$$|\omega_{nm} + \omega| \ll |\omega_{nm} - \omega| \quad (\omega_{nm} \text{ neg.}) \quad (9.50)$$

Whenever operator F is Hermitian, we also have

$$|F_{mn}|^2 = |F_{nm}|^2. \quad (9.51)$$

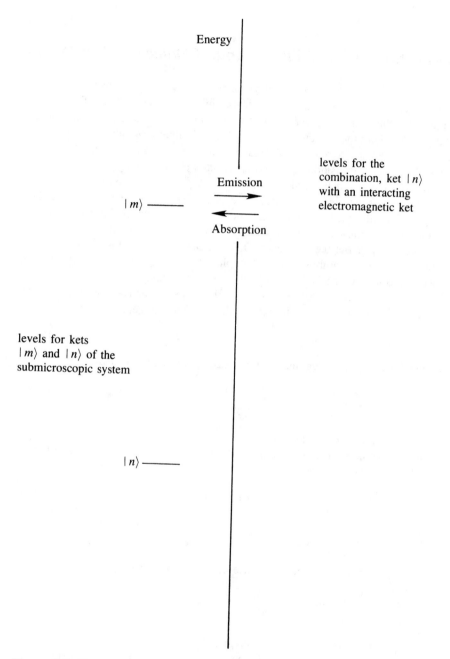

Figure 9.1. Energies of the states involved in electromagnetically induced transitions between kets $|m\rangle$ and $|n\rangle$.

9.6
Transitions To or From a Band of Molecule-Field States

Each state of a system of charged particles works with the electromagnetic field to form combination states extending over the higher, accessible, states of the system. Figure 9.1 shows how the energy of ket $|n\rangle$ plus that of each significant photon state cluster about the energy of ket $|m\rangle$. The field interacts with the charged particles in the bare state and in the corresponding combination states to cause transitions. The rate of transition in either direction depends directly on the density of states about the pertinent level.

Suppose that the given submicroscopic system is initially or finally in ket $|m\rangle$ as Figure 9.1 indicates. Let a lower ket into which or from which the system evolves be labeled $|n\rangle$. This lower particle state combines with photon states to form a band of density $\rho(E_n)$ about the mth level. Quantity $\rho(E_n)$ is defined as in (8.72), with E_n the energy of the combination, system plus electromagnetic state. Furthermore, consider the perturber to be the $H^{(1)}$ of (9.35) and (9.36).

In general, whether result (9.47) or (9.49) is used depends on whether the mth level is lower or higher than the nth level. For the arrangement in Figure 9.1,

$$\omega_{nm} = \omega_n - \omega_m \simeq -\omega, \tag{9.52}$$

formula (9.49) applies. Integrating this expression over the band yields the probability

$$P = \frac{2}{\hbar^2} \int_{\text{band}} |F_{nm}|^2 \frac{1 - \cos(\omega_{nm} + \omega)t}{(\omega_{nm} + \omega)^2} \rho(E_n)\, dE_n \tag{9.53}$$

for the transition to occur in time t.

Since E_n is the energy of the combination, system plus photon state, and since ω_m is here constant, we have

$$E_n = \hbar\omega_n + \hbar\omega \tag{9.54}$$

and

$$dE_n = \hbar\, d(\omega_n + \omega) = \hbar\, d(\omega_n - \omega_m + \omega) = \hbar\, d(\omega_{nm} + \omega). \tag{9.55}$$

Substituting (9.55) into (9.53) and differentiating with respect to time leads to

$$\frac{dP}{dt} = \frac{2}{\hbar} \int_{\text{band}} |F_{nm}|^2 \frac{\sin(\omega_{nm} + \omega)t}{\omega_{nm} + \omega} \rho(E_n)\, d(\omega_{nm} + \omega). \tag{9.56}$$

The principal contribution to the integral occurs in a very narrow region where approximation (9.52) is good. If the band is uniform in this region, expressions $|F_{nm}|$ and $\rho(E_n)$ are approximately constant there and can be factored out of the integral. Furthermore, the limits of integration can be extended to $-\infty$ and $+\infty$ without introducing appreciable error. Then letting

$$\omega_{nm} + \omega = 0, \tag{9.57}$$

we obtain

$$\frac{dP}{dt} = \frac{2}{\hbar} \mid F_{nm} \mid^2 \rho(E_n) \int_{-\infty}^{\infty} \frac{\sin \theta t}{\theta} \, d\theta$$

$$= \frac{2\pi}{\hbar} \mid F_{nm} \mid^2 \rho(E_n). \tag{9.58}$$

In the last step, the result from Example 8.5 has been introduced. Here dP/dt is the time rate of transition, F_{nm} is the matrix element $\langle n \mid F \mid m \rangle$, and $\rho(E_n)$ is the density of the pertinent combination states over the energy E_n.

Note that Equation (9.58) is similar to (8.77); so it is a form of the golden rule of Fermi. In Table 9.1, key relations leading up to this equation have been collected.

Table 9.1
Formulas for Transitions
Induced by a Sinusoidal Perturber

$$H^{(1)}{}_{jk} = \eta(t)(F_{jk}e^{i\omega t} + F_{kj}{}^* e^{-i\omega t})$$

$$\eta(t) = \text{unit step function}$$

$$\omega = \text{photon angular frequency}$$

$$\omega_m = \text{angular frequency for ket } \mid m \rangle$$

$$\omega_n = \text{angular frequency for ket } \mid n \rangle$$

$$\omega_{nm} = \omega_n - \omega_m$$

$$P_{mn} = \frac{2}{\hbar^2} \mid F_{nm} \mid^2 \frac{1 - \cos(\omega_{nm} - \omega)t}{(\omega_{nm} - \omega)^2}$$

$$\rho(E_n) = \text{density of combination states}$$

$$\text{for } \hbar\omega_n + \hbar\omega = E_n$$

$$\frac{dP}{dt} = \frac{2\pi}{\hbar} \mid F_{nm} \mid^2 \rho(E_n)$$

Example 9.1

What dimensions appear in formula (9.58)?

Constant $\hbar$ may be expressed as energy $\times$ time. Since F is measured in energy units, matrix element F_{nm} over the particle system is also. Furthermore, density $\rho(E_n)$ is in states per unit energy for the compound, system plus field. On the other hand, number 2π is dimensionless. Combining these as in (9.58) yields

$$\left[\frac{dP}{dt} \right] = \frac{1}{\text{energy} \times \text{time}} \text{ energy}^2 \; \frac{1}{\text{energy}} = \frac{1}{\text{time}} \; .$$

Multiplying this rate per particle by the number of reactant (absorbing or emitting) particles exposed to the field per unit volume gives us the rate of change in the concentration dC/dt. The dimensions are then

$$\left[\frac{dC}{dt} \right] = \frac{1}{\text{time}} \; \frac{\text{particles}}{\text{volume}} = \frac{\text{concentration}}{\text{time}} \; .$$

9.7
Rates of Induced Absorption or Emission

A band of radiation induces processes of absorption or emission at rates that we can now determine.

When polarized electromagnetic radiation travels in one direction without diverging or converging, the state density over energy is given by formula (9.31). Matrix element F_{nm} is given by (9.20). Substituting these into (9.58) yields the temporal rate of transition

$$\frac{dP}{dt} = \frac{2\pi}{\hbar} \left(\frac{\hbar q A_x^0}{2\mu} \right)^2 \left| \langle n | \frac{\partial}{\partial x} | m \rangle \right|^2 \frac{2u_x(\nu)}{(A_x^0 \omega)^2 \, \epsilon_0 \hbar}$$

$$= \frac{q^2}{2\epsilon_0 \mu^2 \omega^2} \left| \langle n | \frac{\partial}{\partial x} | m \rangle \right|^2 u_x(\nu). \tag{9.59}$$

Here q is the charge on the interacting particle, μ its mass, ϵ_0 the permittivity of space, ω the angular frequency of the radiation, $u_x(\nu) \, d\nu$ the energy density in physical space of the radiation, while the x axis points in the direction of polarization and the z axis points in the direction of propagation.

On multiplying both numerator and denominator of the final expression in (9.59) by $\hbar^2$ and introducing the relationships

$$\hbar\omega = E, \tag{9.60}$$

$$\frac{\hbar}{i} \frac{\partial}{\partial x} = p_x, \tag{9.61}$$

we obtain

$$\frac{dP}{dt} = \frac{q^2}{2\epsilon_0 \mu^2 E^2} \left| \langle n | p_x | m \rangle \right|^2 u_x(\nu). \tag{9.62}$$

A plane polarized wave traveling in the opposite direction yields a similar formula. Superposing two oppositely traveling waves to obtain a standing wave yields two terms which can be combined to form

$$\frac{dP_{x^\pm}}{dt} = \frac{q^2}{2\epsilon_0\mu^2 E^2} \mid \langle n \mid p_x \mid m \rangle \mid^2 u_{x^\pm}(\nu), \tag{9.63}$$

where $u_{x^\pm}(\nu)\, d\nu$ is the energy density in physical space for the standing wave in the infinitesimal band of frequencies $d\nu$. A standing wave polarized in the y direction whose components travel along the x axis similarly yields the transition rate

$$\frac{dP_{y^\pm}}{dt} = \frac{q^2}{2\epsilon_0\mu^2 E^2} \mid \langle n \mid p_y \mid m \rangle \mid^2 u_{y^\pm}(\nu), \tag{9.64}$$

while a standing wave polarized in the z direction with components moving along the y axis yields the transition rate

$$\frac{dP_{z^\pm}}{dt} = \frac{q^2}{2\epsilon_0\mu^2 E^2} \mid \langle n \mid p_z \mid m \rangle \mid^2 u_{z^\pm}(\nu). \tag{9.65}$$

Consequently, a standing wave polarized in the direction of the particle momentum **p** and varying in a perpendicular direction, with energy density over volume of $u_p(\nu)\, d\nu$, induces transitions at the rate

$$\frac{dP}{dt} = \frac{q^2}{2\epsilon_0\mu^2 E^2} \mid \langle n \mid \mathbf{p} \mid m \rangle \mid^2 u_p(\nu). \tag{9.66}$$

In an isotropic field in which relative phases, polarizations, and movements of the constituent standing waves are random, the energy density over frequency and volume is independent of the orientation,

$$u_{x^\pm}(\nu) = u_{y^\pm}(\nu) = u_{z^\pm}(\nu) = u_p(\nu), \tag{9.67}$$

and the net energy density is 3 times that for unidirectional variation;

$$u(\nu) = u_{x^\pm}(\nu) + u_{y^\pm}(\nu) - u_{z^\pm}(\nu) = 3u_p(\nu). \tag{9.68}$$

Introducing (9.68) into (9.66) yields

$$\frac{dP}{dt} = \frac{q^2}{6\epsilon_0\mu^2 E^2} \mid \langle n \mid \mathbf{p} \mid m \rangle \mid^2 u(\nu). \tag{9.69}$$

Thus, the rate at which a submicroscopic particle of charge q and mass μ is induced to go from its mth state to its nth state, energy

$$E = h\nu \tag{9.70}$$

lower, or from its nth state to its mth state, energy (9.70) higher, by isotropic electromagnetic radiation in a continuous band about frequency ν is proportional to the radiation energy density $u(\nu)$ over frequency and volume, if higher-order effects are negligible. Indeed, Equation (9.69) tells us that

$$\text{Induced Rate}_{m \to n} = B_{mn}u(\nu) \tag{9.71}$$

$$\text{Induced Rate}_{n \to m} = B_{nm}u(\nu) \tag{9.72}$$

where

$$B_{mn} = B_{nm} = \frac{q^2}{6\epsilon_0\mu^2 E^2} \mid \langle n \mid \mathbf{p} \mid m \rangle \mid^2. \tag{9.73}$$

The constant of proportionality B_{mn} varies directly with the square of the transition momentum

$$\mid q\langle n \mid \mathbf{p} \mid m \rangle \mid \tag{9.74}$$

and inversely with the square of the energy difference E between the initial and final states of the particle. Quantity B_{mn} is called the *second Einstein coefficient*. The probability per unit time for spontaneous decay A_{mn} will be studied later; it is called the first Einstein coefficient.

9.8
Role Played by the Transition Dipole Moment

We have seen how the first-order perturbation of a charged particle by electromagnetic radiation causes transitions. According to (9.69), the rate varies with the square of the pertinent matrix element for momentum. Now this element can be changed to a geometric form by a simple transformation.

Consider a submicroscopic system containing the particle of mass μ and charge q. Erect a nonrotating frame on the center of mass of the system and let radius vector $\mathbf{r}$ be drawn from the origin of this frame to the charged particle. Let the Hamiltonian operator in absence of the radiation field be H.

The commutator of this H with x, acting on a function ψ, is

$$[H, x]\psi = (Hx - xH)\psi$$

$$= \left(-\frac{\hbar^2}{2\mu} \nabla^2 x + Vx + x\frac{\hbar^2}{2\mu} \nabla^2 - xV \right)\psi$$

$$= -\frac{\hbar^2}{2\mu} \left(2\frac{\partial}{\partial x} \right)\psi = \frac{\hbar}{i\mu} \frac{\hbar}{i} \frac{\partial}{\partial x} \psi = \frac{\hbar}{i\mu} p_x\psi. \tag{9.75}$$

Similar results are obtained with y and z. Composing these yields

$$[H, \mathbf{r}] = \frac{\hbar}{i\mu} \mathbf{p}, \tag{9.76}$$

whence

$$\mathbf{p} = \frac{i\mu}{\hbar}[H, \mathbf{r}] = \frac{i\mu}{\hbar}(H\mathbf{r} - \mathbf{r}H). \tag{9.77}$$

Next, substitute this result into the matrix element for momentum and simplify:

$$\langle n \mid \mathbf{p} \mid m \rangle = \frac{i\mu}{\hbar} (\langle n \mid H\mathbf{r} \mid m \rangle - \langle n \mid \mathbf{r}H \mid m \rangle)$$

$$= \frac{i\mu}{\hbar} (\langle H \mid n \mid \mid \mathbf{r} \mid m \rangle - \langle n \mid \mid \mathbf{r} \mid H \mid m \rangle)$$

$$= \frac{i\mu}{\hbar} (E_n \langle n \mid \mathbf{r} \mid m \rangle - E_m \langle n \mid \mathbf{r} \mid m \rangle)$$

$$= - \frac{i\mu E}{\hbar} \langle n \mid \mathbf{r} \mid m \rangle. \tag{9.78}$$

In the second equality, the Hermiticity of H has been used; in the third equality, the eigenvalue equation and the reality of energy E_n have been employed; in the fourth equality, symbol E has been introduced for the energy difference $E_m - E_n$, as in (9.70).

The final form in (9.78) converts (9.73) to

$$B_{mn} = B_{nm} = \frac{q^2}{6\epsilon_0 \hbar^2} \mid \langle n \mid \mathbf{r} \mid m \rangle \mid^2$$

$$= \frac{2\pi\alpha c}{3\hbar} \mid \langle n \mid \mathbf{r} \mid m \rangle \mid^2. \tag{9.79}$$

When q is the charge e on an electron, then α is the fine structure constant, defined as

$$\alpha = \frac{e^2}{4\pi\epsilon_0 \hbar c}. \tag{9.80}$$

In the approximation that the center of positive charge is at the origin, expression $q \langle n \mid \mathbf{r} \mid m \rangle$ represents the transition dipole moment due to the electron moving between kets $\mid m \rangle$ and $\mid n \rangle$. Substituting the x-component of (9.78) into (9.62) yields

$$\frac{dP}{dt} = \frac{q^2}{2\epsilon_0 \hbar^2} \mid \langle n \mid x \mid m \rangle \mid^2 u_x(\nu) \tag{9.81}$$

for the transition rate per particle exposed to a beam polarized in the x direction.

These formulas apply not only to a single particle transition but also to any multiparticle transition that can be modeled as a single particle shift. Thus, they may be applied to a vibrational transition. Quantity q then has to be chosen so that $q \langle n \mid \mathbf{r} \mid m \rangle$ yields the pertinent physical transition moment. Expression $q \langle m \mid \mathbf{r} \mid m \rangle$ would be the dipole moment before and $q \langle n \mid \mathbf{r} \mid n \rangle$ the dipole moment after the m to n transition.

Example 9.2

What does the coefficient on the right side of (9.79) equal when q is the charge of an electron?

Inserting the accepted values of the fundamental constants gives us

$$\frac{2\pi\alpha c}{3\hbar} = \frac{(6.2832)(7.2973 \times 10^{-3})(2.99792 \times 10^8 \text{ m s}^{-1})}{3(1.05459 \times 10^{-34} \text{ J s})}$$

$$= 4.3447 \times 10^{40} \text{ m J}^{-1} \text{ s}^{-2}.$$

9.9
The Bouguer-Beer Law

Whenever radiation passes through an absorbing medium it induces transitions and is thereby attenuated. Its intensity falls following a law to be derived here.

As before, consider a medium containing atoms or molecules of a given species. Each submicroscopic unit presumably contains an interacting particle of mass μ and charge q. This particle may exist in various states. From (9.81), the rate per particle at which transition occurs from the mth to the nth state, or vice versa, is given by

$$\text{Induced Rate} = \frac{q^2}{2\epsilon_0\hbar^2} \ | \langle n | x | m \rangle |^2 \ u_x(\nu) \qquad (9.82)$$

when the inducing wave is polarized in the x direction and is traveling in the z direction. The separation between the mth state and the nth state is

$$E = h\nu. \qquad (9.83)$$

The density over frequency of the physical density of energy in the wave, u_x, cannot be measured directly. A person can determine the effects of the photons falling on unit area of a detector in unit time, however, and thus get a number that is proportional to an integral of u_x at the surface of the detector. If effectively monochromatic radiation is employed, or if an essentially monochromatic part of the radiation is taken, this number is proportional to u_x at the chosen frequency and is designated the *intensity I* for that frequency.

An emitting or absorbing medium adds energy to or subtracts energy from a passing beam, gradually augmenting or decreasing its intensity with the distance traveled through the medium. Consider a uniform beam moving through a homogeneous material in which the concentration of absorbing systems is C. Let z be the distance the beam has traveled inside the material when it reaches the layer under consideration, as Figure 9.2 shows.

Formula (9.82) gives the rate at which a given atom or molecule absorbs a photon of frequency ν. Multiplying this by $E = h\nu$ gives the time rate of energy change per submicroscopic unit. Then multiplying by the concentration C gives the rate of energy change in the beam per unit volume of absorber:

$$-\frac{d\int u_x(\nu)\,d\nu}{dt} = \frac{\pi q^2\nu}{\epsilon_0\hbar} \ | \langle n | x | m \rangle |^2 \ C \ u_x(\nu). \qquad (9.84)$$

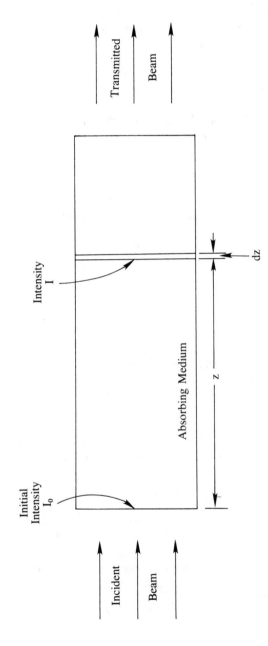

Figure 9.2. Passage of a beam through an absorbing medium with reduction of its intensity from I_0 to I in distance z.

When the density $u_x(\nu)$ is approximately constant through the band where appreciable absorption occurs, the absorption follows a bell-shaped curve. If the width of this curve at half maximum is $\Delta\nu$, then in the numerator on the left of (9.84), we may employ

$$-d \int_{band} u_x(\nu)\, d\nu \;\simeq\; - du_x(\nu) \int_{\Delta\nu} d\nu \;=\; - du_x(\nu)\Delta\nu, \tag{9.85}$$

where $du_x(\nu)$ is the infinitesimal change at the maximum. With this approximation, (9.84) can be rewritten in the form

$$-\frac{du_x(\nu)\,\Delta\nu}{dt} \;=\; \frac{\pi q^2 \nu}{\epsilon_0 \hbar \Delta\nu} \;|\langle n\,|\,x\,|\,m\rangle\,|^2\; C\, u_x(\nu)\,\Delta\nu. \tag{9.86}$$

Dividing both sides of (9.86) by c, the velocity at which the electromagnetic wave moves through the material, replacing $c\, dt$ by dz, $u_x(\nu)\,\Delta\nu$ by (constant) I, and canceling the constant leads to

$$-\frac{dI}{dz} = kCI \tag{9.87}$$

where

$$k = \frac{\pi q^2 \nu}{\epsilon_0 c \hbar\, \Delta\nu} \;|\langle n\,|\,x\,|\,m\rangle\,|^2. \tag{9.88}$$

Quantity I is called the intensity of the radiation at frequency ν, while C is the concentration of the absorbing material, k the absorption coefficient, and z the path length within the material.

Thus, the spatial rate of decrease in intensity along a beam that does not converge or diverge varies directly with the local concentration and the local intensity. The dependence of intensity of light on the length of path was discovered by Pierre Bouguer in 1729; the dependence on concentration of absorbing material by August Beer in 1852.

Example 9.3

How does the intensity of a beam of monochromatic light vary with the distance of its travel through a homogeneous material?

Rearrange (9.87),

$$\frac{dI}{I} = - kC\, dz,$$

and integrate to get

$$\ln \frac{I}{I_0} = - kCz$$

or

$$I = I_0 \, e^{-kCz}.$$

9.10
The Principle of Detailed Balance

So far we have considered the induction of a transition from one ket to another by an interacting electromagnetic wave. However, transition to a lower-energy ket can also take place spontaneously. An excited charged particle can produce a photon (or photons) without being induced to by an external field. The loss of energy takes the particle down to the lower level.

Since in spontaneous decay there is no apparent causative interaction, one is forced to take an indirect approach. We will presume that the rate of such a decay is independent of that of the accompanying induced decay. And, we will determine the overall rate under equilibrium conditions with the principle to be propounded next.

Consider the chemical equilibrium between reactants A, B and products E, F:

$$A + B \rightleftharpoons E + F. \tag{9.89}$$

A person might suppose that adding a catalyst C could increase the relative amounts of E and F. Schematically, at equilibrium we might then have

$$
\begin{array}{ccc}
A + B & \rightleftharpoons & E + F \\
+ & & + \\
C & & C \\
& \searrow \quad D + B \quad \nearrow &
\end{array}
$$

Such a shift produced by another substance C is not observed. An equilibrium is not altered by providing a parallel pathway from reactants to products. At thermal equilibrium, each elementary reaction and its reverse balance each other; they proceed at the same rate. This law is called the *principle of detailed balance*. In statistical mechanics, as discussed by Richard C. Tolman, it is also called the principle of *microscopic reversibility*.

Here we are concerned with the elementary reaction

$$|m\rangle \rightarrow |n\rangle + h\nu, \tag{9.91}$$

where $h\nu$ represents the photon produced. The balancing reaction at equilibrium is

$$|n\rangle + h\nu \rightarrow |m\rangle. \tag{9.92}$$

At thermal equilibrium, under ordinary conditions, the relative populations in the mth and nth states are determined by the Boltzmann distribution law. With N_m the number of systems in the mth state per unit volume, N_n the number in the nth state per unit volume, and $h\nu$ the energy of the mth state above the nth state, we have

$$\frac{N_m}{N_n} = e^{-h\nu/kT}, \tag{9.93}$$

where T is the absolute temperature prevailing and k is the Boltzmann constant.

On the other hand, the density of states in the radiation field is determined by the Bose-Einstein distribution law. Let us consider how next.

9.11
Equilibrium Energy Density in a Radiation Field

The density over frequency of the physical density of energy in an electromagnetic field can be calculated when the different contributions to the field are in thermal equilibrium with each other. Such conditions can be realized by keeping the field from spreading.

Now, an electromagnetic field can be confined to a given region by perfectly conducting walls. Let us suppose that the walls form a rectangular box with edges a, b, c and volume

$$V = abc. \tag{9.94}$$

Also, let us place Cartesian axes along the edges, with the origin at the back lower corner.

For equilibrium, each polarized monochromatic component of the field assumes a standing-wave form. At the postulated walls, the corresponding electric field vanishes, so nodes in the electric wave occur there and the distance between opposite confining walls equals an integral number of half wavelengths. Between the walls at $x = 0$ and $x = a$, we have

$$a = n_x \frac{\lambda_x}{2}. \tag{9.95}$$

The corresponding wave number is

$$k_x = \frac{1}{\lambda_x} = \frac{n_x}{2a}, \qquad n_x = 1, 2, 3, \ldots. \tag{9.96}$$

Similarly, for y and z components we have

$$k_y = \frac{n_y}{2b}, \qquad n_y = 1, 2, 3, \ldots, \tag{9.97}$$

$$k_z = \frac{n_z}{2c}, \qquad n_z = 1, 2, 3, \ldots \tag{9.98}$$

The allowed wave numbers plot in k space as Figure 9.3 shows. Each small rectangular parallelepiped representing a state with a given polarization has the volume

$$\frac{1}{2a} \frac{1}{2b} \frac{1}{2c} = \frac{1}{8abc} = \frac{1}{8V}. \tag{9.99}$$

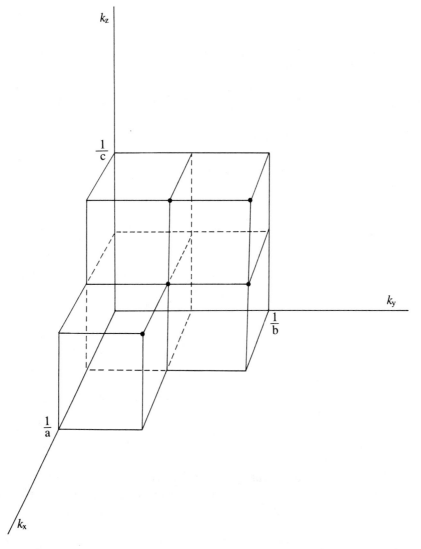

Figure 9.3. Wave numbers for $(n_x, n_y, n_z) = (1, 1, 1)$, $(2, 1, 1)$, $(1, 2, 1)$, $(1, 1, 2)$, $(1, 2, 2)$ and the cells assigned to each of these states.

Since each wave here has two independent polarizations, the volume in k space per state is one-half of (9.99):

$$\frac{1}{16V}.$$ (9.100)

In the continuum approximation, all points a given distance from the origin in the first octant of k space correspond to the same k. The volume covered when k increases by dk is

$$\frac{1}{8}(4\pi k^2\, dk) = \frac{1}{2}\pi k^2\, dk$$ (9.101)

Dividing this by the volume per state, (9.100), yields the number of states corresponding to the increase:

$$dg = \frac{\frac{1}{2}\pi k^2\, dk}{1/(16V)} = 8\pi V k^2\, dk.$$ (9.102)

Einstein's photoelectric equation is

$$E = h\nu$$ (9.103)

where E is the energy of a photon of frequency ν. But since

$$k = \frac{1}{\lambda} = \frac{\nu}{c},$$ (9.104)

we have

$$k^2\, dk = \frac{1}{c^3}\nu^2\, d\nu.$$ (9.105)

Substituting (9.105) into (9.102) yields

$$dg = \frac{8\pi V}{c^3}\nu^2\, d\nu.$$ (9.106)

Photons are indistinguishable particles which are able to occupy a given state without limit. Consequently, they obey the Bose-Einstein distribution law, for which the number of particles at energy E is

$$dN = \frac{dg}{e^{E/kT} - 1}.$$ (9.107)

Here k is Boltzmann's constant and T the absolute temperature. Substituting (9.106) into (9.107), then multiplying by the energy per photon and dividing by the physical volume V, yields the energy density

$$u(\nu)\, d\nu = \frac{h\nu}{V}\frac{1}{e^{h\nu/kT} - 1}\frac{8\pi V}{c^3}\nu^2\, d\nu$$

$$= \frac{8\pi h \nu^3}{c^3} \frac{1}{e^{h\nu/kT} - 1} \, d\nu \qquad (9.108)$$

at equilibrium.

9.12
Spontaneous Emission

Besides being induced to emit or absorb radiation by an applied field, an excited charged particle in a submicroscopic structure can fall to a lower state spontaneously, with an accompanying emission of radiation. The probability for such a transition is related to the coefficient for induced emission or absorption by a procedure developed by Albert Einstein.

Consider a submicroscopic system for which the state described by $|m\rangle$ lies above the state described by $|n\rangle$. Let the probability that the mth state spontaneously radiates and ends up at the nth state, per unit time, be A_{mn}. The corresponding temporal rate of decay per particle is

$$\text{Spontaneous Rate}_{m \to n} = A_{mn}, \qquad (9.109)$$

while the frequency ν of the radiated energy is given by

$$E = h\nu = E_m - E_n \qquad (9.110)$$

in the approximation that energy is conserved in the elementary process.

But Equations (9.71) and (9.72) describe how radiation of frequency ν, at energy density $u(\nu) \, d\nu$, induces the decay and the reverse process, excitation from the nth to the mth state. At equilibrium the net rate of transition upward equals the net rate of transition downward. If N_n and N_m are the numbers of systems in the nth state and in the mth state, per unit volume, we have

$$N_n B_{nm} u(\nu) = N_m [A_{mn} + B_{mn} u(\nu)]. \qquad (9.111)$$

Equilibrium between the two states is governed by the Boltzmann distribution law. Substituting (9.93) and (9.79) into (9.111), canceling N_n, rearranging,

$$B_{mn}(e^{h\nu/kT} - 1)u(\nu) = A_{mn}, \qquad (9.112)$$

and introducing (9.108) leads to

$$A_{mn} = (e^{h\nu/kT} - 1) \frac{8\pi h \nu^3}{c^3} \frac{1}{e^{h\nu/kT} - 1} B_{mn}$$

$$= \frac{8\pi h \nu^3}{c^3} B_{mn}. \qquad (9.113)$$

Thus, the rate of spontaneous emission is proportional to the coefficient for induced decay. Quantity A_{mn} is called the *first Einstein coefficient*; B_{mn} is the corresponding second Einstein coefficient. Introducing the result in (9.79) yields

$$A_{mn} = \frac{8\pi h\nu^3}{c^3} \frac{q^2}{6\epsilon_0\hbar^2} \mid \langle n \mid \mathbf{r} \mid m \rangle \mid^2$$

$$= \frac{\omega\hbar^2\omega^2 q^2}{3\pi c^3 \epsilon_0 \hbar^3} \mid \langle n \mid \mathbf{r} \mid m \rangle \mid^2. \tag{9.114}$$

In a system containing some excited particles at time t, the rate of decay in concentration units equals the rate of decay per particle times the concentration C of the particles. Thus

$$-\frac{dC}{dt} = A_{mn}C \tag{9.115}$$

and

$$C = C_0 e^{-A_{mn}t} \tag{9.116}$$

in absence of an inducing field. By integration, one finds that the mean life of the excited state is

$$\tau = \frac{1}{A_{mn}}. \tag{9.117}$$

Coefficient A_{mn} is called the *decay constant* for the excited state.

When the charged particle is an electron, the coefficient can be expressed in terms of the fine structure constant

$$\alpha = \frac{e^2}{4\pi\epsilon_0\hbar c}. \tag{9.118}$$

Then, formula (9.114) reduces to

$$A_{mn} = \omega(\hbar\omega)^2 \frac{4\alpha}{3c^2\hbar^2} \mid \langle n \mid \mathbf{r} \mid m \rangle \mid^2. \tag{9.119}$$

Here e is the charge on an electron, $\hbar$ Planck's constant divided by 2π, c the speed of light, and $\hbar\omega$ the energy of the mth state above the nth state.

Example 9.4

Formulate the decay constant for a 2p electron falling to the 1s level in an atom.

In a hydrogenlike atom, the energy levels are given by formula (4.67). In a multielectron atom, the active electron is shielded from the nucleus by the additional orbital electrons. Then the Z in the formula has to be replaced by an effective nuclear charge Z_e, which in general varies from level to level. But for simplicity, we will employ the same Z_e for both levels, getting

$$E_{2p} \simeq -\frac{Z_e^2}{2 \times 4}\,\alpha^2\mu c^2, \qquad E_{1s} \simeq -\frac{Z_e^2}{2 \times 1}\,\alpha^2\mu c^2.$$

Then the energy between the two levels is

$$\hbar\omega \simeq \frac{3}{8}\,Z_e^2\alpha^2\mu c^2.$$

From the calculation in Problem 9.15, the matrix element between the 1s and 2p states for the component of $\mathbf{r}$ in the direction in which the p orbital points is approximately $(3/4)a$, where

$$a = \frac{\hbar}{\alpha c\mu Z_e}.$$

Introducing these expressions into (9.119) and reducing yields

$$A_{mn} \simeq \frac{27}{256}\,\omega Z_e^2\alpha^3 \simeq 0.105\,\omega Z_e^2\alpha^3.$$

9.13
Parity Considerations

In the first-order approximation, both Einstein coefficients are proportional to the square of the magnitude of the transition dipole moment $q\langle n \mid \mathbf{r} \mid m\rangle$. Thus, they are small when this moment vanishes. Now, a simple kind of such vanishing occurs when the initial state and the final state have the same parity.

Parity exists for a mode of motion with a center of symmetry. A state whose wave function is not changed by reflection through the center is said to have even parity, with $P = 1$. A state for which the phase of the wave function is changed by π by such reflection is said to have odd parity, with $P = -1$.

When the initial point of $\mathbf{r}$ is at the center of symmetry, reflection of $\mathbf{r}$ through the center reverses its direction. The final $\mathbf{r}$ is then the negative of the original $\mathbf{r}$. So if both $\langle n \mid$ and $\mid m\rangle$ have the same parity, the integrand of $\langle n \mid \mathbf{r} \mid m\rangle$ is changed in sign by the reflection. To each small positive part of the integral, there is then a canceling negative part and

$$\langle n \mid \mathbf{r} \mid m\rangle = 0 \qquad \text{when} \qquad \Delta P = 0. \qquad (9.120)$$

Here ΔP represents the parity change on going from the mth to the nth state.

On the other hand, if $\langle n \mid$ and $\mid m\rangle$ differ in parity, multiplying by $\mathbf{r}$ produces an integrand of even parity. Then $\langle n \mid \mathbf{r} \mid m \rangle$ may differ from zero. For a transition to be allowed in the first-order approximation, we must have

$$\Delta P = \pm 2 \qquad (9.121)$$

when parity is a valid quantum number.

For vibrational states, function (4.51) has even parity when quantum number v is even and odd parity when quantum number v is odd. Thus, parity considerations tell us that transitions for which

$$\Delta v = \pm 1 \qquad (9.122)$$

may be allowed, while those for

$$\Delta v = \pm 2 \qquad (9.123)$$

are forbidden.

For rotational states, function (4.29) has even parity when l (or the corresponding j) is even and odd parity when l (or the corresponding j) is odd. Thus, parity considerations tell us that transitions for which

$$\Delta l = \pm 1 \qquad \text{or} \qquad \Delta j = \pm 1 \qquad (9.124)$$

may be allowed, while those for which

$$\Delta l = \pm 2 \qquad \text{or} \qquad \Delta j = \pm 2 \qquad (9.125)$$

are forbidden.

However, parity considerations do not rule out changes in v by any odd number, or changes in l or j by any odd number.

9.14
Vibrational Transitions

To the approximation that a mode of vibrational motion is harmonic, the simple step operators of Chapter 5 can be employed in evaluating the transition dipole moment and in determining which transitions are allowed.

Now, the transition dipole moment differs from zero only for an asymmetric mode. Let us consider such a mode and set the x axis along the axis of charge separation. Then let us represent the mode by a particle of mass μ carrying the appropriate fraction of the charge on an electron or proton, with the opposite charge being placed at the origin. We also choose the potential to be parabolic:

$$U = \tfrac{1}{2} f x^2. \qquad (9.126)$$

With q the charge on the model particle, the transition dipole moment is

$$q\langle n | x | m \rangle. \qquad (9.127)$$

Operator x is related to the step operators by the formulas in Table 5.2. In the natural units, we find that

$$\frac{1}{\sqrt{2}} a^+ + \frac{1}{\sqrt{2}} a^- = \frac{1}{2}(x - ip) + \frac{1}{2}(x + ip)$$

$$= x. \qquad (9.128)$$

Also,

$$a^+ | m \rangle = \sqrt{m + 1} | m + 1 \rangle \qquad (9.129)$$

and

$$a^- | m \rangle = \sqrt{m} | m - 1 \rangle. \qquad (9.130)$$

Substituting into (9.127) yields

$$q\langle n | x | m \rangle = \frac{q}{\sqrt{2}} \langle n | (a^+ + a^-) | m \rangle$$

$$= \frac{q}{\sqrt{2}} (\sqrt{m + 1} \langle n | m + 1 \rangle + \sqrt{m} \langle n | m - 1 \rangle). \ (9.131)$$

Since the harmonic-oscillator kets form an orthonormal set, this transition moment vanishes except when

$$n = m \pm 1. \qquad (9.132)$$

Thus, we obtain the selection rule

$$\Delta v = \pm 1. \qquad (9.133)$$

Example 9.5

Formulate the decay constant for a harmonic-oscillator excited state.
For decay of the vth level, we have $m = v$ and $n = v - 1$. Then (9.131) yields

$$| \langle v - 1 | x | v \rangle |^2 = \left(\frac{1}{\sqrt{2}} \sqrt{v} \langle v - 1 | v - 1 \rangle \right)^2 = \frac{v}{2}$$

in squared distance units. From (5.46), the natural distance unit squared is

$$\frac{\hbar}{(f\mu)^{1/2}} = \frac{\hbar^2}{\mu\hbar(f/\mu)^{1/2}} = \frac{\hbar^2}{\mu\hbar\omega'}$$

while the energy loss is

$$E = \hbar\omega.$$

Substituting into (9.119) and reducing leads to

$$A_{mn} = \omega(\hbar\omega)^2 \; \frac{4\alpha'}{3c^2\hbar^2} \; \frac{v}{2} \; \frac{\hbar^2}{\mu\hbar\omega} = \frac{2}{3}\omega \; \frac{E}{\mu c^2} \; \alpha'v.$$

In dimensionless constant α', one substitutes for the e in (9.118) the fraction of electronic charge which is placed on the model particle of mass μ. The photon angular frequency is given by ω; its energy by E.

9.15
Transitions in Axial Angular Momentum

In discussing rotational transitions one generally considers the unique axis for the rotator to be the z axis. Then from Equation (3.118) and Example 4.2, we have

$$J_z \,|\, jm\rangle = m\hbar \,|\, jm\rangle. \tag{9.134}$$

Any interacting wave may have components polarized in the z, x, or y directions. Let us here develop selection rules for these.

As fundamental, we consider the commutation relations

$$[J_z, x] = -\frac{\hbar}{i} y, \tag{9.135}$$

$$[J_z, y] = \frac{\hbar}{i} x, \tag{9.136}$$

$$[J_z, z] = 0, \tag{9.137}$$

which follow from (3.53) and (3.55), or from (3.21). We also recall that J_z is Hermitian.

Multiply (9.137) from the left by bra $\langle b|$, from the right by ket $|a\rangle$, and reduce with (9.134):

$$0 = \langle b| [J_z, z] |a\rangle = \langle b| J_z z - z J_z |a\rangle$$

$$= \langle J_z |b| \,|z|\, a\rangle - \langle b| \,|z|\, J_z |a\rangle$$

$$= m_b\hbar \langle b|z|a\rangle - m_a\hbar \langle b|z|a\rangle$$

$$= (m_b - m_a)\hbar \langle b|z|a\rangle. \tag{9.138}$$

For the overall equality in (9.138) to be satisfied, the matrix element $\langle b|z|a\rangle$ has to be zero, and the dipole-moment transition with the wave polarized in the z direction forbidden, except when

$$m_b = m_a. \tag{9.139}$$

Next, we set

$$x \pm iy = x_\pm. \tag{9.140}$$

Then from (9.135) and (9.136), we get

$$[J_z, x_\pm] = -\frac{\hbar}{i} y \pm i \frac{\hbar}{i} x = \pm \hbar x_\pm. \tag{9.141}$$

Consequently, one can construct the sequence

$$
\begin{aligned}
0 = \langle b | ([J_z, x_\pm] \mp \hbar x_\pm) | a \rangle &= \langle b | (J_z x_\pm - x_\pm J_z \mp \hbar x_\pm) | a \rangle \\
&= \langle J_z | b | | x_\pm | a \rangle - \langle b | | x_\pm | J_z | a \rangle \mp \hbar \langle b | x_\pm | a \rangle \\
&= m_b \hbar \langle b | x_\pm | a \rangle - m_a \hbar \langle b | x_\pm | a \rangle \mp \hbar \langle b | x_\pm | a \rangle \\
&= (m_b - m_a \mp 1) \hbar \langle b | x_\pm | a \rangle.
\end{aligned}
\tag{9.142}
$$

For the overall equality in (9.142) to hold, the matrix elements $\langle b | x | a \rangle$ and $\langle b | y | a \rangle$ must both be zero, and the dipole-moment transition with the waves polarized in the x and y directions forbidden, except when

$$m_b = m_a \pm 1. \tag{9.143}$$

Equations (9.139) and (9.143) yield the selection rule

$$\Delta m = 0, \pm 1 \tag{9.144}$$

for the magnetic quantum number m. The rule applies in the first-order approximation to single-photon emission or absorption. Since the controlling matrix element is $q \langle b | \mathbf{r} | a \rangle$, the process is referred to as an electric-dipole-moment transition.

9.16
Shifts in the Rotational or Azimuthal Quantum Number

Rotational states also involve the quantum number j. Indeed, from (3.117) and (3.115), we have the eigenvalue equation

$$J^2 | jm \rangle = j(j + 1)\hbar^2 | jm \rangle \tag{9.145}$$

complementing (9.134). And (3.116) tells us that

$$m = -j, -j + 1, \ldots, j - 1, j. \tag{9.146}$$

From the parity considerations in Section 9.13, j must change by an odd integer for the electric-dipole-moment transition to be allowed. However, Equation (9.144) lets m change only by 1. And from (9.146), j and $-j$ are the limits on m. Thus, we are led to the selection rule

$$\Delta j = \pm 1 \qquad \text{or} \qquad \Delta l = \pm 1. \qquad (9.147)$$

An operator analysis, evaluating $\langle b \mid [J^2, [J^2, \mathbf{r}]] \mid a \rangle$ in two different ways, produces the same result. Since this is fairly lengthy and detailed, we will not develop it here.

9.17
Synopsis

A perturbing potential that varies with time after being applied is generally more complicated than any discussed in Chapter 8. However, such a potential may advantageously be resolved into orthogonal monochromatic components. A typical component has the form

$$H^{(1)} = \eta(t)[Fe^{i\omega t} + Ge^{-i\omega t}] \qquad (9.148)$$

with

$$G_{jk} = F_{kj}^*. \qquad (9.149)$$

The Hilbert space for the submicroscopic system is presumably not affected by the perturbers. Then the state ket can be expressed as a changing superposition of unperturbed kets:

$$\mid I \rangle = \Sigma c_k(t) \mid k \rangle e^{-i\omega_k t}. \qquad (9.150)$$

The time rate of change of c_j is given by (8.132) as before.

To facilitate reduction we transform to b_j by (8.134). This transformation does not alter the formula for the statistical weight:

$$w_j = c_j^* c_j = b_j^* b_j. \qquad (9.151)$$

And derivative $\dot{b}_j$ is still given by (8.135).

Before $t = 0$, $H^{(1)}{}_{jk}$ is zero; after $t = 0$, we have

$$H^{(1)}{}_{jk'} = F_{jk'} e^{i\omega t} + F_{kj}^* e^{-i\omega t}. \qquad (9.152)$$

The expansion

$$b_j = \Sigma \lambda^l b_j^{(l)} \qquad (9.153)$$

is introduced and the equation for $\dot{b}_j{}^{(1)}$ integrated.

When the system is in the mth state initially and angular frequency ω is not too small, we find that

$$b_n \simeq b_n{}^{(1)} = -\frac{2i}{\hbar} F_{mn}{}^* \frac{e^{\frac{1}{2}i(\omega_{nm} \mp \omega)t}}{\omega_{nm} \mp \omega} \sin \tfrac{1}{2}(\omega_{nm} \mp \omega)t. \qquad (9.154)$$

The negative sign is used when ω_{nm} is positive; the positive sign when ω_{nm} is negative.

For the probability of transition in time t, we get

$$P_{mn} = 2 \, | \, F_{mn} \, |^2 \, \frac{1 - \cos(\omega_{nm} \mp \omega)t}{\hbar^2(\omega_{nm} \mp \omega)^2} . \qquad (9.155)$$

Note the similarity to (8.141).

Kets $|\,m\rangle$ and $|\,n\rangle$ describe unperturbed states of the submicroscopic system. With $|\,m\rangle$ above $|\,n\rangle$ in energy, photon states combine with the $|\,n\rangle$ state to produce a continuum of levels about the level of $|\,m\rangle$. Let the density over energy of these states be $\rho(E_n)$. Then the probability of transition, m to n, is

$$P = \frac{2}{\hbar^2} \int_{\text{band}} | \, F_{nm} \, |^2 \, \frac{1 - \cos(\omega_{nm} + \omega)t}{(\omega_{nm} + \omega)^2} \, \rho(E_n) \, dE_n. \qquad (9.156)$$

For the time rate of transition, we then get

$$\frac{dP}{dt} = \frac{2}{\hbar} \, | \, F_{nm} \, |^2 \, \rho(E_n) \int_{-\infty}^{\infty} \frac{\sin \theta t}{\theta} \, d\theta$$

$$= \frac{2\pi}{\hbar} \, | \, F_{nm} \, |^2 \, \rho(E_n). \qquad (9.157)$$

This result is called the golden rule of Fermi.

Far from its source, an electromagnetic wave is governed by its vector potential $\mathbf{A}$. A planar monochromatic component is represented by

$$\mathbf{A} = \mathbf{A}_x{}^0 \cos\left[\omega\left(t - \frac{z}{c} \right) \right] \hat{\mathbf{x}}. \qquad (9.158)$$

Since

$$\nabla \cdot \mathbf{A} = 0 \qquad \text{and} \qquad q^2 A^2 \simeq 0, \qquad (9.159)$$

the perturbation operator is

$$H^{(1)} = \frac{i\hbar q}{\mu} \, \mathbf{A} \cdot \nabla = F e^{i\omega t} + G e^{-i\omega t} \qquad (9.160)$$

with

$$F = \frac{i\hbar q A_x{}^0}{2\mu} \, e^{-i\omega z/c} \frac{\partial}{\partial x} \qquad (9.161)$$

and

$$G = \frac{i\hbar q A_x^0}{2\mu} e^{i\omega z/c} \frac{\partial}{\partial x}. \tag{9.162}$$

Using (9.161), we find that

$$F_{nm} \simeq \frac{i\hbar q A_x^0}{2\mu} e^{-i\omega z/c} \langle n | \frac{\partial}{\partial x} | m \rangle. \tag{9.163}$$

With (9.158), the average rate of energy flow per unit cross section is

$$\bar{P} = \frac{c}{2} (A_x^0 \omega)^2 \epsilon_0. \tag{9.164}$$

In the composite polarized wave, the energy density over volume in the frequency interval $d\nu$ is

$$u_x(\nu) d\nu = u_x(\nu) \frac{dE}{h} = \frac{1}{2}(A_x^0 \omega)^2 \epsilon_0 \rho(E) dE, \tag{9.165}$$

from which

$$\rho(E) = \frac{2u_x(\nu)}{(A_x^0 \omega)^2 \epsilon_0 h}. \tag{9.166}$$

Substituting (9.163) and (9.166) into (9.157) yields

$$\frac{dP}{dt} = \frac{q^2}{2\epsilon_0 \mu^2 E^2} |\langle n | \mathbf{p} | m \rangle|^2 u_p(\nu). \tag{9.167}$$

In an isotropic field, the net energy density is 3 times that for unidirectional variation and

$$u_p(\nu) = \frac{u(\nu)}{3}. \tag{9.168}$$

To a good approximation, we have

$$E = E_m - E_n. \tag{9.169}$$

And introducing the operator for $\mathbf{p}$ leads to

$$\langle n | \mathbf{p} | m \rangle = -\frac{i\mu}{\hbar} (E_m - E_n) \langle n | \mathbf{r} | m \rangle. \tag{9.170}$$

With these, (9.167) reduces to

$$\frac{dP}{dt} = \frac{q^2}{6\epsilon_0 \hbar^2} |\langle n | \mathbf{r} | m \rangle|^2 u(\nu) = B_{mn} u(\nu). \tag{9.171}$$

Expression B_{mn} is called the second Einstein coefficient.
   Alternatively, we have

$$\frac{dP}{dt} = \frac{q^2}{2\epsilon_0\hbar^2} \mid \langle n \mid x \mid m \rangle \mid^2 u_x(\nu). \tag{9.172}$$

In the Bouguer-Beer law

$$-\frac{dI}{dz} = kCI, \tag{9.173}$$

we then find that

$$k = \frac{\pi q^2 \nu}{\epsilon_0 ch\Delta\nu} \mid \langle n \mid x \mid m \rangle \mid^2. \tag{9.174}$$

The first Einstein coefficient A_{mn} is the rate per particle at which spontaneous decay from m to n occurs. At equilibrium the induced rate of excitation, n to m, equals the sum of the induced and the spontaneous rates of deexcitation, m to n; thus,

$$N_n B_{nm} u(\nu) = N_m[A_{mn} + B_{mn} u(\nu)]. \tag{9.175}$$

Introducing the formula for the equilibrium energy density in a radiation field, together with the Boltzmann distribution law, then yields

$$A_{mn} = \frac{8\pi h\nu^3}{c^3} B_{mn}. \tag{9.176}$$

Discussion Questions

9.1. In what elementary discussions is an imposed electromagnetic field considered to be quantized? To what extent may this quantization be neglected in discussing absorption or emission of radiation?

9.2. How is the electromagnetic field described by potentials? How do these enter the Hamiltonian?

9.3. How is it mathematically possible to resolve a fluctuating electromagnetic field into polarized monochromatic components? Why does such a resolution have physical significance?

9.4. How can a vector potential $\mathbf{A}$ give rise to a transverse electromagnetic wave? Why isn't a nonzero scalar potential needed for such a wave?

9.5. Why doesn't the term

$$-\frac{\hbar^2}{2\mu} \nabla^2$$

in the Hamiltonian operator contribute to perturber $H^{(1)}$? Why doesn't $q\phi$ contribute to $H^{(1)}$?

9.6. When does the perturbation operator for an electromagnetic field reduce to

$$\frac{i\hbar q}{\mu} \mathbf{A} \cdot \nabla?$$

To what does the $\mathbf{A}$ for a polarized, monochromatic, planar wave transform this expression?

9.7. How is the rate at which energy is carried by the field obtained?

9.8. Define (a) the state density $\rho(E)$, (b) the energy density $u_x(\nu)$. Show how these are related.

9.9. What form does a suitable polarized monochromatic perturber assume? Need the resulting perturbation expand the space of the kets? Why?

9.10. How is the statistical weight of a state related to the transformed expansion coefficients? When is each of these coefficients a power series in the perturbation parameter?

9.11. How does the probability for an individual transition vary with the extent to which the conservation-of-energy law is violated?

9.12. Why do we discuss transitions from a given state to a band of states and transitions from a band of states to a given state and not from a band of states to a band of states?

9.13. How do we allow for quantization of the electromagnetic field in the expression for rate of transition?

9.14. Why can $|F_{nm}|^2$ and $\rho(E_n)$ be factored out of the integral for the rate of transition? What is the final expression for this rate?

9.15. How is matrix element F_{nm} for the polarized traveling electromagnetic radiation determined? How does this reduce the expression for dP/dt?

9.16. How is the transition rate induced by a standing wave obtained from the rate induced by a traveling wave? How is the transition rate in an isotropic radiation field obtained?

9.17. For (9.75), evaluate $\nabla^2(x\psi) - x\nabla^2\psi$.

9.18. Explain how the *nm*th matrix element for momentum is related to the corresponding matrix element for position.

9.19. How is the transition dipole moment related to the matrix element for momentum?

9.20. How is the intensity of an electromagnetic wave at a given frequency measured? How is the density over frequency of the physical density of energy in the wave related to this intensity?

9.21. Why is

$$\int_{\text{band}} u_x(\nu)\,d\nu$$

the energy per unit volume in the given band of radiation?

9.22. Explain what the band width $\Delta\nu$ represents.

9.23. How is the Bouguer-Beer law of absorption justified?

9.24. What is spontaneous decay? Does such a process have a reverse?

9.25. How can a catalyst shift the position of an equilibrum?

9.26. What rates are equated in the principle of detailed balance?

9.27. Under what conditions are the equilibrium populations in the *m*th and *n*th states governed by the Boltzmann distribution law?

9.28. Why do photons obey the Bose-Einstein distribution law at room temperature?

9.29. Why is the electromagnetic field considered to be confined in our equilibrium calculation?

9.30. How is the confined electromagnetic field quantized? What is the volume in k space per state?

9.31. How do we determine the density over frequency of the physical density of energy in the equilibrium electromagnetic field?

9.32. How is the rate for spontaneous emission found from the rates for the corresponding induced transitions?

9.33. Justify the exponential decay law. How is the decay constant related to the Einstein coefficients?

9.34. What is parity? What kind of systems exhibit parity?

9.35. Explain what happens to the parity of a system when it undergoes an electric-dipole transition.

9.36. If parity is conserved in an electric-dipole emission, what parity does the resulting photon wave function exhibit?

9.37. How is the selection rule

$$\Delta v = \pm 1$$

deduced from properties of operators?

9.38. How is the selection rule for magnetic quantum number m obtained?

9.39. How do we obtain the selection rule for rotational or azimuthal quantum number j?

Problems

9.1. Reconcile the dimensions in formula (9.31).

9.2. Show when the perturber in Hamiltonian (9.9) can be written in the form $-q\mathbf{r}\cdot\mathbf{E}$.

9.3. Rewrite formula (9.79) for B_{mn} in the natural units of Table 6.4.

9.4. How is the result in Example 9.3 altered when the absorbing medium contains one substance at concentration C_1 and a second substance at concentration C_2?

9.5. In a certain absorption cell, solution A transmits 28.2% and solution B transmits 71.8% of the light at a given frequency. What is the transmittance I/I_0 of a uniform mixture of 23.2% by volume of solution A and 76.8% by volume of solution B in the same cell at the same frequency?

9.6. By integration, determine

$$\langle n \,|\, z \,|\, m \rangle$$

for a rotational transition from $j = 1$, $m = 0$ to the $j = 0$ level in a diatomic rotator.

9.7. Formulate the decay constant for the transition from $j = 1$, $m = 0$ to the $j = 0$ level in the diatomic rotator.

9.8. Calculate the decay constant for the transition of Problem 9.7 in HCl. The wave number for the transition is 20.68 cm^{-1} while the dipole moment of HCl equals the electronic charge times 2.20×10^{-9} cm.

9.9. In a straightforward way, evaluate $\langle b \,|\, [J^2, [J^2, \mathbf{r}]] \,|\, a \rangle$.

— — —

9.10. Express the mean radiation energy density

$$\tfrac{1}{2}(A_x{}^0\omega)^2\epsilon_0$$

in terms of amplitude E^0 of the electric intensity.

9.11. Reconcile the dimensions in formula (9.81).

9.12. Rewrite formula (9.88) in the natural units of Table 6.4.

9.13. How does multiplying an absorbing path length by number r change the transmitted fraction of light of a given wavelength?

9.14. How does multiplying the concentration of absorbing material in a given cell by number r alter the transmittance I/I_0 through the cell?

9.15. By integration, determine

$$\langle n \mid \mathbf{r} \mid m \rangle$$

for transition from the 2p to the 1s level in the hydrogen atom.

9.16. Calculate the mean lifetime for a 2p state of the hydrogen atom. The energy released in the descent to the 1s state is 82,260 cm^{-1}.

9.17. Calculate the mean lifetime for the $v = 1$ to $v = 0$ transition in HCl. The wave number for this vibrational transition is 2990 cm^{-1}, the reduced mass of HCl is 1.628×10^{-27} kg, the fraction of charge on the model particle is 0.17.

9.18. Show that $\mathbf{J} \cdot \mathbf{r}$ is a null operator.

References

Books

Bartolo, B. Di: 1976. "Interaction of Radiation with Atoms and Molecules," in Bartolo, B. Di, Pacheso, D., and Goldberg, V. (editors), *Spectroscopy of the Excited State*, Plenum Press, New York, pp. 1-46. This paper discusses the interaction of radiation with atomic and molecular systems in a fundamental, detailed manner.

Bohm, D.: 1951. *Quantum Theory*, Prentice-Hall, New York, pp. 417-448. These sections of Bohm's text are notable for the discussion accompanying the mathematics.

Cohen-Tannoudji, C., Diu, B., and Laloe, F. (translated by Hemley, S. R., Ostrowsky, N., and Ostrowsky, D.): 1977. *Quantum Mechanics*, Wiley, New York, pp. 1304-1355. In this part of their treatise, the authors consider interaction of an atom with an electromagnetic wave, linear and nonlinear responses, oscillations between two states, and decay of a state, all in useful detail.

Johnson, C. S., Jr., and Pederson, L. G.: 1974. *Problems and Solutions in Quantum Chemistry and Physics*, Addison-Wesley, Reading, Mass., pp. 294-315. Chapter 10 of this collection presents solutions of problems on the interaction of radiation with matter.

Articles

Boulil, B., Henri-Rousseau, O., and Deumie, M.: 1988. "Born-Oppenheimer and Pseudo-Jahn-Teller Effects as Considered in the Framework of the Time-Dependent Adiabatic Approximation," *J. Chem. Educ.* **65**, 395-399.

Cray, M., Shih, M.-L., and Milonni, P. W.: 1982. "Stimulated Emission, Absorption, and Interference," *Am. J. Phys.* **50**, 1016-1021.

Doughty, G. R.: 1981. "A New Look at Stimulated Transition Rates," *Am. J. Phys.* **49**, 1071-1073.

Durrant, A. V.: 1976. "Some Basic Properties of Stimulated and Spontaneous Emission: A Semiclassical Approach," *Am. J. Phys.* **44**, 630-635.

Ginzburg, V. L.: 1983. "The Nature of Spontaneous Radiation," *Soviet Phys. Uspekhi* **26**, 713-719.

Henderson, G.: 1980. "Quantum Dynamics and a Semiclassical Description of the Photon," *Am. J. Phys.* **48**, 604-611.

Hilborn, R. C.: 1982. "Einstein Coefficients, Cross Sections, f Values, Dipole Moments, and All That," *Am. J. Phys.* **50**, 982-986.

Kobe, D. H., and Yang, K.-H.: 1983. "Gauge Invariance in Quantum Mechanics: Zero Electromagnetic Field," *Am. J. Phys.* **51**, 163-168.

Lagendijk, A.: 1976. "Spectral Density in Time-Dependent Perturbation Theory," *Am. J. Phys.* **44**, 1098-1100.

Leubner, C.: 1981. "Illustrating Gauge Invariance in Quantum Mechanics through the Free Electron in a Uniform Time-Varying Electric Field," *Am. J. Phys.* **49**, 738-744.

Nagy-Felsobuki, E. I. von: 1989. "Hückel Theory and Photoelectron Spectroscopy," *J. Chem. Educ.* **66**, 821-824.

Sturm, J. E.: 1990. "Grid of Expressions Related to the Einstein Coefficients," *J. Chem. Educ.* **67**, 32-33.

10

QUANTUM ABRUPTNESS
AND NONLOCALITY

10.1
Introduction

The reader has probably noticed that we do not consider the time-dependent Schrödinger equation to be axiomatic. Indeed, it does not tell one whether any change will take place abruptly, as a jump, or smoothly and continuously. Without doubt, it does describe average behavior in an ensemble. But it can only yield the probability for an abrupt change to occur.

Furthermore, the perturbation theory as developed does not give an accurate detailed picture of how transitions occur. Again, the formulas only describe the average behavior observed in suitable ensembles. To obtain information on the aspects that have so far been neglected, we must study individual processes.

First, we recall that in common radioactive materials, the nuclei are isolated from each other by the electronic structures of the atoms. So each decay particle comes from a single nucleus. Its production presumably involves an abrupt jump of the nucleus from an excited state to a lower state.

Individual unstable particles can be traced, when they move rapidly through a medium, by the excitations and ionizations that they produce. The medium may be a photographic emulsion, a supersaturated vapor, a superheated liquid, or a sensitive crystal. Continuous changes in a track are associated with deceleration of the given particle; forks in a track, with abrupt disintegrations.

In Section 10.2, we will consider how changes occur in the electronic structure of an atom or ion. Not unexpectedly, the results support the view a person gets from studying the nuclear and high-energy transitions.

An abrupt change also occurs when a state of a given system is analyzed for one of its contributors. The effect on the system is represented by a reduction or collapse of the governing wave function or ket. Surprisingly, the reduction can affect widely spaced observations. Furthermore, it seems to propagate at tachyonic speeds. Thus, quantum systems exhibit a nonlocality that classical systems do not.

10.2
Visible Quantum Jumps

In Chapters 2 and 3, symmetry considerations, together with the localizability postulate and the local-neglect-of-interaction postulate, led to a form for the Hamiltonian operator. The time-independent form for this operator was employed in constructing the Schrödinger equation for states.

In Chapters 8 and 9, the time-dependent part of the Hamiltonian operator was considered as a perturbation and formulas governing transitions between states were derived. This theory leaves unanswered the question of how individual transitions occur, however. Do they proceed through a coherent superposition of the states involved or as quantum jumps?

To answer this question, one needs to study individual transitions. But for single atoms or ions, allowed electronic transitions are exceedingly fast and difficult to resolve. Nevertheless, one can use a strong resonance fluorescence to monitor when the atom or ion is in a shelf state.

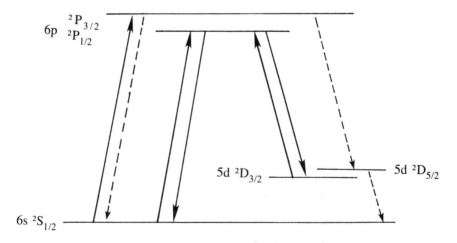

Figure 10.1. Ba^+ levels and transitions used for observing individual quantum jumps. Tunable lasers are employed to excite the $^2P_{1/2}$ level; a filtered barium lamp, the $^2P_{3/2}$ level. The $^2D_{5/2}$ level is the shelf level.

Thus, a person may employ an atom or ion with at least two involved excited states. One of these would be coupled strongly with the ground state, while the other is coupled very weakly. Intense resonance irradiation could cause the strong transition to produce $\sim 10^8$ photons in a one-second interval. Roughly 0.1% of these may be collected and counted. The weak transition may have a mean life of ~ 1 sec. A transition to it, caused by irradiation at the appropriate wavelength, would suppress the strong fluorescence for such times, producing an easily measurable signal.

In practice, a high vacuum apparatus is required. Ions of the chosen material are introduced and clouds of one, two, or three of these may be cooled optically and confined in a tiny radio-frequency ion trap. The small system is then irradiated and the fluorescence measured.

Successful experiments have been run with the Ba^+ and the Hg^+ ions. The levels employed appear in Figures 10.1 and 10.2. For Ba^+, the shelf level is the $5d\ ^2D_{5/2}$ level; for Hg^+, it is the $5d^96s^2\ ^2D_{5/2}$ level. The barium ion yielded results such as appear in Figure 10.3.

If the exciting radiation had driven the ion into a coherent superposition of states, the bright fluorescence signal would have merely dropped in intensity after the lamp was turned on, remaining approximately constant at this lower value. Such action is not observed. The processes appear to occur by quantum jumps.

10.3
Definiteness of Properties

In classical theory all properties of a system are definite at each instant of time, whether or not any one is being adjusted or measured. A classical particle is at some position at each instant in its life with some velocity and momentum. Certain errors

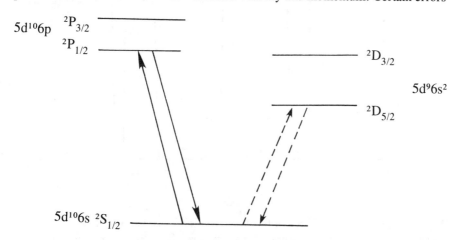

Figure 10.2. Hg^+ levels and transitions involved in studying the periodic shelving of the ion. Here also, the $^2D_{5/2}$ level is the shelf level.

do enter into any preparation or determination. However, in the classical scheme, each of these errors is in principle reducible to zero. An ideal measurement is passive, with no influence on the system.

But in preparing a quantum system, some, not all, of the properties can be set. Thus, a person may fix the energy of a particle in a given potential field, but not simultaneously its position and momentum. Nevertheless, a measurement of an indefinite property can yield a definite value. On repeating the measurement again and again on equivalent systems, the possible values would appear with a distribution determined by the governing ket. Each value would have a certain propensity for manifesting itself.

With such a property, the quantum mechanical measurement acts to alter the system. Instead of providing an increase in one's knowledge of an already preexisting state, it acts to develop a potentiality that is latent. The measuring process is active, not passive, in its effect on the system.

Consider a quantum mechanical system described by the ket $|g\rangle$. Also, consider a measuring process for which we have the operator A and a complete set of orthonormal eigenkets $|1\rangle, |2\rangle, \ldots, |j\rangle, \ldots$ such that

$$A|j\rangle = a_j|j\rangle. \tag{10.1}$$

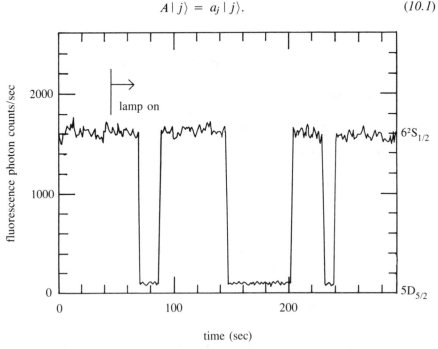

Figure 10.3. A typical record of the 493.3 nm fluorescence from the 6p $^2P_{1/2}$ level of Ba$^+$, showing quantum jumps after the filtered light from a barium lamp, supplying 455.4 nm irradiation, was introduced, according to W. Nagourney, J. Sandberg, and H. Dehmelt (*Phys. Rev. Lett.* **56**, 2797-2799).

On implementation, the measuring process would yield one of the eigenvalues a_1, a_2, ...

Expanding the given ket $|g\rangle$ in terms of the eigenkets leads to

$$|g\rangle = \Sigma c_j |j\rangle. \tag{10.2}$$

If the act of measurement on the system yields an a_k corresponding to a nondegenerate eigenket $|k\rangle$, it would appear to reduce the state ket to $|k\rangle$ appropriately phased. The original composite ket $|g\rangle$ would have apparently degenerated into the pure ket $|k\rangle$ at the time of measurement:

$$|g\rangle \rightarrow |k\rangle. \tag{10.3}$$

We say that $|g\rangle$ *collapses* to $|k\rangle$. The probability for this to occur is given by the probability for $|k\rangle$ to appear in $|g\rangle$. It thus equals the statistical weight of the kth state:

$$w_k = c_k{}^* c_k. \tag{10.4}$$

When the level has a degeneracy of n, the collapse would be to a linear combination of the eigenkets for that level:

$$|g\rangle \rightarrow \sum_k^{k+n-1} d_j |j\rangle. \tag{10.5}$$

Furthermore, the probability for a_k to appear would equal the sum of the weights for the level:

$$w_k = \sum_k^{k+n-1} c_i{}^* c_i. \tag{10.6}$$

So far experiments have supported the existence of the discontinuous change described here. The process appears to be instantaneous in the reference frame of the observer.

Example 10.1

A particle is confined in a large box. At time t, a position measurement reveals the particle to be at point $\mathbf{r}_0$. How does the ket for the particle behave at this time?

Before time t, the ket is one appropriate for confinement in the box. But at time t, this is reduced to a ket describing the particle at point $\mathbf{r}_0$. We represent the process as the collapse

$$|\text{ whole box}\rangle \rightarrow |\mathbf{r}_0\rangle.$$

Ket $|\mathbf{r}_0\rangle$ is represented by a delta function of the coordinates centered on point $\mathbf{r}_0$.

Example 10.2

The box of Example 10.1 is now taken to be rectangular, with Cartesian axes placed along three intersecting edges. Consider the particle to be in a definite energy state initially and describe the X factor of the state function before and after the measurement.

Let A be the length of the x edge. Then from Table 4.1, the initial X factor is

$$X_1(x) = \sqrt{\frac{2}{A}} \, \sin\left(\frac{\pi n_x}{A} x\right)$$

with n_x a positive integer. When the measurement locates the particle at $x = a$, the X factor has collapsed to

$$X_2(x) = [\delta(x - a)]^{1/2}$$

where $\delta(x - a)$ is the one-dimensional Dirac delta centered at $x = a$. The probability density for observing this $X_2(x)$ is $[X_1(a)]^2$.

10.4
Nonlocality in Single Coherent Systems

The classical picture of an atom or molecule has the constituent electrons moving along definite orbits about the nucleus or nuclei. In a collection of such atoms or molecules, all at the same energy state, each member is different. For it is identifiable by the instantaneous relative positions of its electrons and nuclei.

But to get agreement with the measured thermodynamic properties of an assembly of actual atoms or molecules, one has to consider those in the same quantum state as being indistinguishable. These empirical results tell us that the electrons are not localized within the structures as long as the structures are effectively free of each other.

On the other hand, one may bombard a target material with a beam of particles. While passing through some atom or molecule, a particle from the beam may strike an electron and knock it out. The electron and the particle in the beam are localized by the interaction. Recall Section 2.10. If the collision occurs at the point located by radius vector $\mathbf{r}_0$, then the ket for the electron presumably collapses onto $| \mathbf{r}_0 \rangle$ at the time of collision.

A coherent system is a system described by a single ket. This ket is represented by a nonlocal state function subject to the collapse just described.

10.5
The EPR Paradox

But does the collapse actually occur? Are quantum mechanical properties really indefinite? Maybe the theory is just incomplete, as Albert Einstein thought. In support of this view, he together with Boris Podolsky and Nathan Rosen developed the following paradox.

Two systems, 1 and 2, are prepared individually. They proceed to interact for a time and then separate spontaneously.

Suppose that with respect to operator A, the normalized eigenfunctions of the first system are

$$\phi_1(1), \ \phi_2(1), \ \phi_3(1), \ \ldots \tag{10.7}$$

with eigenvalues $a_{11}, a_{12}, a_{13}, \ldots$, while those of the second system are

$$\psi_1(2), \ \psi_2(2), \ \psi_3(2), \ \ldots \tag{10.8}$$

with eigenvalues $a_{21}, a_{22}, a_{23}, \ldots$. We presume that the eigenfunctions in sets (10.7) and (10.8) are labeled so that when the systems interact, the functions pair in order, $\phi_1(1)$ with $\psi_1(2)$, $\phi_2(1)$ with $\psi_2(2)$, $\phi_3(1)$ with $\psi_3(2)$, and so on.

But from how independent choices combine, we know that independent probabilities combine multiplicatively. So to satisfy postulate (IV'), the eigenfunctions in each pair multiply in forming a possible overall state function. The interaction thus results in the superposition

$$\Psi = \sum_j c_j \phi_j(1) \psi_j(2). \tag{10.9}$$

As the systems separate, no outside influence acts to cause form (10.9) to break down; the process is spontaneous. After the systems are completely separated, we suppose that a measurement on the first system yields the eigenvalue a_{1k}. Then from (10.3), Ψ has collapsed to its kth term and the measurement on the second system must yield a_{2k}.

For the same two systems, a person can also construct superposition (10.9) for a second operator B. The corresponding measurement, after the interaction, on the first system may then yield eigenvalue b_{1l}. The Ψ would again have collapsed and the measurement on the second system must now yield b_{2l}.

Einstein, Podolsky, and Rosen considered the two different measurements to be one for momentum and one for position. Since the manipulations were carried out in a region isolated from the second system, yet the momentum and position for this system are obtainable at a given t, it appears that they are ready and waiting there to be measured. But how can this be? Indeed, according to quantum mechanics, a system in an eigenstate with respect to operator A is generally not in an eigenstate with respect to operator B.

Example 10.3

Describe experiments that may be used to study the EPR effect.

(a) Proton pairs in a singlet state. One may bombard a hydrogen containing target with low-energy protons and measure the spin correlations between the recoil protons and the deflected ones. Two Stern-Gerlach devices may be employed.

(b) Low-energy photon pairs emitted in transitions. A typical experiment involves a calcium atom in the excited $4p^2[^1S_0]$ state emitting two photons. The photons are analyzed by spatially separated polarizers.

(c) High-energy photon pairs from positron-electron annihilation. In this experiment the photons produced fly off to Compton polarimeters.

In these various experiments, the analyzers may be meters apart and the measurement directions may be chosen after the particles are separated, when there is no way for a conventional signal to pass from one to the other, alerting it to the choice that has been made.

10.6
Correlations Between Incoherent Product Systems

Many authors have discussed experiments like those proposed by Einstein, Podolsky, and Rosen. Particularly significant is the work of John S. Bell, in 1964, and of others expanding on his approach. They have derived inequalities that can be tested in the laboratory.

Consider the EPR analysis to be carried out repeatedly upon two small systems or particles. For one sequence, the two systems are produced and allowed to interact in a small source region. For another sequence, they are virtually present initially in a composite system located in the small source region. Either way the two systems break apart and are studied after they have become widely separated. At observation posts on each side of the source region the value of a transverse property is measured. We suppose that with the given particles this can exhibit only two values equal in magnitude but opposite in sign (as for instance, spin with $S = \frac{1}{2}$). Let us choose units so that these values are $+1$ and -1.

Let unit vectors $\mathbf{a}$, $\mathbf{a}'$, . . . define the directions, perpendicular to the path of the first particle, chosen at the first position, while unit vectors $\mathbf{b}$, $\mathbf{b}'$, . . . define the directions, perpendicular to the path of the second particle, chosen at the second position. One may find that the results at either measuring position are random. But when one multiplies each result for the first position by the corresponding result for the second position and averages, one generally obtains a value dependent on the relative directions.

For directions $\mathbf{a}$ and $\mathbf{b}$, the average is designated $P(\mathbf{a}, \mathbf{b})$. Since the limits on the individual values are $+1$ and -1, the limits on the product are the same and for the average,

$$-1 \leqslant P(\mathbf{a}, \mathbf{b}) \leqslant 1. \tag{10.10}$$

With common setups, the quantity in parenthesis is represented by the angle between $\mathbf{a}$ and $\mathbf{b}$ when the vectors are drawn from a common origin.

We will suppose that the two measurements in a run are separated by a spacelike interval, so that no signal can pass from one observation post to the other and so affect the correlation. We will also consider that each pair of particles observed forms an incoherent product system in agreement with the concept of locality. According to this, what is done at one post cannot affect the corresponding measurement at the other.

We let ψ represent the state in which the interaction left the two systems in a run, $\epsilon(\psi, \mathbf{a})$ the corresponding probability for obtaining $+1$ at the first observation post using orientation $\mathbf{a}$, $\eta(\psi, \mathbf{b})$ the corresponding probability for obtaining $+1$ at the second observation post with orientation $\mathbf{b}$, and $d\mu(\psi)$ the probability for obtaining the ψ state.

Then at the first post, $1 - \epsilon(\psi, \mathbf{a})$ equals the probability for observing -1, and the average for a given ψ is

$$+1[\epsilon(\psi, \mathbf{a})] - 1[1 - \epsilon(\psi, \mathbf{a})] = 2\epsilon(\psi, \mathbf{a}) - 1. \qquad (10.11)$$

Similarly at the second post, $1 - \eta(\psi, \mathbf{b})$ equals the probability for observing -1, and the average for the given ψ is

$$+1[\eta(\psi, \mathbf{b})] - 1[1 - \eta(\psi, \mathbf{b})] = 2\eta(\psi, \mathbf{b}) - 1. \qquad (10.12)$$

To get the overall average, $P(\mathbf{a}, \mathbf{b})$, multiply the product of (10.11) and (10.12) by the probability for the ψ state and integrate:

$$P(\mathbf{a}, \mathbf{b}) = \int d\mu(\psi)[2\epsilon(\psi, \mathbf{a}) - 1][2\eta(\psi, \mathbf{b}) - 1]. \qquad (10.13)$$

Similar expressions are obtained for $P(\mathbf{a}, \mathbf{b}')$, $P(\mathbf{a}', \mathbf{b})$, and $P(\mathbf{a}', \mathbf{b}')$.

From the difference between $P(\mathbf{a}, \mathbf{b})$ and $P(\mathbf{a}, \mathbf{b}')$, together with the limit

$$|2\epsilon(\psi, \mathbf{a}) - 1| \leqslant 1, \qquad (10.14)$$

we find that

$$|P(\mathbf{a}, \mathbf{b}) - P(\mathbf{a}, \mathbf{b}')| \leqslant \int d\mu(\psi) |2\eta(\psi, \mathbf{b}) - 2\eta(\psi, \mathbf{b}')|. \qquad (10.15)$$

In like manner, one obtains

$$|P(\mathbf{a}', \mathbf{b}) + P(\mathbf{a}', \mathbf{b}')| \leqslant \int d\mu(\psi) |2\eta(\psi, \mathbf{b}) + 2\eta(\psi, \mathbf{b}') - 2|. \qquad (10.16)$$

Because the η's vary from 0 to 1, we also find that

$$|2\eta(\psi, \mathbf{b}) - 2\eta(\psi, \mathbf{b}')| + |2\eta(\psi, \mathbf{b}) + 2\eta(\psi, \mathbf{b}') - 2| \leqslant 2. \qquad (10.17)$$

See Example 10.4.

Combining (10.15), (10.16), (10.17), with

$$\int d\mu(\psi) = 1, \qquad (10.18)$$

leads to

$$|P(\mathbf{a}, \mathbf{b}) - P(\mathbf{a}, \mathbf{b}')| + |P(\mathbf{a}', \mathbf{b}) + P(\mathbf{a}', \mathbf{b}')| \leqslant 2. \qquad (10.19)$$

In a similar way, we also obtain

$$|-P(\mathbf{a}, \mathbf{b}) + P(\mathbf{a}, \mathbf{b}') + P(\mathbf{a}', \mathbf{b}) + P(\mathbf{a}', \mathbf{b}')| \leqslant 2. \qquad (10.20)$$

Since (10.19) and (10.20) are generalizatons of results derived by Bell, they are called *Bell inequalities*. Remember, these inequalities presume that the locality assumption is valid. According to this assumption, any property exhibited by a particle at a point depends only on the state in which the particle has been prepared and on the action of the measuring device at that point. The nearly simultaneous action of a measuring device on a separate particle at a spatially removed point would have no effect.

Example 10.4

What upper limit can one put on the expression

$$|u - v| + |u + v - 1|$$

as long as

$$0 \leqslant u \leqslant 1 \quad \text{and} \quad 0 \leqslant v \leqslant 1?$$

For a particular range of u and v, one may order the enclosed terms so that each result is positive, remove the absolute value signs, and limit the result. Thus when

$$u \geqslant v \quad \text{and} \quad u + v \geqslant 1,$$

we find that

$$u - v + u + v - 1 = 2u - 1 \leqslant 1$$

When

$$u \geqslant v \quad \text{and} \quad u + v \leqslant 1,$$

we similarly obtain

$$u - v + 1 - u - v = 1 - 2v \leqslant 1.$$

On the other hand, when

$$u \leqslant v \quad \text{and} \quad u + v \geqslant 1,$$

the necessary arrangement yields

$$v - u + u + v - 1 = 2v - 1 \leqslant 1.$$

When

$$u \leqslant v \quad \text{and} \quad u + v \leqslant 1,$$

we similarly get

$$v - u + 1 - u - v = 1 - 2u \leqslant 1.$$

So in general, we have

$$|u - v| + |u + v - 1| \leqslant 1$$

as long as

$$0 \leqslant u \leqslant 1 \quad \text{and} \quad 0 \leqslant v \leqslant 1.$$

Example 10.5

Consider the spin ket that yields the eigenvalue $+\frac{1}{2}\hbar$ along a given axis. How may this ket be projected onto an inclined axis?

Choose units so that $\hbar = 2$ and the eigenvalue for the eigenket is $+1$. Projecting this value onto an axis inclined at angle α, as Figure 10.4 shows, then yields (1) $\times \cos \alpha$. But this may come from a superposition of the eigenket for spin $+1$ and the one for spin -1 along the inclined axis.

Let w_+ equal the statistical weight of the $+1$ eigenket and w_- the statistical weight of the -1 eigenket along this axis. For the spin along the axis we have

$$\bar{s} = (1) \cos \alpha = w_+(+1) + w_-(-1)$$

whence

$$w_+ - w_- = \cos \alpha.$$

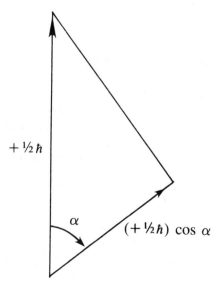

Figure 10.4. Projection of a spin angular momentum onto an axis inclined at angle α.

But the sum of the weights must equal 1:

$$w_+ + w_- = 1.$$

Solving these simultaneous equations yields

$$w_+ = \tfrac{1}{2} + \tfrac{1}{2}\cos\alpha = \cos^2(\tfrac{1}{2}\alpha)$$

and

$$w_- = \tfrac{1}{2} - \tfrac{1}{2}\cos\alpha = \sin^2(\tfrac{1}{2}\alpha).$$

So a superposition representing the projection is

$$|0, +1\rangle = \cos(\tfrac{1}{2}\alpha)|\alpha, +1\rangle + \sin(\tfrac{1}{2}\alpha)|\alpha, -1\rangle$$

where $|\alpha, +1\rangle$ and $|\alpha, -1\rangle$ are the eigenkets for the spin $+1$ and for spin -1 along the inclined axis.

10.7
Correlations Between Spatially Separated Coherent Fermions

But if nonlocality prevails and the spatially separated systems form a coherent whole, the orientation chosen at the first observation post does affect the result obtained at the second observation post. As a result, the correlations are different and for some relative configurations, the Bell inequalities are violated.

Consider a zero-spin system that disintegrates into two indistinguishable, paired, spin-$\tfrac{1}{2}\hbar$ particles. A suitable state function is

$$\frac{1}{\sqrt{2}}[f(\mathbf{r}_1)g(\mathbf{r}_2) + f(\mathbf{r}_2)g(\mathbf{r}_1)]\frac{1}{\sqrt{2}}[|1^{\mathbf{a}}_\uparrow\rangle|2^{\mathbf{a}}_\downarrow\rangle - |1^{\mathbf{a}}_\downarrow\rangle|2^{\mathbf{a}}_\uparrow\rangle], \qquad (10.21)$$

where $f(\mathbf{r}_j)$ is the spatial factor for motion in the first region, $g(\mathbf{r}_j)$ is the spatial factor for motion in the second region, $\mathbf{r}_j$ locates the jth particle, $|j^{\mathbf{a}}_\uparrow\rangle$ is the spin ket for the jth particle with the spin unit in the $+$ or $-$ direction along $\mathbf{a}$. This form is symmetric in the spatial coordinates, antisymmetric in the spin coordinates. So when the two particles are interchanged, the complete state function changes sign in accordance with the Pauli exclusion principle.

Projecting the spin eigenkets for the second system along $\mathbf{b}$, as Example 10.5 shows, gives

$$|2^{\mathbf{a}}_\uparrow\rangle = \cos\tfrac{1}{2}(\mathbf{a}, \mathbf{b})|2^{\mathbf{b}}_\uparrow\rangle + \sin\tfrac{1}{2}(\mathbf{a}, \mathbf{b})|2^{\mathbf{b}}_\downarrow\rangle, \qquad (10.22)$$

where $(\mathbf{a}, \mathbf{b})$ represents the angle between $\mathbf{a}$ and $\mathbf{b}$. Consequently, the spin factor in (10.21) can be rewritten in the form

$$\frac{1}{\sqrt{2}}[\cos\tfrac{1}{2}(\mathbf{a}, \mathbf{b})|1^{\mathbf{a}}_\uparrow\rangle|2^{\mathbf{b}}_\uparrow\rangle + \sin\tfrac{1}{2}(\mathbf{a}, \mathbf{b})|1^{\mathbf{a}}_\uparrow\rangle|2^{\mathbf{b}}_\downarrow\rangle$$

$$- \cos \tfrac{1}{2}(\mathbf{a}, \mathbf{b}) \,|\, 1^{\underline{a}}_{+} \rangle \,|\, 2^{\underline{b}}_{+} \rangle - \sin \tfrac{1}{2}(\mathbf{a}, \mathbf{b}) \,|\, 1^{\underline{a}}_{+} \rangle \,|\, 2^{\underline{b}}_{-} \rangle]. \qquad (10.23)$$

The different spin-ket products in (10.23) are orthogonal and normalized. The orbital factor is also assumed to be normalized.

Now, the product of the bra of (10.23) with (10.23) itself yields

$$\tfrac{1}{2}[\cos^2 \tfrac{1}{2}(\mathbf{a}, \mathbf{b}) + \sin^2 \tfrac{1}{2}(\mathbf{a}, \mathbf{b}) + \cos^2 \tfrac{1}{2}(\mathbf{a}, \mathbf{b}) + \sin^2 \tfrac{1}{2}(\mathbf{a}, \mathbf{b})] = 1. \; (10.24)$$

The terms in (10.24) represent the probabilities for observing

$$|\, 1^{\underline{a}}_{+} \rangle \,|\, 2^{\underline{b}}_{-} \rangle, \; |\, 1^{\underline{a}}_{+} \rangle \,|\, 2^{\underline{b}}_{+} \rangle, \; |\, 1^{\underline{a}}_{-} \rangle \,|\, 2^{\underline{b}}_{+} \rangle, \; |\, 1^{\underline{a}}_{-} \rangle \,|\, 2^{\underline{b}}_{-} \rangle, \qquad (10.25)$$

respectively. Multiplying each by the corresponding spin product, with $\hbar = 2$, and adding yields

$$\begin{aligned}
P(\mathbf{a}, \mathbf{b}) &= \tfrac{1}{2}[(-1) \cos^2 \tfrac{1}{2}(\mathbf{a}, \mathbf{b}) + (+1) \sin^2 \tfrac{1}{2}(\mathbf{a}, \mathbf{b}) \\
&\quad + (-1) \cos^2 \tfrac{1}{2}(\mathbf{a}, \mathbf{b}) + (+1) \sin^2 \tfrac{1}{2}(\mathbf{a}, \mathbf{b})] \\
&= -\cos^2 \tfrac{1}{2}(\mathbf{a}, \mathbf{b}) + \sin^2 \tfrac{1}{2}(\mathbf{a}, \mathbf{b}) \\
&= -\cos (\mathbf{a}, \mathbf{b}). \qquad\qquad (10.26)
\end{aligned}$$

With the four arrangements of (10.20), we would have

$$-P(\mathbf{a}, \mathbf{b}) + P(\mathbf{a}, \mathbf{b}') + P(\mathbf{a}', \mathbf{b}) + P(\mathbf{a}', \mathbf{b}')$$
$$= \cos (\mathbf{a}, \mathbf{b}) - \cos (\mathbf{a}, \mathbf{b}') - \cos (\mathbf{a}', \mathbf{b}) - \cos (\mathbf{a}', \mathbf{b}'). \qquad (10.27)$$

Example 10.6

In an EPR-type experiment with spin-$\tfrac{1}{2}\hbar$ particles, analysis is by Stern-Gerlach devices oriented as Figure 10.5 indicates. Determine the angle ϑ at which the quantity

$$|-P(\mathbf{a}, \mathbf{b}) + P(\mathbf{a}, \mathbf{b}') + P(\mathbf{a}', \mathbf{b}) + P(\mathbf{a}', \mathbf{b}')|$$

would be as large as possible.

For the arrangement in Figure 10.5, we have the angles

$$(\mathbf{a}, \mathbf{b}) = 3\vartheta, \quad (\mathbf{a}, \mathbf{b}') = \vartheta, \quad (\mathbf{a}', \mathbf{b}) = \vartheta, \quad (\mathbf{a}', \mathbf{b}') = \vartheta.$$

Then

$$\begin{aligned}
-P(\mathbf{a}, \mathbf{b}) &+ P(\mathbf{a}, \mathbf{b}') + P(\mathbf{a}', \mathbf{b}) + P(\mathbf{a}', \mathbf{b}') \\
&= -\cos 3\vartheta + \cos \vartheta + \cos \vartheta + \cos \vartheta \\
&= -\cos 3\vartheta + 3\cos \vartheta.
\end{aligned}$$

The extremum occurs where the derivative of this expression is zero. We have

$$\frac{d}{d\vartheta}(-\cos 3\vartheta + 3\cos \vartheta) = 3\sin 3\vartheta - 3\sin \vartheta = 0,$$

whence

$$\sin \vartheta = \frac{1}{\sqrt{2}}$$

and

$$\vartheta = 45°.$$

Example 10.7

Consider spin-$\frac{1}{2}\hbar$ particles being paired, separated, and analyzed in an EPR-type experiment. If the measurements are by Stern-Gerlach devices oriented as Figure 10.5 shows, with $\vartheta = 45°$, by how much is the pertinent Bell inequality violated?

We choose units so that $\frac{1}{2}\hbar = 1$, $-\frac{1}{2}\hbar = -1$. Inserting the given angles into (10.27) then yields

$$|-P(\mathbf{a}, \mathbf{b}) + P(\mathbf{a}, \mathbf{b}') + P(\mathbf{a}', \mathbf{b}) + P(\mathbf{a}', \mathbf{b}')|$$

$$= |\cos 135° - \cos 45° - \cos 45° - \cos 45°|$$

$$= \left| -\frac{1}{\sqrt{2}} - \frac{1}{\sqrt{2}} - \frac{1}{\sqrt{2}} - \frac{1}{\sqrt{2}} \right| = \frac{4}{\sqrt{2}} = 2\sqrt{2} > 2.$$

The sum is larger than 2 by the factor $\sqrt{2}$. Thus, the quantum correlation functions violate Bell inequality (10.20).

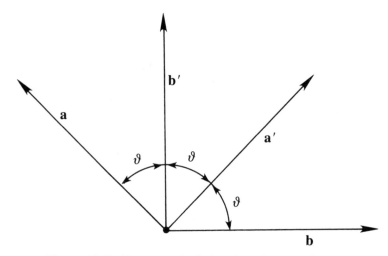

Figure 10.5. The Stern-Gerlach orientations employed.

This experiment has been carried out with protons, as described in Example 10.3(a). The results support the quantum mechanical result above, when one allows for the errors inherent in the measurements.

10.8
A Bell Inequality for Pairs of Pass-or-Fail Filters

Each device employed in an EPR experiment may merely allow particles with a certain polarization to pass, rather than measure plus or minus a unit quantity. Passage may be recorded as $+1$, absorption as 0. Thus, a person could monitor the emission of photon pairs using spatially separated polarization filters. The argument of Section 10.6 must be altered correspondingly.

Again we consider two distinct systems which after interaction are studied in two spatially separated regions. Within each region a transverse property is measured. We suppose that with the given particles the possible values are $+1$ or 0 unit.

As before, base vectors $\mathbf{a}$, $\mathbf{a}'$, . . . define the directions chosen perpendicular to the path of the first particle, at the first position, while base vectors $\mathbf{b}$, $\mathbf{b}'$, ... define the directions chosen perpendicular to the path of the second particle, at the second position. Also, $P(\mathbf{a}, \mathbf{b})$ is constructed as the weighted average of the product of the value found for the first region with the value found for the second region, using directions $\mathbf{a}$ and $\mathbf{b}$, respectively.

As in Section 10.6, we assume that *locality* prevails, that the value obtained in either region is independent of what is being measured in the other region. On the other hand, this value does depend on the state the particles left the region of interaction and the direction chosen at the point of measurement.

As before, function ψ describes the state of the two-particle system. We let $\epsilon(\psi, \mathbf{a})$ be the probability for obtaining $+1$ in the first region with orientation $\mathbf{a}$ and state ψ, while $\eta(\psi, \mathbf{b})$ is the probability for obtaining $+1$ in the second region using orientation $\mathbf{b}$ and state ψ. Furthermore, $d\mu(\psi)$ is the probability for obtaining the ψ state. The resulting mean correlation product is

$$P(\mathbf{a}, \mathbf{b}) = \int d\mu(\psi)\, \epsilon(\psi, \mathbf{a})\, \eta(\psi, \mathbf{b}). \qquad (10.28)$$

We presume that emission from the interaction region is symmetric over time, with respect to the transverse directions, at each observation post. Then for a given orientation $\mathbf{a}$ of the measuring device at the first post, there is 50% probability that a particular particle will pass and

$$\overline{\epsilon(\psi, \mathbf{a})} = \int d\mu(\psi)\, \epsilon(\psi, \mathbf{a}) = \tfrac{1}{2}. \qquad (10.29)$$

For a given orientation $\mathbf{b}$ at the second observation post, we similarly have

$$\overline{\eta(\psi, \mathbf{b})} = \int d\mu(\psi)\, \eta(\psi, \mathbf{b}) = \tfrac{1}{2}. \qquad (10.30)$$

The integral in (10.28) yields the average of the product of ϵ and η:

$$P(\mathbf{a}, \mathbf{b}) = \overline{\epsilon(\psi, \mathbf{a})\, \eta(\psi, \mathbf{b})}. \tag{10.31}$$

But if the distribution of $\epsilon(\psi, \mathbf{a})$ over $\mu(\psi)$ is the same as that of $\eta(\psi, \mathbf{b})$ over $\mu(\psi)$, then

$$\overline{\epsilon(\psi, \mathbf{a})\, \eta(\psi, \mathbf{b})} = \overline{\epsilon(\psi, \mathbf{a})} \ \overline{\eta(\psi, \mathbf{b})} = (\tfrac{1}{2})(\tfrac{1}{2}) = \tfrac{1}{4}. \tag{10.32}$$

On the other hand, the distribution might differ so much that one probability is zero wherever the other is different from zero and

$$\overline{\epsilon(\psi, \mathbf{a})\, \eta(\psi, \mathbf{b})} = 0. \tag{10.33}$$

Consequently, for all possible situations we would have

$$|\, P(\mathbf{a}, \mathbf{b}) - P(\mathbf{a}, \mathbf{b}')\,| \leqslant 0.25. \tag{10.34}$$

This is the *Bell inequality* for the pass-or-fail system.

10.9
Correlations Between Spatially Separated Coherent Photons

As long as a diverging multi-photon system forms a coherent whole, nonlocality prevails and the pertinent Bell inequality may be violated. For pairs of photons the comparison may be with (10.34).

Consider a submicroscopic system that emits two photons, simultaneously or in quick succession, with no net change in spin or parity. As long as neither photon interacts with an independent system, the two may be described by (4.98), by (4.106), or by (4.108).

Then when a measurement in the first region reveals a photon polarized along $\mathbf{a}$, perpendicular to the axis of travel, the complete ket collapses to

$$|\,\mathbf{a}, \mathbf{k}_1\rangle\, |\,\mathbf{a}, \mathbf{k}_2\rangle. \tag{10.35}$$

The photon in the second region is also polarized along $\mathbf{a}$.

To calculate the probability that it would be observed polarized along $\mathbf{b}$, we have to project its electric intensity in that direction. But this projects as a vector. Since $\mathbf{a}$ and $\mathbf{b}$ have been chosen to be unit vectors, we have

$$\mathbf{a} \cdot \mathbf{b} = \cos(\mathbf{a}, \mathbf{b}). \tag{10.36}$$

By (9.21) and (9.22), the magnetic intensity is proportional to the electric intensity. So the Poynting vector (9.23) varies as

$$\cos^2(\mathbf{a}, \mathbf{b}) = \tfrac{1}{2}[1 + \cos 2(\mathbf{a}, \mathbf{b})]. \tag{10.37}$$

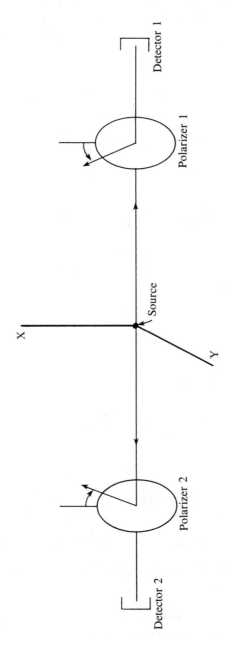

Figure 10.6. EPR experiment with photon pairs.

The photon intensity is proportional to the Poynting vector. So in our experiment, the probability for observing a photon polarized along **b** is (10.37) when one is observed polarized along **a** in the other region. See Figure 10.6 for the setup.

In the symmetric case, when all orientations in the first region are equally likely, the probability for observing a photon along **a** is ½. Consequently, the joint probability is

$$P(\mathbf{a}, \mathbf{b}) = (\tfrac{1}{2})(\tfrac{1}{2})[1 + \cos 2(\mathbf{a}, \mathbf{b})]$$

$$= \tfrac{1}{4}[1 + \cos 2(\mathbf{a}, \mathbf{b})]. \tag{10.38}$$

With the two arrangements of (10.34), we would have

$$P(\mathbf{a}, \mathbf{b}) - P(\mathbf{a}, \mathbf{b}') = \tfrac{1}{4}(\cos 2(\mathbf{a}, \mathbf{b}) - \cos 2(\mathbf{a}, \mathbf{b}'). \tag{10.39}$$

Example 10.8

In an EPR experiment with photon pairs, angles as in Figure 10.5 were employed with $\vartheta = 22.5°$. If the polarizers were 100% efficient, by how much would Bell inequality (10.34) be violated?

From the given ϑ, we have

$$(\mathbf{a}, \mathbf{b}) = 67.5°, \qquad (\mathbf{a}, \mathbf{b}') = 22.5°.$$

Substitute these into (10.39),

$$P(\mathbf{a}, \mathbf{b}) - P(\mathbf{a}, \mathbf{b}') = \tfrac{1}{4}(\cos 135° - \cos 45°)$$

$$= \tfrac{1}{4}\left(-\frac{\sqrt{2}}{2} - \frac{\sqrt{2}}{2}\right) = -\frac{\sqrt{2}}{4} = -0.3536.$$

and the result into the left side of (10.34):

$$|P(67.5°) - P(22.5°)| = 0.3536.$$

This is greater than the upper limit allowed by (10.34) by over 0.1000.

In practice, polarizers are not 100% efficient. A typical experiment yields

$$|P(67.5°) - P(22.5°)| = 0.300 \pm 0.008.$$

The difference between this and 0.3536 can be explained by the apparatus inefficiency.

10.10
Altering Coherences

In a run of an EPR experiment, evidence that the particles form a coherent whole comes from the observed correlations between the widely spaced measurements.

The coherence is introduced by sufficient interaction between the particles in the source region. In a diffraction experiment, evidence for coherence among the different paths comes from the pattern produced by the particles on a screen. This coherence may similarly be produced at the source.

On the other hand, the coherence may be introduced by the way the measurement is made. In Figure 10.7, the essence of a Hanbury-Brown-Twiss (HBT) experiment is diagrammed. Sources 1 and 2 emit photons of the same energy $h\nu$ that are incoherent. The radiation is detected at points A and B, that may be astronomically separated from the sources. At various times a photon is received at A coincident with a photon received at B. The resulting product signals display the familiar diffraction pattern caused by coherent interference.

Interestingly, each detected photon originated partly in each of the two sources; it is assembled from parts contributed by the different sources. Particles like photons and electrons cannot consistently be described as blobs traveling from one point to another.

Chemical systems are incoherent assemblages of coherent molecules or ions. Reacting materials are describable by density matrices. But each product molecule is

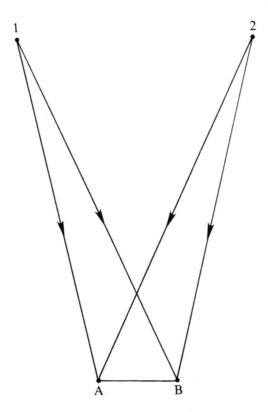

Figure 10.7. Coincident signals at A and B from incoherent sources 1 and 2 are coherent. Thus, interference effects are observed.

a coherent system. Synthesis of a molecule thus involves introducing coherence by interaction. The driving force comes from the greater stability of the coherent system. Conversely, excess energy may lead to decomposition of a substance. This excess energy is introduced by the intervention of independent molecules and so by incoherent means; the result is an incoherent breakup of the given molecules.

10.11
The Implied Action at a Distance

In a run of an EPR experiment, neither particle possesses a definite transverse property (spin or polarization) until a measurement is performed on one of the particles and the wave vector collapses. It appears that such a change of state is instantaneous in the reference frame in which the source is at rest. Alternatively, this frame may be taken as the one in which the center of mass of $\Psi(\mathbf{r}_1, \mathbf{r}_2, t)$ is at rest.

At a particular time t in this frame, a setting of either measuring device picks out one possible component of Ψ. So the orthogonal components are eliminated. The other measuring device may then determine a component of the reduced state function, as we have seen.

When the time between these operations is short enough, inertial frames can be found in which the effect precedes the cause. However, this development does not imply that the causality principle is violated, because the existence of $\Psi(\mathbf{r}_1, \mathbf{r}_2, t)$ as a spacelike structure already picks out the inertial frame in which the center of mass is at rest as a preferred frame.

A difficulty with conventional collapse theory is that it is not time symmetric (or, more precisely, CPT invariant), so some authors have suggested that the accompanying process involves both retarded and advanced de Broglie waves.

In the transaction model, diagrammed in Figure 10.8, an emitter E sends out a conventional retarded wave $\Psi(\mathbf{r}, t)$ in all possible spatial directions. Absorber A responds to the signal by sending back in time the advanced wave $\Psi(\mathbf{r}, t)$ to the emitter.

The advanced wave travels across the same spatial interval and through the same attenuating media as the retarded wave, but in reverse. If we let $\mathbf{r}$ be drawn from the emitter to the absorber, then a *unit* emitted wave is reduced to $\Psi(\mathbf{r}, t)$ at the absorber. And a *unit* echo wave would be reduced to $\Psi^*(\mathbf{r}, t)$ at the emitter. Since the echo leaving the absorber is of strength $\Psi(\mathbf{r}, t)$, rather than unity, the echo wave received back at the emitter is of strength $\Psi^*(\mathbf{r}, t)\,\Psi(\mathbf{r}, t)$. Thus, $\Psi^*\Psi$ measures the probability density at the absorber for the transaction.

The process is completed with satisfaction of the boundary conditions at the emitter and absorber loci. Thus, the transactional interpretation does not alter the existing formalism of quantum mechanics. However, it does offer an explanation for axiom (IV′) and for axiom (II).

In a run of the EPR experiment, the transaction involves one emitter and two absorbers on a straight line with the emitter. The correlation at a distance is imposed by satisfaction of the boundary conditions at the emitter.

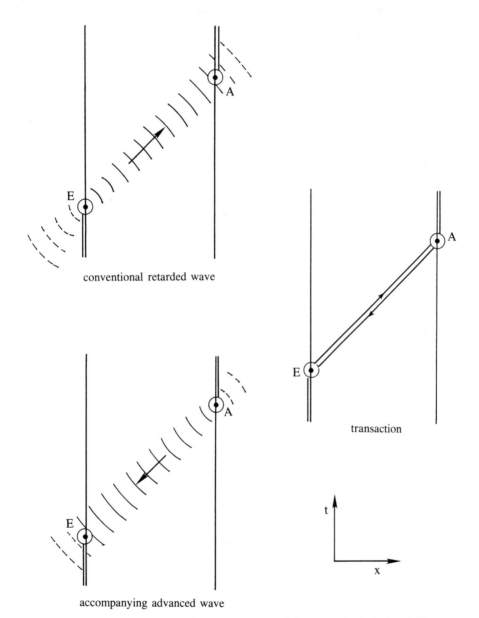

conventional retarded wave

accompanying advanced wave

transaction

Figure 10.8. Basic elements of the transaction model, according to John G. Cramer (*Rev. Mod. Phys.* **58**, 647-687).

Example 10.9

Construct planar retarded and advanced monochromatic de Broglie waves. Let the spatial coordinate be x and the temporal coordinate t. Also let k be the magnitude of the wavevector and ω the angular frequency.

In the *retarded* wave, a given phase moves forward in time. So for the monochromatic wave traveling in the positive direction, left to right, we have

$$\Psi_1 = Ae^{ikx}e^{-i\omega t}.$$

But for the monochromatic wave traveling in the negative direction, right to left, the sign on wavevector k is changed and

$$\Psi_2 = Ae^{ik(-x)}e^{-i\omega t}.$$

In the *advanced* wave, a given phase travels backward in time. So such a wave is obtained from the corresponding retarded wave by reversing the sign on t. For the monochromatic wave traveling in the positive direction, we find

$$\Psi_3 = Ae^{ikx}e^{-i\omega(-t)}.$$

For the monochromatic wave traveling in the negative direction, we similarly find

$$\Psi_4 = Ae^{ik(-x)}e^{-i\omega(-t)}.$$

When the phases are chosen so that $A = A^*$, we would have

$$\Psi_4 = \Psi_1^*.$$

Example 10.10

Construct spherical retarded and advanced monochromatic de Broglie waves. Let the radial coordinate measured from the center be r, while the temporal coordinate is t. Let k be the magnitude of the wavevector and ω the angular frequency.

We consider the wavelets to travel radially, so the amplitude factor has the form

$$Af(k, \vartheta, \varphi)/r.$$

In the *retarded* wave, a given phase moves forward in time. The monochromatic wave traveling radially outward then has the form

$$\Psi_1 = A\,\frac{f}{r}\,e^{ikr}e^{-i\omega t}.$$

To reverse the direction of motion, we change the sign on the wavevector. The negative sign is then moved to coordinate r to give

$$\Psi_2 = A \frac{f}{r} e^{ik(-r)}e^{-i\omega t}.$$

For the *advanced* wave, a given phase travels backward in time. The formula for such a wave is obtained from that for the corresponding retarded wave by reversing the sign on coordinate t. The monochromatic wave traveling radially outward is therefore

$$\Psi_3 = A \frac{f}{r} e^{ikr}e^{-i\omega(-t)},$$

while the monochromatic wave traveling radially inward is

$$\Psi_4 = A \frac{f}{r} e^{ik(-r)}e^{-i\omega(-t)}.$$

With $Af = A^*f^*$, we would have

$$\Psi_4 = \Psi_1^*.$$

The various possible phase shifts at an absorber can be introduced through an imaginary component in Af.

10.12
General Comments

We have seen how the symmetry of the space-time continuum, the localizability of each particle, and the negligible effects of interactions for infinitesimal changes in time led to construction of suitable commutator relations and forms for the key operators. Interactions over finite changes are expressed by potentials in the Hamiltonian.

Any symmetry in the potentials is reflected in the forms assumed by the eigenfunctions. Indeed, these are found to belong to primitive symmetry species of the covering group for the system.

Consistent with these developments, violations of the Bell inequalities occur in a manner that respects the symmetry of each individual coherent system. In the EPR effect, the coherence and symmetry are maintained over large distances.

Furthermore, by introducing advanced as well as retarded de Broglie waves, in the transaction model, one can give individual processes a symmetry over time that otherwise they would not have.

Discussion Questions

10.1. In determining the nature of the transition process, why do we want to observe the behavior of a single submicroscopic system?

10.2. If only a small fraction of the photons emitted can be detected, how can individual transitions be observed?

10.3. In Figure 10.3, why doesn't the photon count per second fall to zero when the ion is in the shelf state?

10.4. How do the experimental results preclude the appearance of a coherent superposition of states?

10.5. Explain the passive nature of a measurement on a classical system.

10.6. When is a measurement on a quantum system (a) passive, (b) active?

10.7. What happens to the state function when one measures an indefinite property, obtaining a definite value? Describe such a measurement.

10.8. How does an atom or molecule in a definite state exhibit the nonlocality of its electrons?

10.9. In an EPR experiment, why should the properties in one of the separated regions be independent of what measurement is being carried out in the other region?

10.10. How are units chosen so that a particle with $S = \frac{1}{2}$ exhibits an angular momentum of $+1$ or -1 with respect to a given direction?

10.11. Show how Inequality (10.15) follows from Equation (10.13).

10.12. How do the results in Example 10.4 apply to Inequality (10.17)?

10.13. Define the correlation probability $P(\mathbf{a}, \mathbf{b})$ for fermions.

10.14. Why are the Bell inequalities important?

10.15. What causes the two-particle state function to collapse in an EPR experiment?

10.16. How does one obtain the Bell inequality for pass or fail testing?

10.17. Explain how the planar polarization of a photon is projected onto a different plane.

10.18. How does the preceding result lead to a quantum theoretical $P(\mathbf{a}, \mathbf{b})$ for photons?

10.19. What is the source of the coherence in a run of the EPR experiment?

10.20. What is the source of the coherence in a conventional diffraction experiment?

10.21. How may coherence be introduced in the measurement process?

10.22. How is coherence introduced in the synthesis of a molecule?

10.23. How is coherence eliminated in the decomposition of a molecule?

10.24. Show that the conventional de Broglie planar wave is a retarded wave.

10.25. Show how the retarded planar de Broglie wave can be transformed to an advanced wave.

10.26. How does the transfer of a particle from an emitter to an absorber proceed, according to the transaction model?

Problems

10.1. Determine the limits on

$$|u - v| + |u + v - 1|$$

when

$$0 \leqslant u \leqslant A \quad \text{and} \quad 0 \leqslant v \leqslant A.$$

10.2. An EPR experiment is being run with pairs of spin-$\frac{1}{2}\hbar$ particles with zero total spin. The Stern-Gerlach devices are oriented as Figure 10.5 shows with $\vartheta = 30°$.

By how much would the Bell inequality be violated if the conditions were ideal?

10.3. From (10.21), the spin ket for $S = 0$, $M_S = 0$ is

$$| 0, 0^a \rangle = \frac{1}{\sqrt{2}} [| 1^a_{\uparrow} \rangle | 2^a_{\downarrow} \rangle - | 1^a_{\downarrow} \rangle | 2^a_{\uparrow} \rangle].$$

Similarly, the spin kets for $S = 1$ and $M_S = 1, 0, -1$ with respect to axis a are

$$| 1, 1^a \rangle = | 1^a_{\uparrow} \rangle | 2^a_{\uparrow} \rangle,$$

$$| 1, 0^a \rangle = \frac{1}{\sqrt{2}} [| 1^a_{\uparrow} \rangle | 2^a_{\downarrow} \rangle + | 1^a_{\downarrow} \rangle | 2^a_{\uparrow} \rangle],$$

$$| 1, -1^a \rangle = | 1^a_{\downarrow} \rangle | 2^a_{\downarrow} \rangle.$$

If the first particle is observed at r_1 with $+1$ spin unit in direction **a**, what would the spin of the second particle be along a parallel **a** at position r_2? Express the resulting collapsed state as a superposition of $| 0, 0^a \rangle$ and $| 1, 0^a \rangle$.

10.4. An EPR experiment is being run with pairs of photons with zero total spin. The polarizers are oriented as Figure 10.5 shows with $\vartheta = 20°$. By how much would one expect the pertinent Bell inequality to be violated under ideal conditions?

10.5. The system described in Problem 10.4 is being studied with a varying ϑ. At what angle would the violation of the Bell inequality disappear?

— — —

10.6. Introduce the appropriate approximations into the expressions for $P(\mathbf{a}, \mathbf{b})$, $P(\mathbf{a}, \mathbf{b}')$, $P(\mathbf{a}', \mathbf{b})$, $P(\mathbf{a}', \mathbf{b}')$ and derive inequality (10.20).

10.7. Determine how the spin ket for angular momentum $\frac{1}{2}\hbar$ along the z axis projects onto a line inclined at angle (a) $30°$, (b) $45°$, (c) $90°$ from the z axis.

10.8. After interacting, paired spin-$\frac{1}{2}\hbar$ particles are separated and observed with Stern-Gerlach devices deployed as Figure 10.5 indicates. If $\vartheta = 20°$, by how much is the pertinent Bell inequality violated over a series of runs?

10.9. A two-photon decay with $\Delta S = 0$ is being studied with polarizers and detectors as in Figure 10.6, angles as in Figure 10.5. If $\vartheta = 25°$, by how much is the Bell inequality violated?

10.10. Employ the setup in Problem 10.9 with a varying ϑ. At what angle would one expect the greatest deviation from the limit imposed by the pertinent Bell inequality?

10.11. At the maximum obtained in Problem 10.10, by how much is the Bell inequality violated?

References

Books

Bunge, M.: 1985. *Treatise on Basic Philosophy*, vol. 7, part I, *Formal and Physical Sciences*, Reidel, Dordrecht, pp. 191-219. Here Bunge discusses in considerable depth the nature of quantum entities, quantum mechanical measurements, separability, and the Bell inequalities.

d'Espagnet, B.: 1976. *Conceptual Foundations of Quantum Mechanics*, 2nd edn., Benjamin, Reading, Mass., pp. 76-293. In this book various Bell inequalities are developed, the question of hidden variables raised, and the observed nonseparability in nature discussed.

Greenberger, D. M. (editor): 1986. *New Techniques and Ideas in Quantum Measurement Theory*, Ann. N.Y. Acad. Sci. vol. 480, pp. 1-632. This annal deals with fundamental problems in quantum theory. Included are discussions of space-time, macroscopic effects, photon experiments, neutron and electron interferometry, and magnetic flux effects, by many different authors.

Schlegel, R.: 1980. *Superposition and Interaction*, University of Chicago Press, Chicago, pp. 139-196. In Chapters 6 and 7 Schlegel treats the locality and measurement problems in an interesting, readable manner.

Articles

Balian, R.: 1989. "On the Principles of Quantum Mechanics and the Reduction of the Wave Packet," *Am. J. Phys.* **57**, 1019-1027.

Ballentine, L. E., and Jarrett, J. P.: 1987. "Bell's Theorem: Does Quantum Mechanics Contradict Relativity?," *Am. J. Phys.* **55**, 696-701.

Beauregard, O. C. de: 1983. "Running Backwards the Mermin Device: Causality in EPR Correlations," *Am. J. Phys.* **51**, 513-516.

Benatti, F., Ghirardi, G. C., Rimini, A., and Weber, T.: 1987. "Quantum Mechanics with Spontaneous Localization and the Quantum Theory of Measurement," *Nuovo Cimento* **100B**, 27-41.

Bennett, C. L.: 1987. "Precausal Quantum Mechanics," *Phys. Rev. A* **36**, 4139-4148.

Bergquist, J. C., Hulet, R. G., Itano, W. M., and Wineland, D. J.: 1986. "Observation of Quantum Jumps in a Single Atom," *Phys. Rev. Lett.* **57**, 1699-1702.

Bitbol, M.: 1988. "The Concept of Measurement and Time Symmetry in Quantum Mechanics," *Phil. Sci.* **55**, 349-375.

Blatt, R., and Zoller, P.: 1988. "Quantum Jumps in Atomic Systems," *Eur. J. Phys.* **9**, 250-256.

Bussey, P. J.: 1988. "Comments on 'When is a Quantum Measurement?' ", *Am. J. Phys.* **56**, 569-570.

Cini, M., De Maria, M., Mattioli, G., and Nicolo, F.: 1979. "Wave Packet Reduction in Quantum Mechanics: A Model of a Measuring Apparatus," *Found. Phys.* **9**, 479-500.

Cohen-Tannoudji, C., and Dalibard, J.: 1986. "Single-Atom Laser Spectroscopy. Looking for Dark Periods in Fluorescence Light," *Europhys. Lett.* **1**, 441-448.

Cramer, J. G.: 1986. "The Transactional Interpretation of Quantum Mechanics," *Rev. Mod. Phys.* **58**, 647-687.

Dehmelt, H.: 1990. "Less is More: Experiments with an Individual Atomic Particle at Rest in Free Space," *Am. J. Phys.* **58**, 17-27.

Dicke, R. H.: 1981. "Interaction-Free Quantum Measurements: A Paradox?", *Am. J. Phys.* **49**, 925-930.

Dicke, R. H.: 1989. "Quantum Measurements, Sequential and Latent," *Found. Phys.* **19**, 385-395.

Dotson, A. C.: 1986. "Bell's Theorem and the Features of Physical Properties," *Am. J. Phys.* **54**, 218-221.

Fine, A.: 1989. "Correlations and Efficiency: Testing the Bell Inequalities," *Found. Phys.* **19**, 453-478.

Fox, J. R.: 1983. "Nonreduction of the State Vector in Quantum Measurement," *Am. J. Phys.* **51**, 49-53.

Franson, J. D.: 1989. "Bell Inequality for Position and Time," *Phys. Rev. Lett.* **62**, 2205-2208.

Friedrichs, K. O.: 1979. "Remarks on the Notion of State in Quantum Mechanics," *Found. Phys.* **9**, 515-524.

Garrett, A. J. M.: 1990. "Bell's Theorem, Inference, and Quantum Transactions," *Found. Phys.* **20**, 381-402.

Gillespie, D. T.: 1986. "Untenability of Simple Ensemble Interpretations of Quantum Measurement Probabilities," *Am. J. Phys.* **54**, 889-894.

Greenberger, D. M., and YaSin, A.: 1989. " 'Haunted' Measurements in Quantum Theory," *Found. Phys.* **19**, 679-704.

Griffiths, R. B.: 1987. "Correlations in Separated Quantum Systems: A Consistent History Analysis of the EPR Problem," *Am. J. Phys.* **55**, 11-17.

Hall, M. J. W.: 1989. "Quantum Mechanics and the Concept of Joint Probability," *Found. Phys.* **19**, 189-207.

Harrison, D.: 1982. "Bells' Inequality and Quantum Correlations," *Am. J. Phys.* **50**, 811-816.

Helliwell, T. M., and Konkowski, D. A.: 1983. "Causality Paradoxes and Nonparadoxes: Classical Superluminal Signals and Quantum Measurements," *Am. J. Phys.* **51**, 996-1003.

Hellman, G.: 1987. "EPR, Bell, and Collapse: A Route Around 'Stochastic' Hidden Variables," *Philosophy of Science* **54**, 558-576.

Herbert, N.: 1975. "Cryptographic Approach to Hidden Variables," *Am. J. Phys.* **43**, 315-316.

Itano, W. M., Heinzen, D. J., Bollinger, J. J., and Wineland, D. J.: 1990. "Quantum Zeno Effect," *Phys. Rev. A* **41**, 2295-2300.

Mattuck, R. D.: 1982. "Bell's Inequality and 'Ghost-Like Action-at-a-Distance' in Quantum Mechanics," *Eur. J. Phys.* **3**, 113-118.

Mattuck, R. D.: 1982. "Ghost-Like Action-at-a-Distance in Quantum Mechanics: An Elementary Introduction to the Einstein, Podolsky, Rosen Paradox," *Eur. J. Phys.* **3**, 107-112.

Muynck, W. M. de: 1986. "On the Relation between the Einstein-Podolsky-Rosen Paradox and the Problem of Nonlocality in Quantum Mechanics," *Found. Phys.* **16**, 973-1002.

Nagourney, W., Sandberg, J., and Dehmelt, H.: 1986. "Shelved Optical Electron Amplifier: Observation of Quantum Jumps," *Phys. Rev. Lett.* **56**, 2797-2799.

Ou, Z. Y., and Mandel, L.: 1988. "Violation of Bell's Inequality and Classical Probability in a Two-Photon Correlation Experiment," *Phys. Rev. Lett.* **61**, 50-53.

Pegg, D. T.: 1982. "Time-Symmetric Electrodynamics and the Kocher-Commins Experiment," *Eur. J. Phys.* **3**, 44-49.

Peres, A.: 1984. "The Classic Paradoxes of Quantum Theory," *Found. Phys.* **14**, 1131-1145.

Peres, A.: 1984. "What is a State Vector?", *Am. J. Phys.* **52**, 644-650.

Peres, A.: 1986. "When is a Quantum Measurement?", *Am. J. Phys.* **54**, 688-692.

Rastall, P.: 1983. "The Bell Inequalities," *Found. Phys.* **13**, 555-570.

Robinson, A. L.: 1982. "Loophole Closed in Quantum Mechanics Test," *Science* **219**, 7 Jan. 40-41.

Robinson, A. L: 1986. "Quantum Jumps Seen in a Single Ion," *Science* **234**, 3 Oct., 24-25.

Seipp, H. P.: 1986. "Bell's Theorem and an Explicit Stochastic Local Hidden-Variable Model," *Found. Phys.* **16**, 1143-1152.

Singh, I., and Whitaker, M. A. B.: 1982. ''Role of the Observer in Quantum Mechanics and the Zeno Paradox,'' *Am. J. Phys.* **50**, 882-887.

Stapp, H. P.: 1979. ''Whiteheadian Approach to Quantum Theory and the Generalized Bell's Theorem,'' *Found. Phys.* **9**, 1-25.

Stapp, H. P.: 1985. ''Bell's Theorem and the Foundations of Quantum Physics,'' *Am. J. Phys.* **53**, 306-317.

Stedman, G. E.: 1985. ''Lecture Demonstration of the Incompatibility of Quantum Predictions with those of a Local Realistic Theory,'' *Am. J. Phys.* **53**, 1143-1149.

Verovnik, I., and Likar, A.: 1988. ''A Fluctuation Interferometer,'' *Am. J. Phys.* **56**, 231-234.

Villars, C. N.: 1984. ''Observables, States and Measurements in Quantum Physics,'' *Eur. J. Phys.* **5**, 177-183.

11

Quantum Spinors

11.1
Introduction

The theory we have developed in the preceding chapters is consistent with Galilean relativity; in a sense, it is a relativistic theory. Furthermore, the possible translational and rotational symmetries of a particle, together with the property of acting at a point, implied that any given particle may have spin. This spin may be either integral $\hbar$ or half-integral $\hbar$ in magnitude.

A system that appears unchanged by a Galilean transformation does not have an invariant wave function, however. Rather, the phase of the function transforms as Equations (2.14) and (2.108) describe.

But under extreme conditions (high pertinent energies, for instance), or where high accuracy is achieved observationally, deviations from the theory become evident. A reason is that, so far, we have not considered the proper connection between time and space. Instead, time has been treated as a 1-dimensional continuum separate from the 3-dimensional continuum of space.

In 1905, Albert Einstein challenged the idea that time is universal and absolute. Indeed, he was led to replace Galilean relativity, which Newtonian mechanics respects, with the relativity that Maxwellian electromagnetism suggests. The invariance of $x^2 + y^2 + z^2$ was replaced with the invariance of $x^2 + y^2 + z^2 - c^2t^2$. Hermann Minkowski, in 1908, noted that this new invariance implies the existence of a single space-time continuum in which the square of the displacement of one point from another may be positive, zero, or negative.

With this Minkowski geometry, the classical invariance of the physical space about a point event has been replaced by invariance of the light hypercone whose vertex is at the event. As a consequence, a person may base a properly relativistic quantum mechanics upon properties on such hypercones. In the formulation, the state concept becomes altered, as we will see.

A particle that travels at the fundamental speed c is governed by an entity whose magnitude is invariant along a nondiverging ray on such a hypersurface. The function itself is a spinor with a vanishing spinorial derivative. A particle that travels more slowly is governed by spinorial functions that are coupled together on such hypersurfaces.

Example 11.1

What are the bases for the myth that time is separate from space, universal and absolute? Why is this idea wrong?

In everyday life a person projects each moment that he or she experiences instantaneously on other objects throughout the universe, without thought. Thus, one imagines that each external object experiences the same time as he or she does.

If the universe were homogeneous and if all sections were at rest with respect to each other, there is no reason why they should not be subject to the same flow of time. But when the bodies or objects move with respect to each other, spatial changes are also involved.

Furthermore, there is no way of communicating instantaneously among the moving bodies. There is an upper limit to the velocity at which signals without precursors can be sent. So the visualized projection is not realizable with any feasible instruments.

Einstein's theory takes into account only what is in principle possible. And it leads to a simple structure of space-time that is supported by numerous experiments.

11.2
The Nature of Space and Time

In the past, space has been surveyed using material standards—measuring sticks, rods, or tapes. Such a standard was presumed to be fixed in length, the same to an observer whether he was moving or at rest with respect to the tool. So the distance between two spatially separated points was presumed to be definite.

Time has been measured with clocks. A suitable one counts the cycles, or given fractional cycles, of a standard repetitive mechanism. Similar mechanisms may, in principle, be placed at desired positions throughout the universe. Different clocks at rest with respect to each other may then be synchronized by very slowly transporting another clock among the positions. A clock transported at a high speed between two given clocks need not give the same result as one transported more slowly.

By itself, space appears to be homogeneous and uniform. Small accessible regions are approximately Euclidean. Indeed, when these are surveyed accurately, definite deviations from the Pythagorean theorem are not found. Furthermore, the time t of a given clock, essentially free of forces, appears to increase smoothly and uniformly.

A reference frame with respect to which all force-free bodies are unaccelerated is called *inertial*. Now, the hypothesis that the space of each such frame is Euclidean and that the corresponding time is uniform and homogeneous does not necessarily lead to invariant distances and times. If Cartesian coordinate axes are erected in two different inertial frames with the origins at the same point event, and if the coordinates of a second such event are x, y, z at time t in the first frame, x', y', z' at time t' in the second frame, one finds that

$$x^2 + y^2 + z^2 - c^2t^2 = x'^2 + y'^2 + z'^2 - c^2t'^2 = s^2. \qquad (11.1)$$

The common magnitude squared is designated s^2 by definition. A possible relative orientation of the spatial axes is depicted in Figure 11.1.

The vector **s** drawn from the common origin to the second point event is said to have components x, y, x, ct in the first inertial frame and x', y', z', ct' in the second inertial frame. The transformation from one inertial frame to another leaves this vector invariant. However, quantity s^2 may be positive, zero, or negative.

A particle may travel from the common origin to the second point event at constant velocity. If its speed in the first frame is v and in the second frame v', then (11.1) can be rewritten in the form

$$v^2t^2 - c^2t^2 = v'^2t'^2 - c^2t'^2 = s^2. \qquad (11.2)$$

But if a point travels so that

$$v = c, \qquad (11.3)$$

then s^2 is zero and

$$v' = c. \qquad (11.4)$$

Thus, a disturbance that travels at speed c in one inertial frame travels at this same speed in all other inertial frames.

Because of this unique role, we call c the *fundamental speed* of the space-time continuum. Insofar as physicists can tell, light travels at this universal speed in free space.

In spreading out from a point event, a wave front traveling at speed c traces out half a hypercone. The other half would be developed by a wave front coming in symmetrically from infinity to the given vertex at speed c. The entire hypersurface is called the light hypercone (or simply the *light cone*) based on the point event. The hypercone based on the origin is depicted in Figure 11.2.

11.3
First and Second Rank Spinors

Each position on the positive-time half of the light cone with vertex at the origin can be represented by a 2-dimensional complex vector. And each position on the negative-time half corresponds to a different 2-dimensional complex vector. The vectors may conveniently be represented by two-element matrices. A product of such

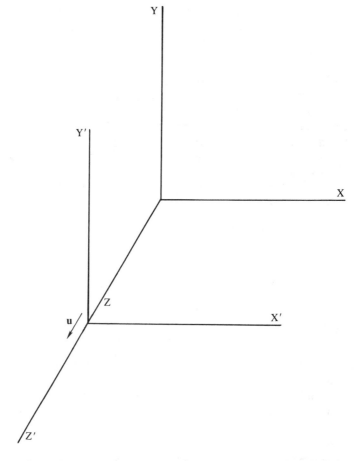

Figure 11.1. Orientation of axes for two inertial frames where the relative motins are in the z and the $-z'$ directions.

a matrix with its Hermitian adjoint is a 2 × 2 complex matrix that represents all points in space-time. As a consequence, either complex matrix can represent any physical vector in the Minkowski continuum.

From (11.1), the square of the radius vector in the Minkowski continuum is

$$s^2 = x^2 + y^2 + z^2 - c^2t^2. \tag{11.5}$$

Let us factor the first two terms and the last two,

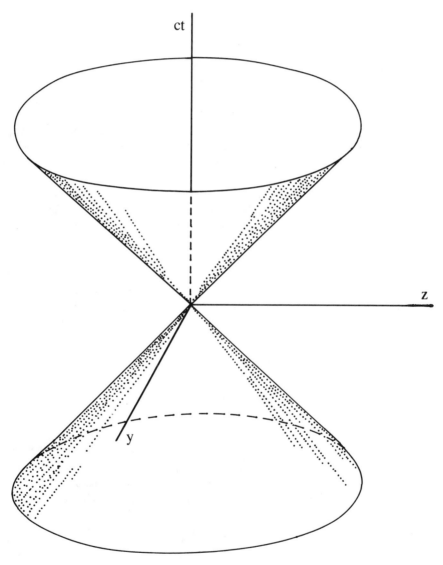

Figure 11.2. Section of the light cone based on the origin.

$$s^2 = (x + iy)(x - iy) + (z + ct)(z - ct), \qquad (11.6)$$

and set

$$x + iy = x_+, \qquad (11.7)$$

$$x - iy = x_-, \qquad (11.8)$$

$$z + ct = z_+, \qquad (11.9)$$

$$z - ct = z_-. \qquad (11.10)$$

Then

$$s^2 = x_+ x_- + z_+ z_-. \qquad (11.11)$$

On the light cone with vertex at the origin, s is zero and

$$x_+ x_- = -z_+ z_-. \qquad (11.12)$$

Since the left side of (11.12) is positive, quantities z_+ and $-z_-$ are either both positive or both negative. When they are both positive, $ct + z$ and $ct - z$ have to be positive, and t itself has to be positive. When they are both negative, $-ct - z$ and $-ct + z$ have to be positive, and t has to be negative.

Let us first consider the situation when both are positive and the event is on the future half of the light cone. Setting the value of z_+ equal to a^2, that of $-z_-$ equal to b^2, and factoring each into a number and its complex conjugate yields

$$z_+ = a^2 = S^{1*}S^1, \qquad (11.13)$$

$$-z_- = b^2 = S^2 S^{2*}, \qquad (11.14)$$

with the phases of S^1 and S^2 undetermined. Equation (11.12) then becomes

$$x_+ x_- = S^{1*}S^1 S^2 S^{2*}. \qquad (11.15)$$

On the left side, we have a number times its complex conjugate. We recognize similar factors on the right and set

$$x_+ = S^{1*}S^2 = ab\, e^{i\alpha}, \qquad (11.16)$$

$$x_- = S^1 S^{2*} = ab\, e^{-i\alpha}. \qquad (11.17)$$

The coordinates determine only the difference α in phase angle, not the absolute phase of either S^2 or S^1. See Figure 11.3.

The 2-dimensional complex vector with components S^1 and S^2 is called the *upper-index first-rank spinor* for the radius vector drawn from the origin to the point (x, y, z, ct) on the light cone. The corresponding matrix is

$$\mathbf{S}^{\cdot} = \begin{pmatrix} S^1 \\ S^2 \end{pmatrix}. \qquad (11.18)$$

The Hermitian adjoint of this is

$$\mathbf{S}^{\cdot\dagger} = (S^{1*} \quad S^{2*}).\tag{11.19}$$

Next, let us consider points on the past half of the light cone, where both z_+ and $-z_-$ are negative. Recalling again that a positive number can be factored into a complex number and its conjugate, we write

$$z_- = S_1^* S_1,\tag{11.20}$$

$$-z_+ = S_2 S_2^*.\tag{11.21}$$

Equation (11.12) is now satisfied when

$$x_+ x_- = S_1^* S_1 S_2 S_2^*.\tag{11.22}$$

As before, we separate the right side into a number times its complex conjugate and identify the former with x_+, the latter with x_-:

$$x_+ = S_1^* S_2,\tag{11.23}$$

$$x_- = S_1 S_2^*.\tag{11.24}$$

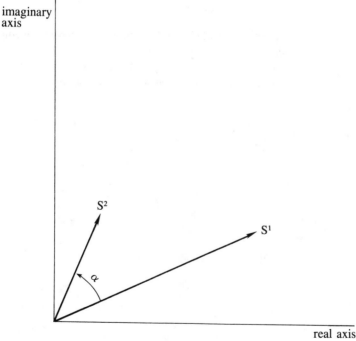

Figure 11.3. Components of spinor $\mathbf{S}^{\cdot}$, for a given (x, y, z, ct) on the light cone, plotted in the complex plane.

The 2-dimensional complex vector having components S_1 and S_2 is the *lower-index first-rank* spinor for the radius vector drawn from the origin to point (x, y, z, ct). We write

$$\mathbf{S.} = \begin{pmatrix} S_1 \\ S_2 \end{pmatrix} . \tag{11.25}$$

This matrix representation of the spinor has the Hermitian adjoint

$$\mathbf{S.}^\dagger = (S_1{}^* \quad S_1{}^*). \tag{11.26}$$

The *second-rank* upper-index spinor corresponding to (11.18) and (11.19) is obtained by multiplying the matrices:

$$\mathbf{M}^{\cdot\cdot} = \mathbf{S}^{\cdot}\mathbf{S}^{\dagger} = \begin{pmatrix} S^1 \\ S^2 \end{pmatrix} (S^{1*} \quad S^{2*}) = \begin{pmatrix} S^1 S^{1*} & S^1 S^{2*} \\ S^2 S^{1*} & S^2 S^{2*} \end{pmatrix}$$

$$= \begin{pmatrix} z_+ & x_- \\ x_+ & -z_- \end{pmatrix} = \begin{pmatrix} z + ct & x - iy \\ x + iy & -z + ct \end{pmatrix} . \tag{11.27}$$

In the fourth step, Equations (11.13), (11.14), (11.16), (11.17) have been employed; in the last step, Equations (11.7)–(11.10).

Similarly, the second-rank lower-index spinor corresponding to (11.25) and (11.26) is constructed by multiplying the matrices and reducing:

$$\mathbf{M..} = \mathbf{S.}\mathbf{S.}^{\dagger} = \begin{pmatrix} S_1 \\ S_2 \end{pmatrix} (S_1{}^* \quad S_2{}^*) = \begin{pmatrix} S_1 S_1{}^* & S_1 S_2{}^* \\ S_2 S_1{}^* & S_2 S_2{}^* \end{pmatrix}$$

$$= \begin{pmatrix} z_- & x_- \\ x_+ & -z_+ \end{pmatrix} = \begin{pmatrix} z - ct & x - iy \\ x + iy & -z - ct \end{pmatrix} . \tag{11.28}$$

A person can also form the *mixed* second-rank spinors

$$\mathbf{M}^{\cdot}. = \mathbf{S}^{\cdot}\mathbf{S.}^{\dagger} \tag{11.29}$$

and

$$\mathbf{M.}^{\cdot} = \mathbf{S.}\mathbf{S}^{\cdot\dagger}. \tag{11.30}$$

Note that the final square matrices representing $\mathbf{M}^{\cdot\cdot}$ and $\mathbf{M..}$ can be formulated whether or not condition (11.12) is satisfied. So they can be constructed for each point in space-time; the point need not be on the light cone with vertex at the origin.

11.4
Spinorial Transformations

Each linear change in a first-rank spinor is effected by a 2×2 square matrix multiplying the column matrix for the original spinor. The same process is also represented by the Hermitian adjoint of the product. Since a second-rank spinor can be resolved into a first-rank spinor times the Hermitian adjoint of such a spinor, it transforms as this product does. Thus, it would be premultiplied by the square matrix and postmultiplied by the Hermitian adjoint of this square matrix. The net transformation preserves s^2 whenever the operating matrix is unimodular. Furthermore, such a transformation of the tensor is valid whether or not the variables x, y, z, ct are on the light cone for the origin.

Shifting from one inertial system to another having the same origin mixes the elements of a spinor in a homogeneously linear manner. The transformed elements are

$$(S^1)' = \alpha S^1 + \beta S^2 \qquad (11.31)$$

and

$$(S^2)' = \gamma S^1 + \delta S^2. \qquad (11.32)$$

Coefficients α, β, γ, δ are the parameters that generate the transformation.

In (11.20), (11.21), (11.23), and (11.24)

$$S_1{}^*, \quad S_1, \quad S_2, \quad S_2{}^* \qquad (11.33)$$

play the same role as

$$-S^2, \quad S^2{}^*, \quad -S^1{}^*, \quad S^1, \qquad (11.34)$$

respectively, play in (11.13), (11.14), (11.16), and (11.17). So Equations (11.31), (11.32) imply that

$$(S_2)'^* = \alpha S_2{}^* - \beta S_1{}^* \qquad (11.35)$$

and

$$-(S_1)'^* = \gamma S_2{}^* - \delta S_1{}^*, \qquad (11.36)$$

whence

$$(S_1)' = \delta^* S_1 - \gamma^* S_2 \qquad (11.37)$$

and

$$(S_2)' = -\beta^* S_1 + \alpha^* S_2. \tag{11.38}$$

Equations (11.31) and (11.32) can be written in matrix form as

$$(S\dot{})' = AS\dot{} \tag{11.39}$$

where

$$A = \begin{pmatrix} \alpha & \beta \\ \gamma & \delta \end{pmatrix} \tag{11.40}$$

while (11.37) and (11.38) can be rewritten as

$$(S.)' = BS. \tag{11.41}$$

where

$$B = \begin{pmatrix} \delta^* & -\gamma^* \\ -\beta^* & \alpha^* \end{pmatrix}. \tag{11.42}$$

Since the Hermitian adjoint of (11.39) is

$$(S^\dagger)' = S^{\cdot\dagger} A^\dagger, \tag{11.43}$$

we have

$$(M\ddot{})' = (S\dot{})'(S^\dagger)' = AS\dot{}S^{\cdot\dagger}A^\dagger = AM\ddot{}A^\dagger. \tag{11.44}$$

Substituting (11.27) into (11.44) yields

$$\begin{pmatrix} z_+{'} & x_-{'} \\ x_+{'} & -z_-{'} \end{pmatrix} = \begin{pmatrix} \alpha & \beta \\ \gamma & \delta \end{pmatrix} \begin{pmatrix} z_+ & x_- \\ x_+ & -z_- \end{pmatrix} \begin{pmatrix} \alpha^* & \gamma^* \\ \beta^* & \delta^* \end{pmatrix}. \tag{11.45}$$

The determinant of (11.45) is

$$-z_+{'}z_-{'} - x_+{'}x_-{'} = (\alpha\delta - \beta\gamma)(-z_+z_- - x_+x_-)(\alpha^*\delta^* - \beta^*\gamma^*). \tag{11.46}$$

Formula (11.11) reduces this to

$$-s'^2 = (\alpha\delta - \beta\gamma)(-s^2)(\alpha^*\delta^* - \beta^*\gamma^*). \tag{11.47}$$

Equations (11.46) and (11.47) can be applied to any radius vector, whether it lies on the light cone based on the origin or not. We need merely require that the fundamental invariance

$$s' = s \tag{11.48}$$

hold. Then (11.47) reduces to

$$1 = (\alpha\delta - \beta\gamma)(\alpha^*\delta^* - \beta^*\gamma^*). \tag{11.49}$$

If we assign zero phase angle to each factor on the right of (11.49), we obtain

$$\alpha\delta - \beta\gamma = 1. \tag{11.50}$$

Equation (11.50) is known as the *unimodular condition*.

Substituting (11.31), (11.32) into the primed form of (11.16) and employing (11.16), (11.17), (11.13), (11.14) to reduce the result yields the transformation equation

$$
\begin{aligned}
x_+{}' &= (\alpha S^1 + \beta S^2)^*(\gamma S^1 + \delta S^2) \\
&= \alpha^*\delta S^{1*}S^2 + \beta^*\gamma S^{2*}S^1 + \alpha^*\gamma S^{1*}S^1 + \beta^*\delta S^{2*}S^2 \\
&= \alpha^*\delta x_+ + \beta^*\gamma x_- + \alpha^*\gamma z_+ - \beta^*\delta z_-.
\end{aligned}
\tag{11.51}
$$

The complex conjugate of (11.51) is

$$x_-{}' = \beta\gamma^* x_+ + \alpha\delta^* x_- + \alpha\gamma^* z_+ - \beta\delta^* z_-. \tag{11.52}$$

Similarly treating (11.13) and (11.14) leads to

$$
\begin{aligned}
z_+{}' &= (\alpha S^1 + \beta S^2)^*(\alpha S^1 + \beta S^2) \\
&= \alpha^*\beta x_+ + \alpha\beta^* x_- + \alpha^*\alpha z_+ - \beta^*\beta z_-
\end{aligned}
\tag{11.53}
$$

and

$$
\begin{aligned}
-z_-{}' &= (\gamma S^1 + \delta S^2)^*(\gamma S^1 + \delta S^2) \\
&= \gamma^*\delta x_+ + \gamma\delta^* x_- + \gamma^*\gamma z_+ - \delta^*\delta z_-.
\end{aligned}
\tag{11.54}
$$

Example 11.2

How are x, y, z, and ct transformed when the phase angle of S^2 minus that of S^1 increases symmetrically by ϕ?

For points on the future half of the light cone with vertex at the origin, we have the spinor components

$$S^1 = a\, e^{iA},$$
$$S^2 = b\, e^{iB}.$$

The difference in phase angle changes as stated when A decreases by $\phi/2$, B increases by $\phi/2$. We then have

$$(S^1)' = a\, e^{i(A-\phi/2)} = e^{-i\phi/2} S^1,$$
$$(S^2)' = b\, e^{i(B+\phi/2)} = e^{i\phi/2} S^2.$$

Putting these relationships into (11.39) yields

$$\begin{pmatrix} (S^1)' \\ (S^2)' \end{pmatrix} = \begin{pmatrix} e^{-i\phi/2} & 0 \\ 0 & e^{i\phi/2} \end{pmatrix} \begin{pmatrix} S^1 \\ S^2 \end{pmatrix}.$$

The elements in operator **A** are

$$\alpha = e^{-i\phi/2}, \qquad \beta = 0, \qquad \gamma = 0, \qquad \delta = e^{i\phi/2}.$$

Since

$$\alpha\delta - \beta\gamma = e^{-i\phi/2}e^{i\phi/2} - 0 = 1,$$

these satisfy the unimodular condition. Therefore, these elements can be employed in (11.45) to transform the coordinates of any event from one inertial frame to another.

Equations (11.51), (11.53), (11.54) now become

$$x_+' = e^{i\phi}x_+ = (\cos \phi + i \sin \phi)x_+,$$

$$z_+' = z_+,$$

$$z_-' = z_-.$$

The first of this set of three equations is

$$x' + iy' = (\cos \phi + i \sin \phi)x - (\sin \phi - i \cos \phi)y,$$

whence

$$x' = x \cos \phi - y \sin \phi,$$

$$y' = x \sin \phi + y \cos \phi.$$

From the other two equations of the set, we obtain

$$z' = z$$

$$ct' = ct.$$

The transformation results from rotating the coordinate axes clockwise through angle ϕ in the xy plane. In matrix form, the final equations become

$$\begin{pmatrix} x' \\ y' \\ z' \\ ct' \end{pmatrix} = \begin{pmatrix} \cos \phi & -\sin \phi & 0 & 0 \\ \sin \phi & \cos \phi & 0 & 0 \\ 0 & 0 & 1 & 0 \\ 0 & 0 & 0 & 1 \end{pmatrix} \begin{pmatrix} x \\ y \\ z \\ ct \end{pmatrix}.$$

Example 11.3

How are x, y, z, and ct transformed when the ratio of amplitude of S^2 to that of S^1 increases by factor e^κ due to reciprocal changes in b and a?

Points on the future half of the light cone expanding from the origin have the spinor components

$$S^1 = a \, e^{iA},$$
$$S^2 = b \, e^{iB}.$$

These change as prescribed when

$$(S^1)' = e^{-\kappa/2}a \, e^{iA} = e^{-\kappa/2} \, S^1,$$
$$(S^2)' = e^{\kappa/2}b \, e^{iB} = e^{\kappa/2} \, S^2;$$

that is, when

$$\begin{pmatrix} (S^1)' \\ (S^2)' \end{pmatrix} = \begin{pmatrix} e^{-\kappa/2} & 0 \\ 0 & e^{\kappa/2} \end{pmatrix} \begin{pmatrix} S^1 \\ S^2 \end{pmatrix}.$$

Operator **A** of (11.39) and (11.40) now has the elements

$$\alpha = e^{-\kappa/2}, \qquad \beta = 0, \qquad \gamma = 0, \qquad \delta = e^{\kappa/2}.$$

Since

$$\alpha\delta - \beta\gamma = e^{-\kappa/2}e^{\kappa/2} - 0 = 1,$$

the unimodular condition is satisfied and the transformation preserves the radius vector in the second-rank spinor. The parameters obtained reduce Equations (11.51), (11.53), (11.54) to

$$x_+' = x_+,$$
$$z_+' = e^{-\kappa}z_+ = (\cosh \kappa - \sinh \kappa)z_+,$$
$$z_-' = e^{\kappa}z_- = (\cosh \kappa + \sinh \kappa)z_-,$$

whence

$$x' = x,$$
$$y' = y,$$
$$z' = z \cosh \kappa - ct \sinh \kappa,$$
$$ct' = -z \sinh \kappa + ct \cosh \kappa.$$

In matrix form, we have

$$
\begin{pmatrix} x' \\ y' \\ z' \\ ct' \end{pmatrix} = \begin{pmatrix} 1 & 0 & 0 & 0 \\ 0 & 1 & 0 & 0 \\ 0 & 0 & \cosh \kappa & -\sinh \kappa \\ 0 & 0 & -\sin \kappa & \cosh \kappa \end{pmatrix} \begin{pmatrix} x \\ y \\ z \\ ct \end{pmatrix}
$$

These relationships constitute the Lorentz transformation equations for going from one inertial frame to another moving as Figure 11.1 depicts. They apply to any event in the space-time continuum.

Example 11.4

How many independent parameters does a spinorial transformation preserving **s** possess?

Spinorial transformation (11.44) is effected by matrix **A**, which contains 4 complex elements. These involve 8 real numbers. For **s** to be preserved, the elements have to satisfy the unimodular condition

$$
\alpha\delta - \beta\gamma = 1.
$$

If the pairs of numbers in α, β, and γ are chosen arbitrarily this condition fixes the pair in δ. Thus, we have 6 independent parameters.

11.5
Minkowskian Vectors

Any ordered set of four expressions or operators that transform as x, y, z, and ct do can replace the components of the radius vector in **M˙˙** and in **M..** without altering the transformation properties. When the expressions are numbers, the set is said to form a *4-vector*.

Consider a 4-vector **V** with the Cartesian components V_x, V_y, V_z, V_{ct}. Substituting these numbers for x, y, z, ct in the final matrix of (11.27) yields

$$
\mathbf{N}^{\cdot\cdot} = \begin{pmatrix} V_z + V_{ct} & V_x - iV_y \\ V_x + iV_y & -V_z + V_{ct} \end{pmatrix} . \tag{11.55}
$$

Similarly substituting into (11.28) gives us

$$\mathbf{N}_{..} = \begin{pmatrix} V_z - V_{ct} & V_x - iV_y \\ V_x + iV_y & -V_z - V_{ct} \end{pmatrix} \tag{11.56}$$

Since $\mathbf{N}^{..}$ and $\mathbf{N}_{..}$ transform as $\mathbf{M}^{..}$ and $\mathbf{M}_{..}$, respectively, do, they are second-rank spinors.

Wolfgang Pauli introduced the matrices

$$\sigma_x = \begin{pmatrix} 0 & 1 \\ 1 & 0 \end{pmatrix}, \tag{11.57}$$

$$\sigma_y = \begin{pmatrix} 0 & -i \\ i & 0 \end{pmatrix}, \tag{11.58}$$

$$\sigma_z = \begin{pmatrix} 1 & 0 \\ 0 & -1 \end{pmatrix}. \tag{11.59}$$

These, together with the unit matrix

$$\mathbf{1} = \begin{pmatrix} 1 & 0 \\ 0 & 1 \end{pmatrix}, \tag{11.60}$$

allow $\mathbf{M}^{..}$, $\mathbf{M}_{..}$, $\mathbf{N}^{..}$, $\mathbf{N}_{..}$ to be written as

$$\mathbf{M}^{..} = x\sigma_x + y\sigma_y + z\sigma_z + ct\mathbf{1}, \tag{11.61}$$

$$\mathbf{M}_{..} = x\sigma_x + y\sigma_y + z\sigma_z - ct\mathbf{1}, \tag{11.62}$$

$$\mathbf{N}^{..} = V_x\sigma_x + V_y\sigma_y + V_z\sigma_z + V_{ct}\mathbf{1}, \tag{11.63}$$

$$\mathbf{N}_{..} = V_x\sigma_x + V_y\sigma_y + V_z\sigma_z - V_{ct}\mathbf{1}, \tag{11.64}$$

A 4-vector may be expanded as a sum of extensions of unit vectors $\mathbf{e}_1, \mathbf{e}_2, \mathbf{e}_3, \mathbf{e}_4$:

$$\mathbf{s} = x\mathbf{e}_1 + y\mathbf{e}_2 + z\mathbf{e}_3 + ct\mathbf{e}_4. \tag{11.65}$$

But for

$$\mathbf{s} \cdot \mathbf{s} = x^2 + y^2 + z^2 - c^2t^2, \tag{11.66}$$

we require that

$$\mathbf{e}_1 \cdot \mathbf{e}_1 = 1, \quad \mathbf{e}_2 \cdot \mathbf{e}_2 = 1, \quad \mathbf{e}_3 \cdot \mathbf{e}_3 = 1, \quad \mathbf{e}_4 \cdot \mathbf{e}_4 = -1, \tag{11.67}$$

and

$$\mathbf{e}_j \cdot \mathbf{e}_k = 0 \qquad \text{when} \qquad j \neq k. \tag{11.68}$$

Differentiating (11.65) yields the displacement

$$d\mathbf{s} = dx\, \mathbf{e}_1 + dy\, \mathbf{e}_2 + dz\, \mathbf{e}_3 + d(ct)\, \mathbf{e}_4. \tag{11.69}$$

Differentials dx, dy, dz, and $d(ct)$ are components of the displacement. If we have two 4-vectors, $\mathbf{U}$ and $\mathbf{V}$, their dot product is the scalar

$$\mathbf{U} \cdot \mathbf{V} = (U_x\mathbf{e}_1 + U_y\mathbf{e}_2 + U_z\mathbf{e}_3 + U_{ct}\mathbf{e}_4) \cdot (V_x\mathbf{e}_1 + V_y\mathbf{e}_2 + V_z\mathbf{e}_3 + V_{ct}\mathbf{e}_4)$$

$$= U_x V_x + U_y V_y + U_z V_z - U_{ct} V_{ct}. \tag{11.70}$$

Example 11.5

How do the Pauli matrices combine with each other?
Apply the rule for matrix multiplication to the product of each matrix with itself,

$$\sigma_x^2 = \begin{pmatrix} 0 & 1 \\ 1 & 0 \end{pmatrix} \begin{pmatrix} 0 & 1 \\ 1 & 0 \end{pmatrix} = \begin{pmatrix} 1 & 0 \\ 0 & 1 \end{pmatrix},$$

$$\sigma_y^2 = \begin{pmatrix} 0 & -i \\ i & 0 \end{pmatrix} \begin{pmatrix} 0 & -i \\ i & 0 \end{pmatrix} = \begin{pmatrix} 1 & 0 \\ 0 & 1 \end{pmatrix},$$

$$\sigma_z^2 = \begin{pmatrix} 1 & 0 \\ 0 & -1 \end{pmatrix} \begin{pmatrix} 1 & 0 \\ 0 & -1 \end{pmatrix} = \begin{pmatrix} 1 & 0 \\ 0 & 1 \end{pmatrix},$$

and with each of the other matrices,

$$\sigma_x\sigma_y = \begin{pmatrix} 0 & 1 \\ 1 & 0 \end{pmatrix} \begin{pmatrix} 0 & -i \\ i & 0 \end{pmatrix} = \begin{pmatrix} i & 0 \\ 0 & -i \end{pmatrix} = i\sigma_z,$$

$$\sigma_y\sigma_x = \begin{pmatrix} 0 & -i \\ i & 0 \end{pmatrix} \begin{pmatrix} 0 & 1 \\ 1 & 0 \end{pmatrix} = \begin{pmatrix} -i & 0 \\ 0 & i \end{pmatrix} = -i\sigma_z,$$

$$\sigma_y \sigma_z = \begin{pmatrix} 0 & -i \\ i & 0 \end{pmatrix} \begin{pmatrix} 1 & 0 \\ 0 & -1 \end{pmatrix} = \begin{pmatrix} 0 & i \\ i & 0 \end{pmatrix} = i\sigma_x,$$

$$\sigma_z \sigma_y = \begin{pmatrix} 1 & 0 \\ 0 & -1 \end{pmatrix} \begin{pmatrix} 0 & -i \\ i & 0 \end{pmatrix} = \begin{pmatrix} 0 & -i \\ -i & 0 \end{pmatrix} = -i\sigma_x,$$

$$\sigma_z \sigma_x = \begin{pmatrix} 1 & 0 \\ 0 & -1 \end{pmatrix} \begin{pmatrix} 0 & 1 \\ 1 & 0 \end{pmatrix} = \begin{pmatrix} 0 & 1 \\ -1 & 0 \end{pmatrix} = i\sigma_y,$$

$$\sigma_x \sigma_z = \begin{pmatrix} 0 & 1 \\ 1 & 0 \end{pmatrix} \begin{pmatrix} 1 & 0 \\ 0 & -1 \end{pmatrix} = \begin{pmatrix} 0 & -1 \\ 1 & 0 \end{pmatrix} = -i\sigma_y.$$

From the first 3 multiplications, the square of a Pauli matrix is unity. From the other results,

$$\sigma_x \sigma_y + \sigma_y \sigma_x = 0,$$
$$\sigma_y \sigma_z + \sigma_z \sigma_y = 0$$
$$\sigma_z \sigma_x + \sigma_x \sigma_z = 0,$$

and

$$\sigma_x \sigma_y - \sigma_y \sigma_x = 2i\sigma_z,$$
$$\sigma_y \sigma_z - \sigma_z \sigma_y = 2i\sigma_x,$$
$$\sigma_z \sigma_x - \sigma_x \sigma_z = 2i\sigma_y.$$

Thus, the Pauli matrices *anticommute* with each other. From their commutators, we construct

$$\sigma \times \sigma = 2i\sigma.$$

11.6
Spinorial Differentiating Operators

A varying scalar physical property is described by a function of the coordinates and time. An infinitesimal change in the property associated with an infinitesimal change in 4-vector s equals the sum of each Cartesian component of the gradient of the function times the corresponding component of ds. Each component of the gradient can be expressed as a constituent vector operator acting on the scalar function. Such operators can occupy a 2 × 2 matrix as the components of a 4-vector, forming a spinorial operator.

Let x, y, z be the Cartesian coordinates of a typical point in an inertial frame at time t. Let us furthermore consider the scalar $\phi(x, y, z, ct)$. Differentiating this function yields

$$d\phi = \frac{\partial\phi}{\partial x}\,dx + \frac{\partial\phi}{\partial y}\,dy + \frac{\partial\phi}{\partial z}\,dz + \frac{\partial\phi}{\partial(ct)}\,d(ct)$$

$$= \frac{\partial\phi}{\partial x}\,dx + \frac{\partial\phi}{\partial y}\,dy + \frac{\partial\phi}{\partial z}\,dz - \left[-\frac{\partial\phi}{\partial(ct)}\right]d(ct). \qquad (11.71)$$

Following (11.70), we can interpret the result as the dot product between the 4-vector with components

$$\frac{\partial\phi}{\partial x}, \qquad \frac{\partial\phi}{\partial y}, \qquad \frac{\partial\phi}{\partial z}, \qquad -\frac{\partial\phi}{\partial(ct)} \qquad (11.72)$$

and the one with components

$$dx, \qquad dy, \qquad dz, \qquad d(ct). \qquad (11.73)$$

Since ϕ is a scalar, the operators

$$\frac{\partial}{\partial x}, \qquad \frac{\partial}{\partial y}, \qquad \frac{\partial}{\partial z}, \qquad -\frac{\partial}{\partial(ct)} \qquad (11.74)$$

transform as elements (11.72) do.

Substituting operators (11.74) for the corresponding vector components in (11.55) yields

$$\mathbf{D}^{\cdot\cdot} = \begin{pmatrix} \dfrac{\partial}{\partial z} - \dfrac{\partial}{\partial(ct)} & \dfrac{\partial}{\partial x} - i\dfrac{\partial}{\partial y} \\[2ex] \dfrac{\partial}{\partial x} + i\dfrac{\partial}{\partial y} & -\dfrac{\partial}{\partial z} - \dfrac{\partial}{\partial(ct)} \end{pmatrix}, \qquad (11.75)$$

while (11.74) and (11.56) give us

$$\mathbf{D}_{\cdot\cdot} = \begin{pmatrix} \dfrac{\partial}{\partial z} + \dfrac{\partial}{\partial(ct)} & \dfrac{\partial}{\partial x} - i\dfrac{\partial}{\partial y} \\[2ex] \dfrac{\partial}{\partial x} + i\dfrac{\partial}{\partial y} & -\dfrac{\partial}{\partial z} + \dfrac{\partial}{\partial(ct)} \end{pmatrix}. \qquad (11.76)$$

Since (11.55) transforms as a second-rank upper-index spinor, operator (11.75) does. Similarly, operator (11.76) transforms as a second-rank lower-index spinor because (11.56) does.

Introducing the Pauli matrices allows (11.75) and (11.76) to be written as

$$\mathbf{D}^{..} = \sigma_x \frac{\partial}{\partial x} + \sigma_y \frac{\partial}{\partial y} + \sigma_z \frac{\partial}{\partial z} - \mathbf{1} \frac{\partial}{\partial (ct)} , \tag{11.77}$$

$$\mathbf{D}_{..} = \sigma_x \frac{\partial}{\partial x} + \sigma_y \frac{\partial}{\partial y} + \sigma_z \frac{\partial}{\partial z} + \mathbf{1} \frac{\partial}{\partial (ct)} . \tag{11.78}$$

Example 11.6

Show that the product of a second-rank upper-index spinor with a first-rank lower-index spinor is a first-rank upper-index spinor.

Let $\mathbf{N}^{..}$ and $\mathbf{T}_.$ be the given spinors in one inertial farme, while $(\mathbf{N}^{..})'$ and $(\mathbf{T}_.)'$ are these spinors in another inertial frame. They transform following law (11.44),

$$(\mathbf{N}^{..})' = \mathbf{A} \, \mathbf{N}^{..} \, \mathbf{A}^\dagger,$$

and law (11.41)

$$(\mathbf{T}_.)' = \mathbf{B} \, \mathbf{T}_.,$$

with $\mathbf{A}$ given by (11.40), $\mathbf{B}$ by (11.42).

The product of the transformed spinors is

$$(\mathbf{N}^{..})' \, (\mathbf{T}_.)' = \mathbf{A} \, \mathbf{N}^{..} \, \mathbf{A}^\dagger \mathbf{B} \, \mathbf{T}_.,$$

as a result. But

$$\mathbf{A}^\dagger \mathbf{B} = \begin{pmatrix} \alpha^* & \gamma^* \\ \beta^* & \delta^* \end{pmatrix} \begin{pmatrix} \delta^* & -\gamma^* \\ -\beta^* & \alpha^* \end{pmatrix}$$

$$= \begin{pmatrix} \alpha^*\delta^* - \beta^*\gamma^* & -\alpha^*\gamma^* + \alpha^*\gamma^* \\ \beta^*\delta^* - \beta^*\delta^* & -\beta^*\gamma^* + \alpha^*\delta^* \end{pmatrix} = \begin{pmatrix} 1 & 0 \\ 0 & 1 \end{pmatrix} ,$$

since the complex conjugate of (11.50) holds. Therefore, the product above reduces to

$$(\mathbf{N}^{..})' \, (\mathbf{T}_.)' = \mathbf{A} \, \mathbf{N}^{..} \, \mathbf{T}_.,$$

the transformation law for a first-rank upper-index spinor.

11.7
A Wave Function that Propagates as Light Does

A light wave traveling in a certain direction in free space without converging or diverging does not change as it moves along a light cone based on a point in its path. A homogeneous beam of antineutrinos or neutrinos moving similarly is described by a wave function whose spinorial derivative on a pertinent light cone vanishes.

Consider a function $\phi^\cdot$ satisfying the equation

$$\mathbf{D}_{\cdot\cdot}\phi^\cdot = 0. \qquad (11.79)$$

From (11.78) and (11.18), this has the form

$$\left(\sigma_x \frac{\partial}{\partial x} + \sigma_y \frac{\partial}{\partial y} + \sigma_z \frac{\partial}{\partial z} + \mathbf{1} \frac{\partial}{\partial(ct)} \right) \begin{pmatrix} \phi^1 \\ \phi^2 \end{pmatrix} = 0. \qquad (11.80)$$

which is linear, homogeneous, and first-order. Each of the two elements satisfying it can therefore be expressed as a linear combination of exponentials,

$$\phi^j = u^j e^{i(\mathbf{k}\cdot\mathbf{r} - \omega t)}, \qquad (11.81)$$

in which u^j is a constant while $\mathbf{k}$ and $\mathbf{r}$ are the Newtonian 3-vectors

$$\mathbf{k} = k_x \mathbf{e}_1 + k_y \mathbf{e}_2 + k_z \mathbf{e}_3, \qquad (11.82)$$

$$\mathbf{r} = x \mathbf{e}_1 + y \mathbf{e}_2 + z \mathbf{e}_3. \qquad (11.83)$$

Substituting (11.81) into (11.80) yields

$$\left(\sigma_x k_x + \sigma_y k_y + \sigma_z k_z - \mathbf{1} \frac{\omega}{c} \right) \begin{pmatrix} u^1 \\ u^2 \end{pmatrix} e^{i(\mathbf{k}\cdot\mathbf{r} - \omega t)} = 0 \qquad (11.84)$$

or

$$\begin{pmatrix} k_z - \dfrac{\omega}{c} & k_x - ik_y \\ k_x + ik_y & -k_z - \dfrac{\omega}{c} \end{pmatrix} \begin{pmatrix} u^1 \\ u^2 \end{pmatrix} e^{i(\mathbf{k}\cdot\mathbf{r} - \omega t)} = 0. \qquad (11.85)$$

Every term in the solution can be separated into a part in which only u^1 differs from zero and a part in which only u^2 differs from zero.

For each of the upper solutions, (11.85) becomes

$$\begin{pmatrix} \left(k_z - \dfrac{\omega}{c} \right) u^1 \\ (k_x + ik_y) u^1 \end{pmatrix} e^{i(\mathbf{k}\cdot\mathbf{r} - \omega t)} = 0. \qquad (11.86)$$

This is satisfied only when

$$k_z = \frac{\omega}{c}, \qquad k_x = 0, \qquad k_y = 0. \qquad (11.87)$$

For each of the lower solutions, (11.85) becomes

$$\begin{pmatrix} (k_x - ik_y)u^2 \\ \left(-k_z - \dfrac{\omega}{c} \right) u^2 \end{pmatrix} e^{i(\mathbf{k \cdot r} - \omega t)} = 0 \qquad (11.88)$$

whence

$$k_z = -\frac{\omega}{c}, \qquad k_x = 0, \qquad k_y = 0. \qquad (11.89)$$

A given phase of ϕ^j propagates according to the law

$$\mathbf{k \cdot r} - \omega t = \text{constant} \qquad (11.90)$$

or

$$k_x x + k_y y + k_z z - \omega t = \text{constant}. \qquad (11.91)$$

Relationships (11.87), which arise when $u^1 \neq 0$, $u^2 = 0$, reduce (11.91) to

$$\frac{\omega}{c} z - \omega t = \text{constant} \qquad (11.92)$$

or

$$z = ct + \frac{c}{\omega} \text{ (constant)}. \qquad (11.93)$$

Similarly, Equations (11.89), which arise when $u^1 = 0$, $u^2 \neq 0$, reduce (11.91) to

$$z = -ct + \frac{c}{\omega} \text{ (constant)}. \qquad (11.94)$$

Either the upper, or the lower, solutions can be superposed to form a wave packet. The corresponding particle velocity is given by the group velocity

$$v = \frac{d\omega}{dk}. \qquad (11.95)$$

From (11.87), we obtain

$$v_z = \frac{d}{dk}(ck) = c, \qquad v_x = 0, \qquad v_y = 0, \qquad (11.96)$$

and from (11.89),

$$v_z = \frac{d}{dk}(-ck) = -c, \qquad v_x = 0, \qquad v_y = 0. \qquad (11.97)$$

The upper solutions describe particles moving at speed c in the direction of increasing z; the lower solutions, similar particles moving in the opposite direction at speed c. According to Einstein's relativity theory, such particles have no rest mass. Consequently, they may be

   (a) electron neutrinos or antineutrinos,
   (b) muon neutrinos or antineutrinos,
   (c) tau neutrinos or antineutrinos.

11.8
Disturbances that Propagate More Slowly

In Galilean relativity the space observed by a surveyor in one inertial frame is the same as the space observed by a surveyor in another frame moving with respect to the first frame. But in Einsteinian relativity the two spaces are different; only the light cone based on a common origin is invariant. As a consequence the gradient in physical space describing the analytical variation of a state function has to be replaced by derivatives on the past half and on the future half of the light cone. In the simplest theory these are coupled in an antisymmetric manner. Eliminating either of the spinors leads to a second-order equation that is a generalization of the Schrödinger equation.

We have seen how the lower-index spinorial derivative can be physically significant when it vanishes, as in (11.79). When it does not vanish, as in

$$\mathbf{D}_{..}\phi^{\cdot} = -A\chi_{.}, \qquad (11.98)$$

where A is a parameter, a new spinor $\chi_{.}$ is defined. This can be subjected to the condition

$$\mathbf{D}^{..}\chi_{.} = A\phi^{\cdot} \qquad (11.99)$$

that is similar to (11.98) except for a change in sign.

Let $\mathbf{D}^{..}$ act on (11.98). Then reduce the right side with (11.99) to obtain

$$\mathbf{D}^{..}\mathbf{D}_{..}\phi^{\cdot} = -A\mathbf{D}^{..}\chi_{.} = -A^2\phi^{\cdot}, \qquad (11.100)$$

a differential equation in which the function is directed onto the future half of the origin's light cone and the differentiating operator is a scalar. Eliminating $\phi^{\cdot}$ from (11.98) and (11.99) similarly gives us

$$\mathbf{D}_{..}\mathbf{D}^{..}\chi_{.} = -A^2\chi_{.}, \qquad (11.101)$$

a differential equation in which the function is directed onto the past half of the light cone and the differentiating operator is another scalar.

When what is called space depends on the choice of inertial frame, the conventional gradient is not physically significant by itself, but with the light cone still invariant, spinorial derivatives are physically significant. So these replace the del operator of Schrödinger theory. Indeed, Equations (11.100) and (11.101) have the same form as the Schrödinger equation

$$\nabla^2 \Psi = -k^2 \Psi \qquad (11.102)$$

with the appropriate spinorial operators, $\mathbf{D}^{\cdot\cdot} \mathbf{D}_{\cdot\cdot}$ and $\mathbf{D}_{\cdot\cdot} \mathbf{D}^{\cdot\cdot}$, replacing ∇^2.

Since (11.100) and (11.101) are linear, homogeneous, second-order differential equations, the elements of their spinorial solutions can be expressed as linear combinations of exponentials. Typical terms are

$$\phi^j = u^j e^{i(\mathbf{k}\cdot\mathbf{r}-\omega t)} \qquad (11.103)$$

and

$$\chi_j = v_j e^{i(\mathbf{k}\cdot\mathbf{r}-\omega t)} \qquad (11.104)$$

where u^j, v_j are constants and $\mathbf{k}$, $\mathbf{r}$ are the 3-dimensional wavevector and position vector, as before.

Differentiating (11.103) or (11.104) with respect to x, y, z, ct leads to multiplication of the expression by ik_x, ik_y, ik_z, $-i\omega/c$, respectively. A result of substituting (11.103) into (11.100) is therefore

$$\begin{pmatrix} k_z + \dfrac{\omega}{c} & k_x - ik_y \\[2mm] k_x + ik_y & -k_z + \dfrac{\omega}{c} \end{pmatrix} \begin{pmatrix} k_z - \dfrac{\omega}{c} & k_x - ik_y \\[2mm] k_x + ik_y & -k_z - \dfrac{\omega}{c} \end{pmatrix} \begin{pmatrix} u^1 \\[2mm] u^2 \end{pmatrix} e^{i(\mathbf{k}\cdot\mathbf{r}-\omega t)}$$

$$= A^2 \begin{pmatrix} u^1 \\[2mm] u^2 \end{pmatrix} e^{i(\mathbf{k}\cdot\mathbf{r}-\omega t)}. \qquad (11.105)$$

Since the exponential is not zero, it can be canceled from (11.105). Since the spinor elements u^1 and u^2 can be chosen arbitrarily, spinor $\mathbf{u}^{\cdot}$ can also be canceled from the equation. Then A^2 is interpreted as A^2 times the unit matrix. Equating elements on both sides of the remaining equality yields one condition

$$k_x^2 + k_y^2 + k_z^2 - \frac{\omega^2}{c^2} = A^2, \qquad (11.106)$$

whence

$$\mathbf{k}\cdot\mathbf{k} - \frac{\omega^2}{c^2} = A^2. \qquad (11.107)$$

Because of the de Broglie relationship

$$\mathbf{k} = \frac{\mathbf{p}}{\hbar} \qquad (11.108)$$

and the Einstein relationship

$$\omega = \frac{E}{\hbar}, \tag{11.109}$$

(11.107) can be rewritten in the form

$$\frac{1}{\hbar^2} \left(\mathbf{p \cdot p} - \frac{E^2}{c^2} \right) = A^2. \tag{11.110}$$

Relative to the inertial frame in which momentum $\mathbf{p}$ is zero, the energy is

$$E = m_0 c^2. \tag{11.111}$$

So (11.110) yields

$$A^2 = - \frac{m_0^2 c^2}{\hbar^2} \tag{11.112}$$

whence

$$A = \frac{i m_0 c}{\hbar}. \tag{11.113}$$

Here m_0 is the rest mass of the particle, c the speed of light in free space, and $\hbar$ Planck's constant divided by 2π.

Substituting terms (11.103) and (11.104) into (11.98), (11.99), carrying out the differentiations, and reducing with (11.113), (11.108), (11.109) leads to

$$\begin{pmatrix} p_z - \dfrac{E}{c} & p_x - i p_y \\[2mm] p_x + i p_y & -p_z - \dfrac{E}{c} \end{pmatrix} \phi = -m_0 c \chi. \tag{11.114}$$

and

$$\begin{pmatrix} p_z + \dfrac{E}{c} & p_x - i p_y \\[2mm] p_x + i p_y & -p_z + \dfrac{E}{c} \end{pmatrix} \chi = m_0 c \phi \tag{11.115}$$

in which p_x, p_y, p_z, E can be interpreted either as the algebraic quantities for a given type of motion or as the corresponding operators.

Introducing the Pauli matrices (11.57)–(11.59) with

$$\sigma = \sigma_x \mathbf{e}_1 + \sigma_y \mathbf{e}_2 + \sigma_z \mathbf{e}_3 \tag{11.116}$$

allows us to simplify (11.114) and (11.115) to

$$- \left(\sigma \cdot \mathbf{p} - \mathbf{1}\frac{E}{c} \right) \phi - m_0 c \chi = 0, \tag{11.117}$$

$$\left(\sigma \cdot \mathbf{p} + \mathbf{1}\frac{E}{c} \right) \chi - m_0 c \phi = 0 \tag{11.118}$$

or

$$\left[\begin{pmatrix} \dfrac{E}{c} - \sigma \cdot \mathbf{p} & 0 \\ 0 & \dfrac{E}{c} + \sigma \cdot \mathbf{p} \end{pmatrix} - \begin{pmatrix} 0 & m_0 c \\ m_0 c & 0 \end{pmatrix}\right] \begin{pmatrix} \phi^{\cdot} \\ \chi_{\cdot} \end{pmatrix} = 0. \quad (11.119)$$

The coupled spinorial differential equations (11.98) and (11.99) do not impose any limitation on wavevector $\mathbf{k}$, except that it belong to a 4-vector of magnitude A, according to (11.107). This quantity differs from zero and the particle travels more slowly than a neutrino when m_0 differs from zero.

Example 11.7

What is the relationship between $\phi^{\cdot}$ and $\chi_{\cdot}$ when the free particle is at rest? When

$$p_x = p_y = p_z = 0,$$

(11.114) and (11.115) reduce to

$$\begin{pmatrix} -\dfrac{E}{c} & 0 \\ 0 & -\dfrac{E}{c} \end{pmatrix} \phi^{\cdot} = \begin{pmatrix} -m_0 c & 0 \\ 0 & -m_0 c \end{pmatrix} \chi_{\cdot}$$

and

$$\begin{pmatrix} \dfrac{E}{c} & 0 \\ 0 & \dfrac{E}{c} \end{pmatrix} \chi_{\cdot} = \begin{pmatrix} m_0 c & 0 \\ 0 & m_0 c \end{pmatrix} \phi^{\cdot}.$$

With

$$E = m_0 c^2,$$

we find that

$$\phi^{\cdot} = \chi_{\cdot}.$$

The upper-index and the lower-index spinor functions have equivalent elements.
Giving the particle a small momentum introduces deviations, but still we would expect that

$$\phi^{\cdot} \simeq \chi_{\cdot}.$$

and

$$u^j \simeq v_j.$$

Example 11.8

The momentum-energy 4-vector, with components p_x, p_y, p_z, E/c, exists whether E is negative or positive. Should we allow negative-energy solutions in quantum mechanics?

If we change the classical relativistic formula (11.111) to

$$E = -m_0c^2,$$

Equation (11.112) and Equations (11.114), (11.115) still hold as written. For a particle at rest, these latter reduce to the matrix equations in Example 11.7, which are satisfied with the negative energy when

$$\phi^. = -\chi.$$

and

$$u^j = -v_j.$$

Giving the particle a small momentum would be expected to leave

$$\phi^. \simeq -\chi.$$

and

$$u^j \simeq -v_j.$$

In prequantum theory, negative-energy solutions are rejected because they cannot be reached by a continuous loss of energy. They are separated from the observed states by an energy gap. But in quantum theory, a system can jump from one energy level to a discretely lower one; so the negative-energy solutions cannot be rejected, out of hand. How they are employed will be considered in Section 12.3.

11.9
The Dirac Equation

Example 11.7 indicates that the elements of $\phi^.$ approach those of $\chi.$ as the velocity of the particle is made small with respect to c. So in dealing with a low-energy particle, a person can cut the number of functions of appreciable amplitude in half by going to an equation in which the dependent variables are $\phi^. + \chi.$ and $\phi^. - \chi.$, hybrid spinors. Since $\phi^.$ and $\chi.$ need contain only one nonzero function each, we may thus end up with only one function appreciably different from zero. This would be comparable to the single function which we have employed as the state function.

Let us proceed by construction, forming the sum and the difference of (11.117) and (11.118), and multiplying the results by c to get

$$(E - m_0c^2)(\phi^\cdot + \chi.) - c\sigma\mathbf{p}(\phi^\cdot - \chi.) = 0, \qquad (11.120)$$

$$(E + m_0c^2)(\phi^\cdot - \chi.) - c\sigma\mathbf{p}(\phi^\cdot + \chi.) = 0. \qquad (11.121)$$

If we now let

$$\alpha = \begin{pmatrix} 0 & \sigma \\ \sigma & 0 \end{pmatrix}, \qquad (11.122)$$

$$\beta = \begin{pmatrix} 1 & 0 \\ 0 & -1 \end{pmatrix}, \qquad (11.123)$$

$$\Psi = \begin{pmatrix} \phi^\cdot + \chi. \\ \phi^\cdot - \chi. \end{pmatrix}, \qquad (11.124)$$

the equations can be rewritten in the form

$$(E - c\alpha\mathbf{p} - \beta m_0c^2)\Psi = 0 \qquad (11.125)$$

or

$$(c\alpha\mathbf{p} + \beta m_0c^2)\Psi = E\Psi. \qquad (11.126)$$

When the particle under consideration is not free, its interactions may be described as in (3.66), (3.77), and (9.4). Then $\mathbf{p}$ has to be replaced with $\mathbf{p} - q\mathbf{A}$ and E with $E - q\phi$. Equation (11.126) becomes

$$[c\alpha\cdot(\mathbf{p} - q\mathbf{A}) + \beta m_0c^2]\Psi = (E - q\phi)\Psi. \qquad (11.127)$$

Here q is the interaction coefficient, $\mathbf{A}$ the vector potential, and ϕ the scalar potential, of the field. A form of (11.127) was constructed by Paul A. M. *Dirac* in 1928.

Equations (11.120) and (11.121) similarly become

$$(E - m_0c^2 - q\phi)(\phi^\cdot + \chi.) - c\sigma\cdot(\mathbf{p} - q\mathbf{A})(\phi^\cdot - \chi.) = 0, \qquad (11.128)$$

$$(E - m_0^2 - q\phi)(\phi^\cdot - \chi.) - c\sigma\cdot(\mathbf{p} - q\mathbf{A})(\phi^\cdot + \chi.) = 0 \qquad (11.129)$$

when the particle is not free.

The matrices defined by (11.122) and (11.123) are known as the Dirac matrices.

Example 11.9

How do the Dirac matrices combine with each other?
Apply the rule for block multiplication of matrices to representative combinations:

$$\alpha_j \alpha_k = \begin{pmatrix} 0 & \sigma_j \\ \sigma_j & 0 \end{pmatrix} \begin{pmatrix} 0 & \sigma_k \\ \sigma_k & 0 \end{pmatrix} = \begin{pmatrix} \sigma_j \sigma_k & 0 \\ 0 & \sigma_j \sigma_k \end{pmatrix} \quad ,$$

$$\alpha_j \beta = \begin{pmatrix} 0 & \sigma_j \\ \sigma_j & 0 \end{pmatrix} \begin{pmatrix} 1 & 0 \\ 0 & -1 \end{pmatrix} = \begin{pmatrix} 0 & -\sigma_j \\ \sigma_j & 0 \end{pmatrix} ,$$

$$\beta \alpha_j = \begin{pmatrix} 1 & 0 \\ 0 & -1 \end{pmatrix} \begin{pmatrix} 0 & \sigma_j \\ \sigma_j & 0 \end{pmatrix} = \begin{pmatrix} 0 & \sigma_j \\ -\sigma_j & 0 \end{pmatrix} ,$$

$$\beta \beta = \begin{pmatrix} 1 & 0 \\ 0 & -1 \end{pmatrix} \begin{pmatrix} 1 & 0 \\ 0 & -1 \end{pmatrix} = \begin{pmatrix} 1 & 0 \\ 0 & 1 \end{pmatrix} .$$

Since the product of each component of σ with the same component is unity, the product of each component of α with itself is unity. Since the components of σ anticommute with each other, the components of α anticommute. Furthermore, each component of α anticommutes with β, and the square of β is unity.

Constructing the commutator of one Dirax matrix with another and reducing with the results from Example 11.5 leads to

$$\alpha_j \alpha_k - \alpha_k \alpha_j = \begin{pmatrix} \sigma_j \sigma_k - \sigma_k \sigma_j & 0 \\ 0 & \sigma_j \sigma_k - \sigma_k \sigma_j \end{pmatrix}$$

$$= \begin{pmatrix} 2i\epsilon_{jkl}\sigma_l & 0 \\ 0 & 2i\epsilon_{jkl}\sigma_l \end{pmatrix}$$

$$= 2i\epsilon_{jkl} \begin{pmatrix} \sigma_l & 0 \\ 0 & \sigma_l \end{pmatrix} \equiv 2i\epsilon_{jkl}\Sigma_l,$$

if we introduce the *permutation symbol*

$\epsilon_{jkl} = 1$ if *jkl* is 123 or a cyclic permutation thereof,

$\epsilon_{jkl} = -1$ if *jkl* is 213 or a cyclic permutation thereof,

$\epsilon_{jkl} = 0$ otherwise,

and the *sigma symbol*

$$\Sigma_l = \begin{pmatrix} \sigma_l & 0 \\ 0 & \sigma_l \end{pmatrix},$$

together with the Einstein *summation convention*, whereby repetition of an index indicates summation over that index.

Example 11.10

Write out the Dirac matrices in full.

Substitute (11.57), (11.58), (11.59), (11.60) into the components of (11.122) and into (11.123):

$$\alpha_x = \begin{pmatrix} 0 & \sigma_x \\ \sigma_x & 0 \end{pmatrix} = \begin{pmatrix} 0 & 0 & 0 & 1 \\ 0 & 0 & 1 & 0 \\ 0 & 1 & 0 & 0 \\ 1 & 0 & 0 & 0 \end{pmatrix},$$

$$\alpha_y = \begin{pmatrix} 0 & \sigma_y \\ \sigma_y & 0 \end{pmatrix} = \begin{pmatrix} 0 & 0 & 0 & -i \\ 0 & 0 & i & 0 \\ 0 & -i & 0 & 0 \\ i & 0 & 0 & 0 \end{pmatrix},$$

$$\alpha_z = \begin{pmatrix} 0 & \sigma_z \\ \sigma_z & 0 \end{pmatrix} = \begin{pmatrix} 0 & 0 & 1 & 0 \\ 0 & 0 & 0 & -1 \\ 1 & 0 & 0 & 0 \\ 0 & -1 & 0 & 0 \end{pmatrix},$$

$$\beta = \begin{pmatrix} 1 & 0 \\ 0 & -1 \end{pmatrix} = \begin{pmatrix} 1 & 0 & 0 & 0 \\ 0 & 1 & 0 & 0 \\ 0 & 0 & -1 & 0 \\ 0 & 0 & 0 & -1 \end{pmatrix}.$$

From (11.124), the Dirac state function is

$$\Psi = \begin{pmatrix} \phi^1 + \chi_1 \\ \phi^2 + \chi_2 \\ \phi^1 - \chi_1 \\ \phi^2 - \chi_2 \end{pmatrix}.$$

11.10
Review

For Schrödinger quantum mechanics, time is universal and absolute, separate from space. Each vector in the space is represented by an invariant displacement. For Einstein relativistic quantum mechanics, on the other hand, time is bound up with space in a single 4-dimensional continuum. A vector in the continuum may be either spacelike or timelike, represented by a displacement in space or in time, or neither, a null vector.

Space may be surveyed and passage of time measured with respect to inertial frames. Let us choose a point in the continuum as origin and erect on it a Cartesian coordinate system for inertial frame Σ and for inertial frame Σ'. The variables locating a second point would be x, y, z, t in Σ and x', y', z', t' in Σ'.

For Galilean relativistic theory, the spatial distance of the point from the origin is invariant and

$$r^2 = x^2 + y^2 + z^2 = x'^2 + y'^2 + z'^2. \tag{11.130}$$

For Einstein relativistic theory (the special theory), we have instead an invariant

$$s^2 = x^2 + y^2 + z^2 - c^2t^2 = x'^2 + y'^2 + z'^2 - c^2t'^2, \tag{11.131}$$

where c is the fundamental speed, determined by experiment to be the speed of light in free space. So points for which $s = 0$ are said to lie on the light cone based on the origin.

The components of 4-vector **s** may be collected as

$$x + iy = x_+, \qquad x - iy = x_-, \tag{11.132}$$

$$z + ct = z_+, \qquad z - ct = z_-. \tag{11.133}$$

Then Equation (11.131) can be rewritten as

$$s^2 = x_+x_- + z_+z_- = x_+'x_-' + z_+'z_-'. \tag{11.134}$$

In the transformation from Σ to Σ', the light cone based on the origin is invariant. On the future half of this cone, we can set

$$z_+ = S^{1}{}^*S^1, \qquad -z_- = S^2 S^{2}{}^*, \qquad (11.135)$$

$$x_+ = S^{1}{}^*S^2, \qquad x_- = S^1 S^{2}{}^*, \qquad (11.136)$$

and construct the first-rank spinor

$$S^{\cdot} = \begin{pmatrix} S^1 \\ S^2 \end{pmatrix}. \qquad (11.137)$$

Similarly, on the past half of the light cone, we set

$$z_- = S_1{}^*S_1, \qquad -z_+ = S_2 S_2{}^*, \qquad (11.138)$$

$$x_+ = S_1{}^*S_2, \qquad x_- = S_1 S_2{}^*, \qquad (11.139)$$

and construct the first-rank spinor

$$S_{\cdot} = \begin{pmatrix} S_1 \\ S_2 \end{pmatrix}. \qquad (11.140)$$

By matrix multiplication, one obtains the second-rank spinors

$$\mathbf{M}^{\cdot\cdot} = \mathbf{S}^{\cdot}\mathbf{S}^{\cdot\dagger} = \begin{pmatrix} z + ct & x - iy \\ x + iy & -z + ct \end{pmatrix} \qquad (11.141)$$

and

$$\mathbf{M}_{\cdot\cdot} = \mathbf{S}_{\cdot}\mathbf{S}_{\cdot}{}^{\dagger} = \begin{pmatrix} z - ct & x - iy \\ x + iy & -z - ct \end{pmatrix}. \qquad (11.142)$$

For a 4-vector with Cartesian components V_x, V_y, V_z, V_{ct}, the second-rank spinors

$$\mathbf{N}^{\cdot\cdot} = \begin{pmatrix} V_z + V_{ct} & V_x - iV_y \\ V_x + iV_y & -V_z + V_{ct} \end{pmatrix} \qquad (11.143)$$

and

$$\mathbf{N}_{\cdot\cdot} = \begin{pmatrix} V_z - V_{ct} & V_x - iV_y \\ V_x + iV_y & -V_z - V_{ct} \end{pmatrix} \qquad (11.144)$$

similarly exist.

The linear transformation

$$(\mathbf{S}^{\cdot})' = \mathbf{A}\,\mathbf{S}^{\cdot} \qquad (11.145)$$

becomes

$$(\mathbf{M}^{\cdot\cdot})' = \mathbf{A}\,\mathbf{M}^{\cdot\cdot}\,\mathbf{A}^{\dagger} \tag{11.146}$$

and

$$(\mathbf{N}^{\cdot\cdot})' = \mathbf{A}\,\mathbf{N}^{\cdot\cdot}\,\mathbf{A}^{\dagger}. \tag{11.147}$$

Equations (11.146) and (11.147) apply throughout the space-time continuum.

Differentiating the scalar function $\phi(x, y, z, ct)$ yields the Cartesian components

$$\frac{\partial \phi}{\partial x}, \qquad \frac{\partial \phi}{\partial y}, \qquad \frac{\partial \phi}{\partial z}, \qquad -\frac{\partial \phi}{\partial (ct)}\;. \tag{11.148}$$

These can be paired with the components of ds

$$dx, \qquad dy, \qquad dz, \qquad d(ct) \tag{11.149}$$

in a dot product to yield $d\phi$.

Components (11.148) may be employed in (11.143) and (11.144). The resulting spinors involve the operators

$$\mathbf{D}^{\cdot\cdot} = \sigma_x \frac{\partial}{\partial x} + \sigma_y \frac{\partial}{\partial y} + \sigma_z \frac{\partial}{\partial z} - \mathbf{1}\,\frac{\partial}{\partial (ct)} \tag{11.150}$$

and

$$\mathbf{D}_{..} = \sigma_x \frac{\partial}{\partial x} + \sigma_y \frac{\partial}{\partial y} + \sigma_z \frac{\partial}{\partial z} + \mathbf{1}\,\frac{\partial}{\partial (ct)} \tag{11.151}$$

acting on ϕ. Here the Pauli matrices are

$$\sigma_x = \begin{pmatrix} 0 & 1 \\ 1 & 0 \end{pmatrix}, \tag{11.152}$$

$$\sigma_y = \begin{pmatrix} 0 & -i \\ i & 0 \end{pmatrix}, \tag{11.153}$$

$$\sigma_z = \begin{pmatrix} 1 & 0 \\ 0 & -1 \end{pmatrix}, \tag{11.154}$$

while the unit matrix is

$$\mathbf{1} = \begin{pmatrix} 1 & 0 \\ 0 & 1 \end{pmatrix}. \tag{11.155}$$

In a Lorentz transformation based on the common origin of Σ and Σ', the origin's light cone is invariant. So derivatives on this cone can be physically significant. But operator (11.150) acting on the lower index spinor function

$$\chi. = \begin{pmatrix} \chi_1 \\ \chi_2 \end{pmatrix} \qquad (11.156)$$

yields an upper index spinor function

$$\mathbf{D}^{\cdot\cdot}\chi. = A\phi^{\cdot} \qquad (11.157)$$

with

$$\phi^{\cdot} = \begin{pmatrix} \phi^1 \\ \phi^2 \end{pmatrix} \qquad (11.158)$$

Coupled with this may be the condition

$$\mathbf{D}_{..}\phi^{\cdot} = -A\chi.. \qquad (11.159)$$

For motion in free space, we find that

$$A = \frac{im_0c}{\hbar}. \qquad (11.160)$$

Multiplying (11.157) by $\hbar/i$ and introducing the operator relations

$$p_x = \frac{\hbar}{i}\frac{\partial}{\partial x}, \qquad p_y = \frac{\hbar}{i}\frac{\partial}{\partial y}, \qquad p_z = \frac{\hbar}{i}\frac{\partial}{\partial z}, \qquad E = -\frac{\hbar}{i}\frac{\partial}{\partial t}, \qquad (11.161)$$

transforms (11.157) to

$$\left(\sigma\cdot\mathbf{p} + \mathbf{1}\,\frac{E}{c} \right)\chi. - m_0c\phi^{\cdot} = 0. \qquad (11.162)$$

Similarly from (11.159), we obtain

$$-\left(\sigma\cdot\mathbf{p} - \mathbf{1}\,\frac{E}{c} \right)\phi^{\cdot} - m_0c\chi. = 0. \qquad (11.163)$$

With the Dirac 4×4 matrices

$$\alpha = \begin{pmatrix} 0 & \sigma \\ \sigma & 0 \end{pmatrix}, \qquad \beta = \begin{pmatrix} 1 & 0 \\ 0 & -1 \end{pmatrix}, \qquad (11.164)$$

and the bispinor

$$\Psi = \begin{pmatrix} \phi^{\cdot} + \chi. \\ \phi^{\cdot} - \chi. \end{pmatrix}, \qquad (11.165)$$

Equations (11.162) and (11.163) combine to yield

$$(c\alpha \cdot \mathbf{p} + \beta m_0 c^2)\Psi = E\Psi. \qquad (11.166)$$

When the particle is not free, (11.166) becomes

$$[c\alpha \cdot (\mathbf{p} - q\mathbf{A}) + \beta m_0 c^2]\Psi = (E - q\phi)\Psi, \qquad (11.167)$$

where $\mathbf{A}$ is the vector potential, ϕ the scalar potential, and q the interaction coefficient (charge).

Discussion Questions

11.1. What is wrong with common-sense notions of (a) space, (b) time?

11.2. How is the space apparent to an inertial observer surveyed?

11.3. How is the time apparent to such an observer measured?

11.4. What is the fundamental invariance of the space-time continuum?

11.5. How does the fundamental invariance allow one speed to be the same in all inertial frames?

11.6. What is a light cone?

11.7. Why are x, y, z, ct considered components of a 4-dimensional vector?

11.8. Explain when the 2-dimensional tensor

$$\begin{pmatrix} z + ct & x - iy \\ x + iy & -z + ct \end{pmatrix}$$

factors into a 2-dimensional vector and its Hermitian adjoint.

11.9. Likewise discuss factoring of the tensor

$$\begin{pmatrix} z - ct & x - iy \\ x + iy & -z - ct \end{pmatrix}$$

into a vector and its Hermitian adjoint.

11.10. What are (a) $\mathbf{S}$', (b) $\mathbf{S}_.$, (c) $\mathbf{S}'\mathbf{S}'^{\dagger}$, (d) $\mathbf{S}'^{\dagger}\mathbf{S}'$, (e) $\mathbf{S}_.\mathbf{S}_.^{\dagger}$, (f) $\mathbf{S}_.^{\dagger}\mathbf{S}_.$?

11.11. Why can a person replace S^1 with S_2^* and S^2 with $-S_1^*$ in the transformation equation

$$(\mathbf{S}')' = \mathbf{A}\,\mathbf{S}'?$$

11.12. How are the elements in $\mathbf{B}$ of

$$(\mathbf{S}_.)' = \mathbf{B}\,\mathbf{S}_.$$

related to the elements in $\mathbf{A}$ of

$$(\mathbf{S}')' = \mathbf{A}\,\mathbf{S}'?$$

11.13. How is the transformation law for a second-rank spinor related to the law for first-rank spinors?

11.14. What relationship must the elements of transformation matrices **A** and **B** satisfy? How is this reduced to the unimodular condition?

11.15. Determine **B**†**A**.

11.16. How may (a) a rotation, (b) a Lorentz transformation be represented by a spinorial transformation?

11.17. How do the components of a 4-vector form second-rank spinors?

11.18. How are these spinors expressed as a linear function of the Pauli matrices?

11.19. Explain how the scalar product of two 4-vectors is formed.

11.20. Show that

$$\begin{pmatrix} \dfrac{\partial}{\partial z} \mp \dfrac{\partial}{\partial(ct)} & \dfrac{\partial}{\partial x} - i\dfrac{\partial}{\partial y} \\[2mm] \dfrac{\partial}{\partial x} + i\dfrac{\partial}{\partial y} & -\dfrac{\partial}{\partial z} \mp \dfrac{\partial}{\partial(ct)} \end{pmatrix}$$

transforms as a second-rank spinor.

11.21. Explain how $\mathbf{D}_{..}\phi^{\cdot}$ transforms.

11.22. What properties does a homogeneous beam of neutrinos possess? Why does such a beam satisfy

$$\mathbf{D}_{..}\,\phi^{\cdot} = 0?$$

11.23. What geometric fields are invariant in (a) Galilean relativity, (b) Einsteinian relativity?

11.24. Why may spinorial derivatives be physically significant?

11.25. How is the significance of A in

$$\mathbf{D}_{..}\,\phi^{\cdot} = -A\chi_{.,}$$

$$\mathbf{D}^{..}\chi_{.} = A\phi^{\cdot}$$

determined? What is this significance?

11.26. Under what circumstances do the elements of $\phi^{\cdot}$ approach those in $\chi_{.}$? Explain.

11.27. What advantage is there in combining the dependent spinors in the hybrid form

$$\Psi = \begin{pmatrix} \phi^{\cdot} + \chi_{.} \\ \phi^{\cdot} - \chi_{.} \end{pmatrix}?$$

11.28 Why can p_x, p_y, p_z, and E be interpreted in the equations either as (1) algebraic quantities or as (b) operators?

11.29. Why does introduction of a field cause $\mathbf{p} - q\mathbf{A}$ to replace $\mathbf{p}$ and $E - q\phi$ to replace E in the Dirac equation?

Problems

11.1. Explain why a person cannot consider z and ct to be real and imaginary parts of one coordinate, as he or she takes x and y to be real and imaginary parts of another coordinate, of a 2-dimensional space.

11.2. Show that a reflection in a plane cannot be represented by a spinorial transformation.

11.3. How are x, y, z, and ct altered when the signs of S^1 and S^2 are reversed?

11.4. What kind of spinor is

$$\begin{pmatrix} (z - ct)T^1 + (x - iy)T^2 \\ (x + iy)T^1 - (z + ct)T^2 \end{pmatrix} \ ?$$

11.5. Show that $\mathbf{M}^{..}\mathbf{M}_{..}$ is a scalar.

11.6. Identify

$$\begin{pmatrix} 2\,\dfrac{\partial}{\partial(S_1 S_1{}^*)} & 2\,\dfrac{\partial}{\partial(S_2 S_1{}^*)} \\[2ex] 2\,\dfrac{\partial}{\partial(S_1 S_2{}^*)} & 2\,\dfrac{\partial}{\partial(S_2 S_2{}^*)} \end{pmatrix}.$$

11.7. Derive Dirac's equation in the form

$$(E + c\alpha \cdot \mathbf{p} + \beta m_0 c^2)\Psi = 0.$$

What is Ψ now?

—   —   —

11.8. How could spinors be employed to preserve the invariance of $x^2 + y^2 + z^2$ in a Galilean transformation?

11.9. Show that an inversion cannot be represented by a spinor transformation.

11.10. If a physical transformation multiplies S^1 by k, what does it do to S^2?

11.11. What physical transformation results when spinor $\mathbf{S}^{.}$ is subjected to a transformation that differs from the identity transformation by

$$\frac{\sigma_z}{2i}\,d\varphi\,?$$

11.12. Obtain the transformation laws for (a) $\mathbf{N}_{..}$, (b) $\mathbf{M}^{.}_{.}$, (c) $\mathbf{M}_{.}^{.}$.

11.13. Show that $\mathbf{D}^{..}\,\mathbf{D}_{..}$ behaves as a scalar operator.

11.14. Prove that when a square matrix $\mathbf{A}$ has a vanishing anticommutator with σ_x, σ_y, and σ_z, so that

$$\mathbf{A}\sigma + \sigma\mathbf{A} = 0,$$

it equals zero.

References

Books

Cartan, E.: 1966. *The Theory of Spinors*, M.I.T. Press, Cambridge, Mass., pp. 41-51, 125-151. This is a translation of a French book that was compiled from notes of Cartan's lectures, gathered and arranged by A. Mercier. Thus, it is quite terse, yet detailed.

Duffey, G. H.: 1980. *Theoretical Physics: Classical and Modern Views*, Krieger, Melbourne, Fla., pp. 554-578. The presentation is like that in this chapter.

Eisele, J. A.: 1969. *Modern Quantum Mechanics with Applications to Elementary Particle Physics*, Wiley-Interscience, New York, pp. 94-101, 180-302, 433-459. The emphasis in this readable book is on matrix methods. While little is given on spinors, the Dirac equation is treated in detail.

Synge, J. L.: 1964. *Relativity: The Special Theory*, 2nd edn., North-Holland, Amsterdam, pp. 102-110. The nature of spinors and their linear transformations are discussed.

Articles

Chapman, T. C., and Leiter, D. J.: 1976. "On the Generally Covariant Dirac Equation," *Am. J. Phys.* **44**, 858-862.

Dumais, J.-F.: 1977. "On Some Properties of γ Matrices," *Am. J. Phys.* **45**, 352-354.

Edmonds, J. D., Jr.: 1978. "Maxwell's Eight Equations as One Quaternion Equation," *Am. J. Phys.* **46**, 430-431.

Fernow, R. C.: 1976. "Expansions of Spin-½ Expectation Values," *Am. J. Phys.* **44**, 560-563.

Frescura, F. A. M., and Hiley, B. J.: 1981. "Geometric Interpretation of the Pauli Spinor," *Am. J. Phys.* **49**, 152-157.

Hestenes, D.: 1979. "Spin and Uncertainty in the Interpretation of Quantum Mechanics," *Am. J. Phys.* **47**, 399-415.

Moyer, D. F.: 1981. "Origins of Dirac's Electron, 1925-1928; Evaluations of Dirac's Electron, 1928-1932; Vindications of Dirac's Electron, 1932-1934," *Am. J.Phys.* **49**, 944-948, 1055-1062, 1120-1125.

Weaver, D. L.: 1976. "Unitary Transformations of the Dirac Equation," *Am. J. Phys.* **47**, 32-35.

12

SIMPLE DIRAC SYSTEMS

12.1
The One-Particle Dirac Equation

In Galilean relativity space is physically separable from time. Furthermore, both space and time are invariant; neither is affected by a transformation from one reference frame to another. So in our fundamental discussions we could employ derivatives over space and time in representing the commutation relations and in constructing the Schrödinger equations.

But in Einsteinian relativity, only the light cone based on the pertinent common origin is invariant. To be physically significant, derivatives must be defined on this cone. The simplest possibilities involve second-rank spinorial operators acting on first-rank spinorial functions.

Thus, we were led to construct Equations (11.98) and (11.99). These combine to yield

$$\mathbf{D}^{..} \mathbf{D}_{..} \phi^{.} = -A^2 \phi^{.} \qquad (12.1)$$

and

$$\mathbf{D}_{..} \mathbf{D}^{..} \chi_{.} = -A^2 \chi_{.}, \qquad (12.2)$$

generalizations of the Schrödinger equation.

Expressions (11.103) and (11.104), describing a free particle in a given momentum state, satisfy (12.1), (12.2) with

$$A^2 = \mathbf{k} \cdot \mathbf{k} - \frac{\omega^2}{c^2} = k^2 - \frac{E^2}{\hbar^2 c^2} = -\frac{m_0^2 c^2}{\hbar^2}. \qquad (12.3)$$

Thus A is the 4-vector form of the k in (11.102).

Each pure-momentum solution allows the substitutions

$$\frac{\partial}{\partial x_j} = ik_j = i\frac{p_j}{\hbar} \qquad (12.4)$$

and

$$\frac{\partial}{\partial t} = -i\omega = -i\frac{E}{\hbar} \qquad (12.5)$$

to be made, as in Schrödinger theory. These lead to Equations (11.114) and (11.115), which can be combined to form the Dirac equation

$$(E - c\boldsymbol{\alpha}\cdot\mathbf{p} - \beta m_0 c^2)\Psi = 0. \qquad (12.6)$$

Matrices α and β, of (11.122) and (11.123), have the properties

$$\alpha_x^2 = \alpha_y^2 = \alpha_z^2 = \beta^2 = 1, \qquad (12.7)$$

$$\alpha_x\alpha_y + \alpha_y\alpha_x = \alpha_y\alpha_z + \alpha_z\alpha_y = \alpha_z\alpha_x + \alpha_x\alpha_z$$

$$= \alpha_x\beta + \beta\alpha_x = \alpha_y\beta + \beta\alpha_y = \alpha_z\beta + \beta\alpha_z = 0. \qquad (12.8)$$

State function Ψ is the hybrid spinor

$$\Psi = \begin{pmatrix} \phi^1 + \chi_1 \\ \phi^2 + \chi_2 \\ \phi^1 - \chi_1 \\ \phi^2 - \chi_2 \end{pmatrix} = \begin{pmatrix} \Psi_1 \\ \Psi_2 \\ \Psi_3 \\ \Psi_4 \end{pmatrix}. \qquad (12.9)$$

Premultiplying Ψ by its Hermitian adjoint yields a probability density ρ:

$$\Psi^\dagger\Psi = (\Psi_1^* \; \Psi_2^* \; \Psi_3^* \; \Psi_4^*) \begin{pmatrix} \Psi_1 \\ \Psi_2 \\ \Psi_3 \\ \Psi_4 \end{pmatrix}$$

$$= \Psi_1^*\Psi_1 + \Psi_2^*\Psi_2 + \Psi_3^*\Psi_3 + \Psi_4^*\Psi_4$$

$$= \rho. \qquad (12.10)$$

When the given particle carries q units of charge and moves in a field of vector potential $\mathbf{A}$ and scalar potential ϕ per unit charge, (12.6) has to be replaced by

$$[E - q\phi - c\boldsymbol{\alpha}\cdot(\mathbf{p} - q\mathbf{A}) - \beta m_0 c^2]\Psi = 0. \qquad (12.11)$$

12.2
Spatially Periodic Translational States

A set of identical particles free to move over all space at a uniform density exhibits a continuum of energy levels. If we subdivide the space into identical cells and require conditions in any one cell to be repeated in all other cells, discrete levels appear, at *both* positive and negative energies. The excess above m_0c^2 of the former checks with the Schrödinger formula when the excess is not large. However, two independent Dirac states are obtained for each given set of quantum numbers n_x, n_y, and n_z.

Let us suppose that particles of rest mass m_0 move with null potential energy through a rectangular parallelepiped having perpendicular edges a, b, and c as in Table 4.1 in Section 4.2. Let us place coordinate axes along a right-handed combination of the edges. We also suppose that the volume is large enough so adding similar blocks to fill all space does not appreciably alter conditions in the given region. With these added, the boundaries are eliminated and the wave function in the given volume is repeated periodically in the three perpendicular directions.

Being free, the particles possess $\phi^{\cdot}$ and χ. obeying (12.1) and (12.2). These spinors can be expressed as linear combinations of exponential terms

$$\phi_j = u^j e^{i(\mathbf{k}\cdot\mathbf{r} - \omega t)}, \tag{12.12}$$

$$\chi_j = v_j e^{i(\mathbf{k}\cdot\mathbf{r} - \omega t)}, \tag{12.13}$$

as in Section 11.8.

The periodic boundary conditions require each constituent wave length, λ_x, λ_y, λ_z, to be a unit fraction of the corresponding edge length. Thus,

$$a = n_x\lambda_x = \frac{2\pi n_x}{k_x} = \frac{n_x}{\bar{k}_x} \quad \text{with} \quad n_x = \ldots, -1, 0, 1, \ldots, \tag{12.14}$$

$$b = n_y\lambda_y = \frac{2\pi n_y}{k_y} = \frac{n_y}{\bar{k}_y} \quad \text{with} \quad n_y = \ldots, -1, 0, 1, \ldots, \tag{12.15}$$

$$c = n_z\lambda_z = \frac{2\pi n_z}{k_z} = \frac{n_z}{\bar{k}_z} \quad \text{with} \quad n_z = \ldots, -1, 0, 1, \ldots. \tag{12.16}$$

Substituting (12.12), (12.13) into (12.1) and (12.2) leads to (11.105) and the related equation. For these to be satisfied, (11.106) must hold, together with (11.109) and (11.112), whence

$$k_x^2 + k_y^2 + k_z^2 - \frac{E^2}{\hbar^2c^2} = -\frac{m_0^2c^4}{\hbar^2c^2} \tag{12.17}$$

or

$$m_0^2c^4 + c^2\hbar^2\bar{k}_x^2 + c^2\hbar^2\bar{k}_y^2 + c^2\hbar^2\bar{k}_z^2 = E^2. \tag{12.18}$$

Introducing quantization conditions (12.14), (12.15), (12.16) yields

$$E^2 = m_0^2 c^4 + c^2 h^2 \left(\frac{n_x^2}{a^2} + \frac{n_y^2}{b^2} + \frac{n_z^2}{c^2} \right)$$

$$= m_0^2 c^4 \left[1 + \frac{h^2}{m_0^2 c^2} \left(\frac{n_x^2}{a^2} + \frac{n_y^2}{b^2} + \frac{n_z^2}{c^2} \right) \right] \qquad (12.19)$$

whence

$$E = \pm m_0 c^2 \left[1 + \frac{h^2}{m_0^2 c^2} \left(\frac{n_x^2}{a^2} + \frac{n_y^2}{b^2} + \frac{n_z^2}{c^2} \right) \right]^{\frac{1}{2}}$$

$$\simeq \pm m_0 c^2 \left[1 + \frac{h^2}{2 m_0 c^2 m_0} \left(\frac{n_x^2}{a^2} + \frac{n_y^2}{b^2} + \frac{n_z^2}{c^2} \right) \right]$$

$$= \pm \left[m_0 c^2 + \frac{h^2}{2 m_0} \left(\frac{n_x^2}{a^2} + \frac{n_y^2}{b^2} + \frac{n_z^2}{c^2} \right) \right]. \qquad (12.20)$$

In the next-to-last step, the radical has been expanded with the binomial theorem, and higher terms have been dropped. Since n_x, n_y, and n_z are the quantum numbers employed in Table 4.1, the kinetic energy in the positive energy state of (12.20) agrees with that obtained in the Schrödinger theory, as long as the approximation is valid. Term $m_0 c^2$ is the rest energy of a particle.

Bringing in de Broglie's relationship

$$\hbar \mathbf{k} = h \mathbf{\acute{K}} = \mathbf{p} \qquad (12.21)$$

and rearranging converts (12.18) to

$$E^2 - m_0^2 c^4 - c^2 p_x^2 - c^2 p_y^2 - c^2 p_z^2 = 0 \qquad (12.22)$$

whence

$$(E - m_0 c^2)(E + m_0 c^2) - c(p_x - i p_y) c(p_x + i p_y) - c p_z c p_z = 0. \qquad (12.23)$$

To obtain information on the corresponding state functions, we substitute (12.12), (12.13) into (12.6) with

$$u^j + v_j = w_j \qquad (12.24)$$

and

$$u^j - v_j = w_{j+2}. \qquad (12.25)$$

Canceling the exponential factor then yields

$$
\begin{pmatrix}
(E - m_0c^2)\,w_1 + 0 - cp_zw_3 - c(p_x - ip_y)\,w_4 \\
0 + (E - m_0c^2)\,w_2 - c(p_x + ip_y)\,w_3 + cp_zw_4 \\
- cp_zw_1 - c(p_x - ip_y)\,w_2 + (E + m_0c^2)\,w_3 + 0 \\
- c(p_x + ip_y)\,w_1 + cp_zw_2 + 0 + (E + m_0c^2)\,w_4
\end{pmatrix}
=
\begin{pmatrix}
0 \\ 0 \\ 0 \\ 0
\end{pmatrix}. \qquad (12.26)
$$

The first equation in (12.26) agrees with (12.23) when

$$w_1 = a(E + m_0c^2), \qquad (12.27)$$

$$w_2 = 0, \qquad (12.28)$$

$$w_3 = acp_z, \qquad (12.29)$$

$$w_4 = ac(p_x + ip_y). \qquad (12.30)$$

The other equations in (12.26) are then satisfied identically. The second equation in (12.26) agrees with (12.23) when

$$w_1 = 0, \qquad (12.31)$$

$$w_2 = a(E + m_0c^2), \qquad (12.32)$$

$$w_3 = ac(p_x - ip_y), \qquad (12.33)$$

$$w_4 = -acp_z. \qquad (12.34)$$

The other equations in (12.26) are satisfied by (12.31)–(12.34) identically. The third equation in (12.26) agrees with (12.23) when

$$w_1 = acp_z. \qquad (12.35)$$

$$w_2 = ac(p_x + ip_y), \qquad (12.36)$$

$$w_3 = a(E - m_0c^2), \qquad (12.37)$$

$$w_4 = 0. \qquad (12.38)$$

Then the other equations in (12.26) are satisfied identically. The fourth equation in (12.26) agrees with (12.23) when

$$w_1 = ac(p_x - ip_y), \qquad (12.39)$$

$$w_2 = -acp_z, \qquad (12.40)$$

$$w_3 = 0, \qquad (12.41)$$

$$w_4 = a(E - m_0c^2). \qquad (12.42)$$

As before, the other equations are satisfied identically.

When the momenta are zero and

$$E = m_0 c^2, \tag{12.43}$$

only (12.27)–(12.30) and (12.31)–(12.34) give nonvanishing state functions. When the momenta are zero and

$$E = -m_0 c^2, \tag{12.44}$$

only (12.35)–(12.38) and (12.39)–(12.42) give nonvanishing state functions.

Since a freely translating particle can be viewed from a frame in which it is at rest, the particle exists in two independent intrinsic states with either sign of the energy. We will see later that these correspond to the two orientations of spin that fundamental particles exhibit. How the negative energy states are to be interpreted will be considered next.

12.3
Antiparticles

Each allowed n_x, n_y, and n_z in (12.20) yields a positive energy and a negative energy of the same magnitude. To each positive level of a free particle corresponds an equally degenerate negative level. Furthermore, there is no upper limit on the quantum numbers, so there is no upper or lower limit on the levels. Introducing a field generally shifts levels, but the absence of limits persists.

Furthermore, transitions can occur among the levels, with gains and losses of quanta of energy. Presumably, a particle could keep falling down the ladder of levels, losing any desired amount of total energy. If the study of energy in thermodynamics tells us anything, it is that such a surfeit of energy is not available. Perpetual motion machines do not exist and cannot be constructed.

What has gone wrong? Might the theory be trying to tell us something about the nature of empty space? Indeed, when Dirac was confronted with this dilemma, he suggested that the negative energy states were all filled with equivalent unobservable particles. The Pauli exclusion principle then prevented the particle with a positive energy from going down into a negative level.

Any one of these unobservable entities can be excited into a positive level. The result would be an observable particle and the hole from which the particle was removed. This hole would carry the opposite charge and spin from the excited particle and so would be observed as its antiparticle.

12.4
The 4-Dimensional Sigma Matrix

A positive energy state of a free particle is a superposition of the state described by (12.27)–(12.30) and the state described by (12.31)–(12.34). A negative energy state is a superposition of the state described by (12.35)–(12.38) and the state described by (12.39)–(12.42). In the inertial frame in which the particle is at rest, a positive energy state exhibits a null w_3 and w_4; a negative energy state, a null w_1 and w_2. Furthermore, the initial expressions and certain superpositions of them are eigenfunctions of very simple operators to be constructed here.

From the 2-dimensional Pauli matrices, let us form

$$\Sigma = \begin{pmatrix} \sigma & 0 \\ 0 & \sigma \end{pmatrix}. \tag{12.45}$$

Then

$$\Sigma_z = \begin{pmatrix} \sigma_z & 0 \\ 0 & \sigma_x \end{pmatrix} = \begin{pmatrix} 1 & 0 & 0 & 0 \\ 0 & -1 & 0 & 0 \\ 0 & 0 & 1 & 0 \\ 0 & 0 & 0 & -1 \end{pmatrix}, \tag{12.46}$$

$$\Sigma_x = \begin{pmatrix} \sigma_x & 0 \\ 0 & \sigma_x \end{pmatrix} = \begin{pmatrix} 0 & 1 & 0 & 0 \\ 1 & 0 & 0 & 0 \\ 0 & 0 & 0 & 1 \\ 0 & 0 & 1 & 0 \end{pmatrix}, \tag{12.47}$$

$$\Sigma_y = \begin{pmatrix} \sigma_y & 0 \\ 0 & \sigma_y \end{pmatrix} = \begin{pmatrix} 0 & -i & 0 & 0 \\ i & 0 & 0 & 0 \\ 0 & 0 & 0 & -i \\ 0 & 0 & i & 0 \end{pmatrix}. \tag{12.48}$$

In a given wavevector state, free particles have the state function

$$\Psi = \begin{pmatrix} w_1 \\ w_2 \\ w_3 \\ w_4 \end{pmatrix} e^{i(\mathbf{k}\cdot\mathbf{r} - \omega t)}, \tag{12.49}$$

according to (12.9), (12.12), (12.13), (12.24), (12.25). The possible states with given energy are superpositions of those for which the w_j's are given by (12.27)–(12.30), (12.31)–(12.34) or by (12.35)–(12.38), (12.39)–(12.42).

When the observer is at rest with respect to the particle and its energy is positive, (12.29) and (12.33) vanish,

$$w_3 = 0, \tag{12.50}$$

(12.30) and (12.34) vanish,

$$w_4 = 0. \tag{12.51}$$

When w_2 is also zero, we find that

$$\Sigma_z \Psi = \begin{pmatrix} 1 & 0 & 0 & 0 \\ 0 & -1 & 0 & 0 \\ 0 & 0 & 1 & 0 \\ 0 & 0 & 0 & -1 \end{pmatrix} \begin{pmatrix} w_1 \\ 0 \\ 0 \\ 0 \end{pmatrix} e^{i(\mathbf{k}\cdot\mathbf{r} - \omega t)}$$

$$= \begin{pmatrix} w_1 \\ 0 \\ 0 \\ 0 \end{pmatrix} e^{i(\mathbf{k}\cdot\mathbf{r} - \omega t)} = +1\Psi, \tag{12.52}$$

Ψ is an eigenfunction of Σ_z with the eigenvalue $+1$. On the other hand, when w_1 is zero, we find that

$$\Sigma_z \Psi = \begin{pmatrix} 1 & 0 & 0 & 0 \\ 0 & -1 & 0 & 0 \\ 0 & 0 & 1 & 0 \\ 0 & 0 & 0 & -1 \end{pmatrix} \begin{pmatrix} 0 \\ w_2 \\ 0 \\ 0 \end{pmatrix} e^{i(\mathbf{k}\cdot\mathbf{r} - \omega t)}$$

$$= \begin{pmatrix} 0 \\ -w_2 \\ 0 \\ 0 \end{pmatrix} e^{i(\mathbf{k}\cdot\mathbf{r} - \omega t)} = -1\Psi, \tag{12.53}$$

Ψ is an eigenfunction of Σ_z with the eigenvalue -1.

When the ratio w_2/w_1 is ± 1, we have

$$\Sigma_x \Psi = \begin{pmatrix} 0 & 1 & 0 & 0 \\ 1 & 0 & 0 & 0 \\ 0 & 0 & 0 & 1 \\ 0 & 0 & 1 & 0 \end{pmatrix} \begin{pmatrix} w_1 \\ \pm w_1 \\ 0 \\ 0 \end{pmatrix} e^{i(\mathbf{k \cdot r} - \omega t)}$$

$$= \begin{pmatrix} \pm w_1 \\ w_1 \\ 0 \\ 0 \end{pmatrix} e^{i(\mathbf{k \cdot r} - \omega t)} = \pm 1 \Psi. \tag{12.54}$$

This Ψ is an eigenfunction of Σ_x with the eigenvalue ± 1. When the ratio w_2/w_1 is $\pm i$, we have

$$\Sigma_y \Psi = \begin{pmatrix} 0 & -i & 0 & 0 \\ i & 0 & 0 & 0 \\ 0 & 0 & 0 & -i \\ 0 & 0 & i & 0 \end{pmatrix} \begin{pmatrix} w_1 \\ \pm i w_1 \\ 0 \\ 0 \end{pmatrix} e^{i(\mathbf{k \cdot r} - \omega t)}$$

$$= \begin{pmatrix} \pm w_1 \\ i w_1 \\ 0 \\ 0 \end{pmatrix} e^{i(\mathbf{kr} - \omega t)} = \pm 1 \Psi. \tag{12.55}$$

The Ψ is now an eigenfunction of Σ_y with the eigenvalue ± 1.

When the energy is negative and the particle is at rest, similar results are obtained. Because

$$\Sigma_j^2 = \begin{pmatrix} \sigma_j & 0 \\ 0 & \sigma_j \end{pmatrix} \begin{pmatrix} \sigma_j & 0 \\ 0 & \sigma_j \end{pmatrix} = \begin{pmatrix} \sigma_j^2 & 0 \\ 0 & \sigma_j^2 \end{pmatrix} = \begin{pmatrix} 1 & 0 \\ 0 & 1 \end{pmatrix}, \tag{12.56}$$

any Ψ is an eigenfunction of Σ_j^2 and of

$$\Sigma^2 = \Sigma \cdot \Sigma = \Sigma_x^{\,2} + \Sigma_y^{\,2} + \Sigma_z^{\,2}. \tag{12.57}$$

The eigenvalue for Σ^2 is 3.

Example 12.1

What are the commutators of Σ_z with α_x, α_y, α_z, and β?

Employ definitions (12.45), (11.122), (11.123) in constructing the commutators, carry out the indicated operations, and reduce making use of the results in Example 11.5:

$$\Sigma_z\alpha_x - \alpha_x\Sigma_z = \begin{pmatrix} \sigma_z & 0 \\ 0 & \sigma_z \end{pmatrix} \begin{pmatrix} 0 & \sigma_x \\ \sigma_x & 0 \end{pmatrix} - \begin{pmatrix} 0 & \sigma_x \\ \sigma_x & 0 \end{pmatrix} \begin{pmatrix} \sigma_z & 0 \\ 0 & \sigma_z \end{pmatrix}$$

$$= \begin{pmatrix} 0 & \sigma_z\sigma_x - \sigma_x\sigma_z \\ \sigma_z\sigma_x - \sigma_x\sigma_z & 0 \end{pmatrix} = \begin{pmatrix} 0 & 2i\sigma_y \\ 2i\sigma_y & 0 \end{pmatrix} = 2i\alpha_y,$$

$$\Sigma_z\alpha_y - \alpha_y\Sigma_z = \begin{pmatrix} \sigma_z & 0 \\ 0 & \sigma_z \end{pmatrix} \begin{pmatrix} 0 & \sigma_y \\ \sigma_y & 0 \end{pmatrix} - \begin{pmatrix} 0 & \sigma_y \\ \sigma_y & 0 \end{pmatrix} \begin{pmatrix} \sigma_z & 0 \\ 0 & \sigma_z \end{pmatrix}$$

$$= \begin{pmatrix} 0 & \sigma_z\sigma_y - \sigma_y\sigma_z \\ \sigma_z\sigma_y - \sigma_y\sigma_z & 0 \end{pmatrix} = \begin{pmatrix} 0 & -2i\sigma_x \\ -2i\sigma_x & 0 \end{pmatrix} = -2i\alpha_x,$$

$$\Sigma_z\alpha_z - \alpha_z\Sigma_z = \begin{pmatrix} \sigma_z & 0 \\ 0 & \sigma_z \end{pmatrix} \begin{pmatrix} 0 & \sigma_z \\ \sigma_z & 0 \end{pmatrix} - \begin{pmatrix} 0 & \sigma_z \\ \sigma_z & 0 \end{pmatrix} \begin{pmatrix} \sigma_z & 0 \\ 0 & \sigma_z \end{pmatrix}$$

$$= \begin{pmatrix} 0 & \sigma_z^{\,2} \\ \sigma_z^{\,2} & 0 \end{pmatrix} - \begin{pmatrix} 0 & \sigma_z^{\,2} \\ \sigma_z^{\,2} & 0 \end{pmatrix} = 0,$$

$$\Sigma_z\beta - \beta\Sigma_z = \begin{pmatrix} \sigma_z & 0 \\ 0 & \sigma_z \end{pmatrix} \begin{pmatrix} 1 & 0 \\ 0 & -1 \end{pmatrix} - \begin{pmatrix} 1 & 0 \\ 0 & -1 \end{pmatrix} \begin{pmatrix} \sigma_z & 0 \\ 0 & \sigma_z \end{pmatrix}$$

$$= \begin{pmatrix} \sigma_z & 0 \\ 0 & -\sigma_z \end{pmatrix} - \begin{pmatrix} \sigma_z & 0 \\ 0 & -\sigma_z \end{pmatrix} = 0.$$

Example 12.2

Determine the commutator of Σ_x and Σ_y.

Construct $[\Sigma_x, \Sigma_y]$, carry out the indicated operations, and use the results in Example 11.5 to reduce:

$$\Sigma_x\Sigma_y - \Sigma_y\Sigma_x = \begin{pmatrix} \sigma_x & 0 \\ 0 & \sigma_x \end{pmatrix} \begin{pmatrix} \sigma_y & 0 \\ 0 & \sigma_y \end{pmatrix} - \begin{pmatrix} \sigma_y & 0 \\ 0 & \sigma_y \end{pmatrix} \begin{pmatrix} \sigma_x & 0 \\ 0 & \sigma_x \end{pmatrix}$$

$$= \begin{pmatrix} \sigma_x\sigma_y - \sigma_y\sigma_x & 0 \\ 0 & \sigma_x\sigma_y - \sigma_y\sigma_x \end{pmatrix} = \begin{pmatrix} 2i\sigma_z & 0 \\ 0 & 2i\sigma_z \end{pmatrix} = 2i\Sigma_z.$$

12.5
Orbital Angular Momentum

The operator acting on Ψ on the left side of (11.126) is neither spherically symmetric nor cylindrically symmetric. As a result, making the potential in (11.127) spherically symmetric or cylindrically symmetric about the origin does not give the energy operator similar symmetry. The energy operator for a particle in a central field does not commute with the operator for orbital angular momentum around any given axis. However, adding the right expression to the energy expression does yield such a commuting operator. This corresponds to adding the spin of the particle to its orbital angular momentum.

Dirac Equation (11.127) or (12.11), for motion in a field of potential

$$q\phi = V(r), \qquad \mathbf{A} = 0, \tag{12.58}$$

is

$$[(c\boldsymbol{\alpha}\cdot\mathbf{p} + \beta m_0 c^2 + V(r)]\Psi = E\Psi. \tag{12.59}$$

Energy operator H is the expression in brackets.

The operator for angular momentum around the z axis caused by motion of the particle through space is

$$L_z = \frac{\hbar}{i} \frac{\partial}{\partial \varphi} = \frac{\hbar}{i} \left(x \frac{\partial}{\partial y} - y \frac{\partial}{\partial x} \right) = x p_y - y p_x. \qquad (12.60)$$

The commutator of L_z with H is

$$[L_z, H] = L_z H - H L_z. \qquad (12.61)$$

Since $\beta m_0 c^2$ is a constant matrix, L_z commutes with it. Since $V(r)$ does not vary with angle φ, L_z also commutes with it. The only expression in H that need not commute is $\mathbf{p}$. Therefore, (12.61) reduces to

$$(L_z, H] = c\boldsymbol{\alpha}\cdot[(x p_y - y p_x)\mathbf{p} - \mathbf{p}(x p_y - y p_x)]. \qquad (12.62)$$

In the brackets, each positive term is paired with a similar negative term by commutation. All these cancel except some from $\mathbf{p}$ involving p_x and p_y.

In particular, we have the nonzero components

$$x p_y p_x - p_x x p_y = p_y (x p_x - p_x x) = p_y \left(x \frac{\hbar}{i} \frac{\partial}{\partial x} - \frac{\hbar}{i} \frac{\partial}{\partial x} x \right)$$

$$= p_y \left(x \frac{\hbar}{i} \frac{\partial}{\partial x} - x \frac{\hbar}{i} \frac{\partial}{\partial x} - \frac{\hbar}{i} \right) = i\hbar p_y, \qquad (12.63)$$

$$- y p_x p_y + p_y y p_x = -p_x (y p_y - p_y y) = -i\hbar p_x. \qquad (12.64)$$

Consequently, (12.62) reduces to

$$[L_z, H] = i\hbar c(\alpha_x p_y - \alpha_y p_x). \qquad (12.65)$$

Because L_z does not commute with H when the potential is spherically symmetric, solutions of the corresponding Dirac equation are generally not eigenfunctions of L_z and no energy state need exhibit a value for this operator. Since L_z is the operator for orbital angular momentum around the z axis, this angular momentum is not conserved.

12.6
Total Angular Momentum

The expression that has to be added onto L_z to form an operator commuting with H for a central field is a kind of constant. So it can be interpreted as defining an intrinsic property of the particle. Such a property is evident even when the particle is free and at rest.

In Section 12.4, we found that the hybrid spinor for a free particle at rest with only one nonzero component is an eigenfunction of Σ_z with the two eigenvalues $+1$

and -1. So we are led to try a constant a times Σ_z. Parameter a will give us the particle's spin angular momentum that can act along the z axis, or along any other given axis.

The commutator of this operator with the energy operator is

$$[a\Sigma_z, H] = a(\Sigma_z H - H\Sigma_z),\tag{12.66}$$

where

$$H = c\boldsymbol{\alpha}\cdot\mathbf{p} + \beta m_0 c^2 + V(r)\tag{12.67}$$

from (12.59). Since $V(r)$ is a varying number times the unit matrix, Σ_z commutes with it. From Example 12.1, Σ_z also commutes with $\beta m_0 c^2$ and with $c\alpha_z p_z$. We are left with

$$a(\Sigma_z H - H\Sigma_z) = ac[(\Sigma_z \alpha_x - \alpha_x \Sigma_z)p_x + (\Sigma_z \alpha_y - \alpha_y \Sigma_z)p_y]$$

$$= i(2a)c(\alpha_y p_x - \alpha_x p_y).\tag{12.68}$$

The trial is successful. We need only set

$$a = \frac{\hbar}{2}\tag{12.69}$$

to have (12.68) equal the negative of (12.65). Thus, the operator

$$J_z = L_z + \frac{\hbar}{2}\Sigma_z\tag{12.70}$$

commutes with H in the spherically symmetric field. Similarly, one finds that

$$J_x = L_x + \frac{\hbar}{2}\Sigma_x\tag{12.71}$$

and

$$J_y = L_y + \frac{\hbar}{2}\Sigma_y\tag{12.72}$$

commute with H.

Expressions (12.71), (12.72) combine to yield

$$J_x J_y = \left(L_x + \frac{\hbar}{2}\Sigma_x \right)\left(L_y + \frac{\hbar}{2}\Sigma_y \right)$$

$$= L_x L_y + \frac{\hbar}{2}(\Sigma_x L_y + L_x \Sigma_y) + \frac{\hbar^2}{4}\Sigma_x \Sigma_y\tag{12.73}$$

and

$$J_y J_x = L_y L_x + \frac{\hbar}{2}(\Sigma_y L_x + L_y \Sigma_x) + \frac{\hbar^2}{4}\Sigma_y \Sigma_x.\tag{12.74}$$

From the representation of L_x, L_y, and L_z as differentiating operators, as in (12.60), we have

$$L_x L_y - L_y L_x = i\hbar L_z. \qquad (12.75)$$

The differentiating operators commute with the sigma operators; so

$$\Sigma_x L_y - L_y \Sigma_x = 0, \qquad (12.76)$$

$$\Sigma_y L_x - L_x \Sigma_y = 0. \qquad (12.77)$$

From Example 12.2

$$\Sigma_x \Sigma_y - \Sigma_y \Sigma_x = 2i\Sigma_z. \qquad (12.78)$$

With (12.75)–(12.78), the difference between (12.73) and (12.74) reduces to

$$J_x J_y - J_y J_x = i\hbar L_z + 0 + \frac{\hbar^2}{4} 2i\Sigma_z$$

$$= i\hbar \left(L_z + \frac{\hbar}{2} \Sigma_z \right) = i\hbar J_z. \qquad (12.79)$$

Thus, the commutation rules for the complete J_k operators are the same as for the L_k ones. Formulas based on these commutation rules are valid. Consequently, the operator

$$J_x^2 + J_y^2 \qquad (12.80)$$

does not alter the quantum numbers of an eigenfunction of J_z on which it acts and

$$J_x^2 + J_y^2 + J_z^2 \qquad (12.81)$$

does not.

A person can form a common set of eigenfunctions for the energy operator H, the z component of the total angular momentum operator J_z, and the total angular momentum squared operator

$$\mathbf{J} \cdot \mathbf{J} = J^2 = J_x^2 + J_y^2 + J_z^2. \qquad (12.82)$$

The corresponding eigenvalues are the energy E, the z component of total angular momentum

$$L_z + \frac{\hbar}{2} \Sigma_z = M_L \hbar + M_S \frac{\hbar}{2}, \qquad (12.83)$$

and the total angular momentum squared

$$J(J + 1)\hbar^2. \qquad (12.84)$$

Example 12.3

Assume that each component of an operator **A** commutes with each component of matrix Σ and reduce the expression

$$\Sigma \cdot \mathbf{A} \, \Sigma \cdot \mathbf{A}.$$

Employ the summation convention, whereby repetition of an index indicates summation over that index. Expand each dot product, rearrange, add and subtract **A·A**, write out representative terms, and manipulate:

$$\Sigma \cdot \mathbf{A} \, \Sigma \cdot \mathbf{A} = \Sigma_j A_j \, \Sigma_k A_k = \Sigma_j \Sigma_k A_j A_k$$

$$= \delta_{jk} A_j A_k + (1 - \delta_{jk}) \, \Sigma_j \Sigma_k A_j A_k$$

$$= \mathbf{A} \cdot \mathbf{A} + \Sigma_1 \Sigma_2 A_1 A_2 + \Sigma_2 \Sigma_1 A_2 A_1 + \ldots$$

$$= \mathbf{A} \cdot \mathbf{A} + \tfrac{1}{2}\Sigma_1\Sigma_2 A_1 A_2 - \tfrac{1}{2}\Sigma_2\Sigma_1 A_1 A_2 + \tfrac{1}{2}\Sigma_2\Sigma_1 A_2 A_1 - \tfrac{1}{2}\Sigma_1\Sigma_2 A_2 A_1 + \ldots$$

$$= \mathbf{A} \cdot \mathbf{A} + \tfrac{1}{2}\Sigma_1\Sigma_2 (A_1 A_2 - A_2 A_1) - \tfrac{1}{2}\Sigma_2\Sigma_1 (A_1 A_2 - A_2 A_1) + \ldots$$

$$= \mathbf{A} \cdot \mathbf{A} + i\Sigma_3 (A_1 A_2 - A_2 A_1) + \ldots$$

$$= \mathbf{A} \cdot \mathbf{A} + i\Sigma \cdot \mathbf{A} \times \mathbf{A}.$$

In the fifth step, the anticommuting nature of the Σ_j's was introduced (this property follows from the similar nature of the σ_j's); in the seventh step, the reduction of the commutators from Example 12.2 was employed; in the last step, the connection of the commutators to the components of $\mathbf{A} \times \mathbf{A}$ was used.

Example 12.4

Reduce the expression

$$\alpha \cdot \mathbf{A} \, \alpha \cdot \mathbf{B}$$

in which operators **A** and **B** commute with Dirac matrix α.

Proceed as in Example 12.3:

$$\alpha \cdot \mathbf{A} \, \alpha \cdot \mathbf{B} = \alpha_j A_j \alpha_k B_k \, \alpha_j \alpha_k A_j B_k$$

$$= \delta_{jk} A_j B_k + (1 - \delta_{jk}) \, \alpha_j \alpha_k A_j B_k$$

$$= \mathbf{A} \cdot \mathbf{B} + \alpha_1 \alpha_2 A_1 B_2 + \alpha_2 \alpha_1 A_2 B_1 + \ldots$$

$$= \mathbf{A} \cdot \mathbf{B} + \tfrac{1}{2}\alpha_1\alpha_2 A_1 B_2 - \tfrac{1}{2}\alpha_2\alpha_1 A_1 B_2 + \tfrac{1}{2}\alpha_2\alpha_1 A_2 B_1 - \tfrac{1}{2}\alpha_1\alpha_2 A_2 B_1 + \ldots$$

$$= \mathbf{A} \cdot \mathbf{B} + \tfrac{1}{2}\alpha_1\alpha_2 (A_1 B_2 - A_2 B_1) - \tfrac{1}{2}\alpha_2\alpha_1 (A_1 B_2 - A_2 B_1) + \ldots$$

$$= \mathbf{A} \cdot \mathbf{B} + i\Sigma_3 (A_1 B_2 - A_2 B_1) + \ldots$$

$$= \mathbf{A} \cdot \mathbf{B} + i\Sigma \cdot \mathbf{A} \times \mathbf{B}.$$

12.7
Operators for a Central Field

When the potential is scalar and is a function merely of distance r from a center, the coordinate r can be separated from the angular coordinates in the energy operator. A convenient angular-momentum expression is

$$\hbar k = \beta(\Sigma \cdot \mathbf{L} + \hbar). \tag{12.85}$$

Indeed, the effect of (12.85) acting twice is multiplication by

$$
\begin{aligned}
\hbar^2 k^2 &= \beta(\Sigma \cdot \mathbf{L} + \hbar)\beta(\Sigma \cdot \mathbf{L} + \hbar) \\
&= \beta^2(\Sigma \cdot \mathbf{L}\ \Sigma \cdot \mathbf{L} + 2\hbar\Sigma \cdot \mathbf{L} + \hbar^2) \\
&= \mathbf{L} \cdot \mathbf{L} + i\Sigma \cdot \mathbf{L} \times \mathbf{L} + 2\hbar\Sigma \cdot \mathbf{L} + \hbar^2 \\
&= \mathbf{L} \cdot \mathbf{L} + i\Sigma \cdot i\hbar\mathbf{L} + 2\hbar\Sigma \cdot \mathbf{L} + \hbar^2 \\
&= L^2 + \hbar\Sigma \cdot \mathbf{L} + \tfrac{1}{4}\hbar^2\Sigma^2 + \tfrac{1}{4}\hbar^2 \\
&= (\mathbf{L} + \tfrac{1}{2}\hbar\Sigma)^2 + \tfrac{1}{4}\hbar^2 \\
&= J(J + 1)\hbar^2 + \tfrac{1}{4}\hbar^2 \tag{12.86}
\end{aligned}
$$

when the operand is a pertinent eigenspinor. In the second step, we have employed the commutativity of β with $\mathbf{L}$ and Σ; in the third step, the result from Example 12.3; in the fourth step, the commutator relations for the components of $\mathbf{L}$; in the fifth step, the value of Σ^2 from (12.57); in the seventh step, result (12.84). Since J is a half positive odd integer, number k can be any positive or negative integer:

$$k = \pm 1, \pm 2, \ldots . \tag{12.87}$$

The component of α in the direction of radius vector $\mathbf{r}$ is

$$\alpha_r = \alpha \cdot \hat{\mathbf{r}} \tag{12.88}$$

where $\hat{\mathbf{r}}$ is $\mathbf{r}/r$. A radial-momentum operator is

$$p_r = \hat{\mathbf{r}} \cdot \mathbf{p} - \frac{i\hbar}{r}. \tag{12.89}$$

With these definitions, we can construct

$$\alpha_r p_r + \frac{i}{r}\alpha_r\beta\hbar k = \alpha \cdot \hat{\mathbf{r}}\ \hat{\mathbf{r}} \cdot \mathbf{p} - \alpha \cdot \hat{\mathbf{r}}\ \frac{i\hbar}{r} + \frac{i}{r}\ \alpha \cdot \hat{\mathbf{r}}\ \Sigma \cdot \mathbf{L} + \frac{i}{r}\ \alpha \cdot \hat{\mathbf{r}}\hbar$$

$$= \alpha \cdot \hat{\mathbf{r}}(\hat{\mathbf{r}} \cdot \mathbf{p} + \frac{i}{r}\ \Sigma \cdot \mathbf{L}) = \alpha \cdot \hat{\mathbf{r}}(\hat{\mathbf{r}} \cdot \mathbf{p} + i\Sigma \cdot \hat{\mathbf{r}} \times \mathbf{p})$$

$$= \alpha \cdot \hat{\mathbf{r}}(\alpha \cdot \hat{\mathbf{r}}\alpha \cdot \mathbf{p}) = \alpha_r^2\ \alpha \cdot \mathbf{p} = \alpha \cdot \mathbf{p}. \tag{12.90}$$

In the third step, the cross product for **L** was introduced; in the fourth step, the result from Example 12.4.

With (12.90), the Dirac equation for a particle in a central field (12.59) can be written in the form

$$\left[c\alpha_r p_r + \frac{i\hbar c}{r} \alpha_r \beta k + \beta m_0 c^2 + V(r) \right] \Psi = E\Psi. \tag{12.91}$$

Operators α_r and β need only be Hermitian matrices satisfying

$$\alpha_r^2 = \beta^2 = 1 \tag{12.92}$$

and

$$\alpha_r \beta + \beta \alpha_r = 0. \tag{12.93}$$

Thus, we may let

$$\beta = \begin{pmatrix} 1 & 0 \\ 0 & -1 \end{pmatrix} \tag{12.94}$$

and

$$\alpha_r = \begin{pmatrix} 0 & -i \\ i & 0 \end{pmatrix} \tag{12.95}$$

In formula (12.4), the x_j axis can be pointed in the direction of any $\hat{\mathbf{r}}$. So we can write

$$\hat{\mathbf{r}} \cdot \mathbf{p} = \frac{\hbar}{i} \frac{\partial}{\partial r} = -i\hbar \frac{\partial}{\partial r}. \tag{12.96}$$

Operator p_r, of (12.89), therefore has the structure

$$p_r = -i\hbar \left(\frac{\partial}{\partial r} + \frac{1}{r} \right). \tag{12.97}$$

The angular and spin parts of Ψ determine quantum number k, but need have no other influence on the radial dependence. Thus, we may factor the radial part out and write

$$\Psi = \begin{pmatrix} r^{-1} F(r) \\ r^{-1} G(r) \end{pmatrix} Y(\vartheta, \varphi, \tau) T(t). \tag{12.98}$$

Substituting (12.98) into (12.91), with (12.94)–(12.97), yields

$$(E - m_0c^2 - V)F + \hbar c \frac{dG}{dr} + \frac{\hbar ck}{r} G = 0, \qquad (12.99)$$

$$(E + m_0c^2 - V)G - \hbar c \frac{dF}{dr} + \frac{\hbar ck}{r} F = 0. \qquad (12.100)$$

12.8
Behavior in the Outlying Regions of a Central Field

The potential that a particle experiences is central when the source is an entity spherically symmetric about the origin. When the entity is relatively compact, a person expects that potential to tend toward a constant as distance r of the particle from the center increases. Wave functions F and G then approach a simple transcendental form asymptotically.

Consider a particle of rest mass m_0 subject to the potential

$$V = V(r) \qquad (12.101)$$

which becomes constant as r increases without limit. Let us choose the zero of energy E such that

$$V = 0 \quad \text{when} \quad r = \infty. \qquad (12.102)$$

Behavior of the particle is governed by (12.99) and (12.100). For convenience, let us divide each of these equations by m_0c^2 and introduce the reduced (dimensionless) energy

$$\epsilon = \frac{E}{m_0c^2} \qquad (12.103)$$

and the reduced (dimensionless) radial distance

$$x = \frac{m_0c}{\hbar} r, \qquad (12.104)$$

obtaining

$$\left(\epsilon - 1 - \frac{V}{m_0c^2} \right) F + \frac{dG}{dx} + \frac{k}{x} G = 0, \qquad (12.105)$$

$$\left(\epsilon + 1 - \frac{V}{m_0c^2} \right) G - \frac{dF}{dx} + \frac{k}{x} F = 0. \qquad (12.106)$$

Wherever x is large enough, (12.102) applies and (12.105), (12.106) reduce to

$$(\epsilon - 1)F + \frac{dG}{dx} = 0. \qquad (12.107)$$

$$(\epsilon + 1)G - \frac{dF}{dx} = 0. \tag{12.108}$$

Multiplying (12.107) by $(\epsilon + 1)$,

$$(\epsilon^2 - 1)F + \frac{d}{dx}(\epsilon + 1)G = 0, \tag{12.109}$$

combining with (12.108), and rearranging leads to the relationship

$$\frac{d^2F}{dx^2} + (\epsilon^2 - 1)F = 0, \tag{12.110}$$

whose solution is

$$F = A\, e^{-(1-\epsilon^2)^{\frac{1}{2}}x} + B\, e^{(1-\epsilon^2)^{\frac{1}{2}}x}. \tag{12.111}$$

Substituting (12.111) into (12.108) yields

$$G = C\, e^{-(1-\epsilon^2)^{\frac{1}{2}}x} + D\, e^{(1-\epsilon^2)^{\frac{1}{2}}x}$$

$$= -\left(\frac{1-\epsilon}{1+\epsilon}\right)^{\frac{1}{2}} A\, e^{-(1-\epsilon^2)^{\frac{1}{2}}x} + \left(\frac{1-\epsilon}{1+\epsilon}\right)^{\frac{1}{2}} B\, e^{(1-\epsilon^2)^{\frac{1}{2}}x}. \tag{12.112}$$

When the particle is free, the energy is greater than the rest energy,

$$\epsilon > 1, \tag{12.113}$$

the exponents in (12.111) and (12.112) are imaginary, and the solutions are sinusoidal. These are acceptable.

But when the particle is bound, the energy is less than the rest energy,

$$\epsilon < 1, \tag{12.114}$$

the exponents in (12.111) and (12.112) are real, and we must have

$$B = D = 0 \tag{12.115}$$

to keep F and G from increasing without limit as x increases. The asymptotic behavior is like that found for the corresponding Schrödinger system.

12.9
Functions F and G for Hydrogen-like Structures

The nature of the radial functions closer to the center is determined by the function $V(r)$. Because of its importance, we will here consider the inverse-r potential that hydrogen-like structures exhibit.

Consider the particle to be an electron of charge $-e$ in the field of a nucleus of charge Ze, so the potential is

$$V = -\frac{Ze^2}{4\pi\epsilon_0 r}.$$ (12.116)

Substituting this into (12.105), (12.106) with

$$\alpha = \frac{e^2}{4\pi\epsilon_0 \hbar c}$$ (12.117)

leads to

$$\left(\epsilon - 1 + \frac{Z\alpha}{x} \right) F + \frac{dG}{dx} + \frac{k}{x} G = 0,$$ (12.118)

$$\left(\epsilon + 1 + \frac{Z\alpha}{x} \right) G - \frac{dF}{dx} + \frac{k}{x} F = 0.$$ (12.119)

Quantity ϵ is the energy in units of the rest energy of the electron $m_0 c^2$, Z is the atomic number on the nucleus, k is the angular quantum number introduced in Section 12.7, α is the fine-structure constant, and x is the radial distance in units of $\hbar/m_0 c$.

Forms that explicitly exhibit the asymptotic behavior found in Section 12.8 are

$$F = f e^{-(1-\epsilon^2)^{1/2} x},$$ (12.120)

$$G = g\, e^{-(1-\epsilon^2)^{1/2} x}.$$ (12.121)

Substituting these into (12.118) and (12.119) changes the differential equations to

$$\left(\epsilon - 1 + \frac{Z\alpha}{x} \right) f - (1 - \epsilon^2)^{1/2} g + \frac{dg}{dx} + \frac{k}{x} g = 0,$$ (12.122)

$$\left(\epsilon + 1 + \frac{Z\alpha}{x} \right) g + (1 - \epsilon^2)^{1/2} f - \frac{df}{dx} + \frac{k}{x} f = 0.$$ (12.123)

Since we expect f and g to be analytic, we write them as power series

$$f = \sum_i x^s a_i x^i,$$ (12.124)

$$g = \sum_i x^s b_i x^i,$$ (12.125)

in which summation occurs over the repeated index i. Substituting these into (12.122) and (12.123) yields the recurrence relations

$$(\epsilon - 1)a_{\nu-1} + Z\alpha a_\nu - (1 - \epsilon^2)^{1/2} b_{\nu-1} + (k + s + \nu)b_\nu = 0,$$ (12.126)

$$(\epsilon + 1)b_{\nu-1} + Z\alpha b_\nu + (1 - \epsilon^2)^{\frac{1}{2}}a_{\nu-1} + (k - s - \nu)a_\nu = 0. \qquad (12.127)$$

Multiplying the first of these simultaneous equations by $(1 - \epsilon^2)^{\frac{1}{2}}$, the second by $\epsilon - 1$ and subtracting produces the connection between the two series:

$$[(1 - \epsilon)Z\alpha + (\nu + s + k)(1 - \epsilon^2)^{\frac{1}{2}}]b_\nu$$
$$= -[(1 - \epsilon^2)^{\frac{1}{2}}Z\alpha + (\nu + s - k)(\epsilon - 1)]a_\nu. \qquad (12.128)$$

If there were no minimum $s + i$, for (12.124) and (12.125), a singularity that would preclude normalizing Ψ would appear at the center. But if the term below a certain one

$$\nu = m' \qquad (12.129)$$

is zero in either series, (12.128) tells us that the corresponding term in the other series is also generally zero and (12.126), (12.127) yield

$$Z\alpha a_{m'} + (k + s + m')b_{m'} = 0, \qquad (12.130)$$

$$Z\alpha b_{m'} + (k - s - m')a_{m'} = 0. \qquad (12.131)$$

For $a_{m'}$ and $b_{m'}$ to differ from zero, the determinant of the coefficients in these simultaneous equations must equal zero and

$$(Z\alpha)^2 = [k + (s + m')][k - (s + m')]. \qquad (12.132)$$

The higher $s + m'$ satisfying (12.132) is

$$s + m' = [k^2 - (Z\alpha)^2]^{\frac{1}{2}}. \qquad (12.133)$$

When $a_{m'-1}$ and $b_{m'-1}$ are zero, the recurrence relations allow all earlier coefficients to be zero also. So we may just as well take $m' = 0$ in (12.133), obtaining

$$s = [k^2 - (Z\alpha)^2]^{\frac{1}{2}} \qquad (12.134)$$

and

$$i = 0, 1, 2, \ldots \qquad (12.135)$$

for (12.124), (12.125).

The factors $\Sigma a_i x^i$ and $\Sigma b_i x^i$ introduce the waviness and the nodes into F and G. Since the degree of each polynomial need only equal the number of radial nodes in the corresponding function, the degree need not be infinite.

If the degree were infinite, and the series did not terminate, functions f and g would alter the asymptotic behavior of F and G. To prevent such alteration, we assume that terms for which

$$i \geqslant n' + 1 \qquad (12.136)$$

are zero. Letting ν in (12.126) and (12.127) be $n' + 1$ then yields

$$(\epsilon - 1)a_{n'} - (1 - \epsilon^2)^{1/2}b_{n'} = 0, \qquad (12.137)$$

$$(\epsilon + 1)b_{n'} + (1 - \epsilon^2)^{1/2}a_{n'} = 0, \qquad (12.138)$$

whence

$$b_{n'} = -\left(\frac{1 - \epsilon}{1 + \epsilon}\right)^{1/2}a_{n'}. \qquad (12.139)$$

The coefficient of $a_{n'}$ in (12.139) agrees with the coefficient of $A \exp[-(1 - \epsilon^2)^{1/2}x]$ in (12.112), as indeed it must.

12.10
Energy Levels of Hydrogen-like Structures

In the Coulomb-field Ψ, $F(r)$ and $G(r)$ appear multiplied by r^{-1}; as a result, exponent s must be positive when i begins at zero, as in (12.134) and (12.135), to keep Ψ normalizable. Similarly, the series for f and g must terminate, and (12.139) hold, so that Ψ exhibit the proper asymptotic behavior and be normalizable.

These conditions enable us to obtain a formula for the energy from (12.128). We set

$$\nu = n', \qquad (12.140)$$

the maximum, substitute (12.139) into (12.128), and reduce, obtaining

$$(1 - \epsilon)Z\alpha + (n' + s + k)(1 - \epsilon^2)^{1/2}$$
$$= (1 + \epsilon)Z\alpha - (n' + s - k)(1 - \epsilon^2)^{1/2}. \qquad (12.141)$$

Then

$$(n' + s)(1 - \epsilon^2)^{1/2} = Z\alpha\epsilon, \qquad (12.142)$$

whence

$$\epsilon = \frac{1}{\left[1 + \dfrac{(Z\alpha)^2}{(n' + s)^2}\right]^{1/2}}. \qquad (12.143)$$

Here ϵ is the energy of the hydrogen-like structure divided by the rest energy of the electron,

$$\epsilon = \frac{E}{m_0c^2}, \qquad (12.144)$$

Z is the atomic number of the nucleus, α the fine-structure constant, $n' + 1$ the number of terms in the polynomial factor in F or G, and s is given by (12.134),

$$s = [k^2 - (Z\alpha)^2]^{\frac{1}{2}}, \qquad (12.145)$$

with k the angular-momentum quantum number of Section 12.7. We also let

$$n' + |k| = n. \qquad (12.146)$$

For structures in which $Z\alpha$ is small, an expansion of ϵ as a power series in $(Z\alpha)^2$ is useful. First, (12.145) is used to express ϵ as a function of s. Then, derivatives of ϵ with respect to s and derivatives of s with respect to $(Z\alpha)^2$ are constructed and evaluated at $(Z\alpha)^2 = 0$. Employing the results in a Taylor series leads to the formula

$$\epsilon = 1 - \frac{(Z\alpha)^2}{2(n' + |k|)^2} - \frac{(Z\alpha)^4}{2(n' + |k|)^4} \left(\frac{n' + |k|}{|k|} - \frac{3}{4} \right) - \cdots$$

$$= 1 - \frac{(Z\alpha)^2}{2n^2} - \frac{(Z\alpha)^4}{2n^4} \left(\frac{n}{|k|} - \frac{3}{4} \right) - \cdots, \qquad (12.147)$$

whence

$$E = m_0 c^2 \left[1 - \frac{(Z\alpha)^2}{2n^2} - \frac{(Z\alpha)^4}{2n^4} \left(\frac{n}{|k|} - \frac{3}{4} \right) - \cdots \right]. \qquad (12.148)$$

The first term, $m_0 c^2$, is the rest energy of the electron. The second term, $-(m_0 c^2 Z^2 \alpha^2)/(2n^2)$, agrees with the Schrödinger formula since n is identified as the principal quantum number. The third term is the main "relativistic" correction term. Note how it depends on k, as well as on n.

Example 12.5

How is k related to azimuthal quantum number l?

Quantum number l is well defined for a Schrödinger particle in a central field. It is not an exact quantum number in Dirac theory, however.

But if the energy were not significantly below the rest energy,

$$\epsilon \simeq 1,$$

we would find that

$$G \ll F.$$

Operator β of (12.94), acting on Ψ, produces

$$\begin{pmatrix} 1 & 0 \\ 0 & -1 \end{pmatrix} \begin{pmatrix} r^{-1}F(r) \\ r^{-1}G(r) \end{pmatrix} Y T = \begin{pmatrix} r^{-1}Y(r) \\ -r^{-1}G(r) \end{pmatrix} Y T.$$

When G is negligible, the change in sign of G can be neglected and we can replace β by the unit matrix:

$$\beta \simeq 1.$$

From (12.85), with (12.57) and result (12.84), we now find that

$$\hbar^2 k = \hbar\Sigma\cdot\mathbf{L} + \hbar^2$$

$$= (\mathbf{L} + \tfrac{1}{2}\hbar\Sigma)^2 - \mathbf{L}^2 - \tfrac{3}{4}\hbar^2 + \hbar^2$$

$$= j(j + 1)\hbar^2 - l(l + 1)\hbar^2 + \tfrac{1}{4}\hbar^2$$

whence

$$k = j(j + 1) - l(l + 1) + \tfrac{1}{4}.$$

Substituting in

$$j = l \pm \tfrac{1}{2},$$

we obtain

$$k = \pm l + \tfrac{1}{2} \pm \tfrac{1}{2},$$

whence

$$l = k - 1$$

or

$$l = -k,$$

depending on whether k is positive or negative.

12.11
Restrictions on the Solutions

According to (12.87), quantum number k may equal any positive or negative integer:

$$k = \pm1, \pm2, \pm3, \ldots . \tag{12.149}$$

Furthermore, number n' is the degree of factors $\Sigma a_i x^i$ and $\Sigma b_i x^i$ in the series for f and g. So n' may be zero or any positive integer. Consequently from (12.146), we have

$$n = 1, 2, 3, \ldots . \tag{12.150}$$

When n' is zero, each series in (12.124) and (12.125) contains only a single term. Then Equations (12.137), (12.138) must reduce to (12.130), (12.131), with m' equal to zero:

$$Z\alpha a_0 + (k + s)b_0 = 0, \tag{12.151}$$

$$Z\alpha b_0 + (k - s)a_0 = 0. \tag{12.152}$$

But (12.151) implies that

$$b_0 = - \frac{Z\alpha}{k + s} a_0. \tag{12.153}$$

Also, according to (12.134), we have

$$s < |k|. \tag{12.154}$$

And from (12.139), the coefficient of a_0 in (12.153) must be negative. So only positive values of k are allowed when n' is zero.

When n' is a positive integer, however, this restriction is not present; both positive and negative k's are allowed.

Exponent s is smallest and F and G most singular at the origin when k is 1. However, no essential difficulty arises till

$$Z\alpha = 1. \tag{12.155}$$

Then, in the S state,

$$s = 0, \tag{12.156}$$

according to (12.134). Since

$$\alpha = \frac{1}{137.04} , \tag{12.157}$$

Equation (12.155) would be satisfied at

$$Z = 137.04. \tag{12.158}$$

For the 1S state, n' is zero and energy ϵ varies with $Z\alpha$ as Figure 12.1 depicts. When $Z\alpha > 1$, $k^2 = 1$, parameter s^2 is negative and energy ϵ is complex. The system apparently becomes unstable as Z increases beyond 137.

However, if one allows for the finite dimensions of the nucleus, suitable solutions are obtained for considerably higher nuclear charges. But at

$$Z \simeq 170, \tag{12.159}$$

instability sets in.

12.12
Concluding Remarks

On going to Einstein relativity, each classical 3-vector is paired with a classical scalar to produce an invariant 4-vector. And the classical gradient operator is paired with differentiation with respect to ct as in (11.77) and (11.78). So the spinorial derivatives, in Equations (11.98) and (11.99), are physically significant. These equations combine to give (12.1) and (12.2), in which the classical Laplacian ∇^2 of the Schrödinger equation is replaced by $\mathbf{D}^{\cdot\cdot}\mathbf{D}_{\cdot\cdot}$.

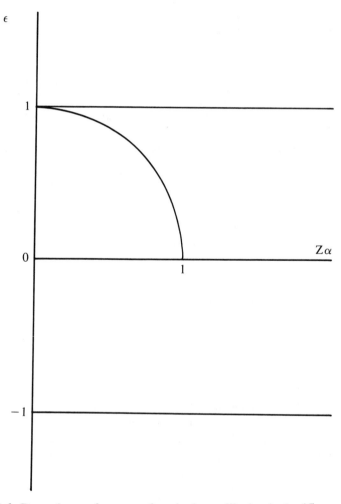

Figure 12.1. Dependence of energy ϵ for a hydrogenlike ion in the 1S state on $Z\alpha$.

As a result, a coherent state of a particle is described by two spinorial functions χ. and ϕ'. Conventionally, these are combined in form (12.9). This obeys the Dirac equation

$$[E - q\phi - c\alpha \cdot (\mathbf{p} - q\mathbf{A}) - \beta m_0 c^2]\Psi = 0. \qquad (12.160)$$

For a central electrostatic field, (12.160) reduces to

$$H\Psi = (c\alpha \cdot \mathbf{p} + \beta m_0 c^2 + V(r))\Psi = E\Psi. \qquad (12.161)$$

The operator for any component of orbital angular momentum does not commute with the Hamiltonian H. However, the operator

$$J_z = L_z + \frac{\hbar}{2} \Sigma_z \qquad (12.162)$$

does. For the common eigenstates, we find that the corresponding eigenvalue is

$$M_L \hbar + M_S \frac{\hbar}{2}. \qquad (12.163)$$

The angular and spin parts of Ψ influence the radial dependence through quantum number k, for which

$$\hbar k = \beta(\Sigma \cdot \mathbf{L} + \hbar). \qquad (12.164)$$

As a consequence, one may factor the bispinor as

$$\Psi = \begin{pmatrix} r^{-1}F(r) \\ r^{-1}G(r) \end{pmatrix} Y(\vartheta, \varphi, \tau)T(t) \qquad (12.165)$$

and obtain

$$F = f\, e^{-(1-\epsilon^2)^{1/2}x}, \qquad (12.166)$$

$$G = g\, e^{-(1-\epsilon^2)^{1/2}x} \qquad (12.167)$$

with

$$\epsilon = \frac{E}{m_0 c^2}, \qquad (12.168)$$

$$x = \frac{m_0 c}{\hbar} r. \qquad (12.169)$$

On seeking solutions for hydrogen-like structures, we employ

$$f = x^s \Sigma a_i x^i, \qquad (12.170)$$

$$g = x^s \Sigma b_i x^i. \qquad (12.171)$$

With the first term in both series constant, we obtain

$$s = [k^2 - (Z\alpha)^2]^{1/2} \qquad (12.172)$$

and

$$\epsilon = \cfrac{1}{\left[1 + \cfrac{(Z\alpha)^2}{(n' + s)^2} \right]^{1/2}} \qquad (12.173)$$

whence

$$\epsilon = 1 - \frac{(Z\alpha)^2}{2n^2} - \frac{(Z\alpha)^4}{2n^4} \left(\frac{n}{|k|} - \frac{3}{4} \right) - \ldots \qquad (12.174)$$

The second term corresponds to the Schrödinger result.

Functions $F(r)$ and $G(r)$ are most singular at $r = 0$ when k is 1 and s is 0. Then

$$Z\alpha = 1. \qquad (12.175)$$

At this stage the binding energy of the electron becomes equal to its rest energy.

Furthermore, no energy is needed to excite the highest negative-energy particle. This particle may then act to cancel one unit of charge on the nucleus. The accompanying hole may be ejected as a positron. Note that the instability here resides in the electronic structures—not in the nuclear structure.

Discussion Questions

12.1. Describe the 4-vector form of the 3-dimensional wavevector **k**. How does this 4-vector enter the equations of motion?

12.2. Why are the Dirac equations matrix equations?

12.3. Explain how the probability density of a particle may be calculated in Dirac theory.

12.4. Are periodic boundary conditions consistent with relativity?

12.5. How do these boundary conditions yield the energy levels for a freely translating particle?

12.6. Why does the Einstein-relativistic energy E include the rest energy?

12.7. How does each given n_x, n_y, n_z give rise to two completely distinct states at each energy level?

12.8. Why is there neither an upper limit nor a lower limit to the energy levels of a Dirac particle?

12.9. Given the absence of a lower limit on E, what keeps one from extracting an infinite amount of energy from any given particle?

12.10. In what sense is empty space not void?

12.11. What is an antiparticle?

12.12. Into what pure states may each energy level of a free particle be resolved?

12.13. For what simple operators do pairs of bispinors for the states of Question 12.12 yield eigenvalues of ± 1?

12.14. Why doesn't the angular momentum operator L_z commute with the energy operator H when the potential is spherically symmetric?

12.15. What has to be added to L_z to obtain an operator commuting with H for a spherically symmetric field?

12.16. Why is the added operator in Question 12.15 considered to define an intrinsic property of the particle itself?

12.17. When is $\mathbf{A} \times \mathbf{A}$ not zero?

12.18. When is angular-momentum operator k an eigenoperator? What are its eigenvalues?

12.19. How can $\boldsymbol{\alpha} \cdot \mathbf{p}$ be expressed in terms of α_r and k?

12.20. Why do we need only two coupled ordinary differential equations to describe the radial behavior in a central field?

12.21. How do we determine the asymptotic behavior of a particle in a central field?

12.22. How do the simultaneous differential equations of Question 12.20 appear when the particle moves in a Coulombic field?

12.23. What role is played by each factor in

$$F = (x^s)\Sigma(a_i x^i)\ e^{-(1-\epsilon^2)^{1/2}x},$$

$$G = (x^s)\Sigma(b_i x^i)\ e^{-(1-\epsilon^2)^{1/2}x}?$$

12.24. Why do we choose

$$s = +\ [k^2 - (Z\alpha)^2]^{1/2}$$

when

$$i = 0, 1, \ldots, n'$$

in F and G?

12.25. Why is there a further restriction on the angular-momentum quantum number k when $n' = 0$?

12.26. How is the energy of a hydrogen-like system obtained from the recurrence relations?

12.27. Why is this result expanded as a power series in $(Z\alpha)^2$?

12.28. Why isn't l an exact quantum number in a Dirac central field? In what sense is l determined by k?

Problems

12.1. For a free Dirac particle, determine the product of the phase velocity with the group velocity.

12.2. Find the commutators of Σ_x with α_x, α_y, α_z.

12.3. Show that α_r and β, of (12.95) and (12.94), satisfy (12.92) and (12.93).

12.4. Derive (12.147) from (12.143).

12.5. Hydrogen-like levels satisfy the formula

$$\text{Level}_n = \text{Level}_\infty - R\mu_r \left[\frac{Z^2}{n^2} + \frac{Z^4\alpha^2}{n^3} \left(\frac{1}{J + \frac{1}{2}} - \frac{3}{4n} + \Lambda_{nlZA} \right) + \ldots \right],$$

in which R is the Rydberg constant. What part does Schrödinger theory yield? What part is not explained by Dirac theory?

12.6. Determine how energy ϵ varies with $Z\alpha$ when n' is zero in an H-like structure.

12.7. What numbers can n', k, l, and j assume when $n = 3$? Label the corresponding states.

— — —

12.8. Continue the expansion of the radical in (12.20) through the next term. Then determine the ratio of this correction term to the preceding term.

12.9. The repeating unit for a translating electron is a cube with edges 1.000 Å long. Assume that the electron is in a completely symmetric state, with $n_x = n_y = n_z = n$. Determine the ratio of the relativistic correction term of Problem 12.8 to the preceding term when $n = 0, 1, 2, 3$.

12.10. Simplify $\Sigma \cdot \mathbf{A} \, \Sigma \cdot \mathbf{B}$ when each component of $\mathbf{A}$ commutes with each component of Σ.

12.11. What matrix converts Σ to α? Simplify $\alpha \cdot \mathbf{A} \, \Sigma \cdot \mathbf{B}$ when each component of $\mathbf{A}$ commutes with each component of Σ.

12.12. How can the simultaneous equations (12.105), (12.106) be put into the form

$$\frac{du_1}{dx} = f_2(x) \, u_2,$$

$$\frac{du_2}{dx} = f_1(x) \, u_1?$$

12.13. Use the equations from Problem 12.12 to show that F and G cannot vanish simultaneously at any point where x is finite.

References

Books

Blinder, S. M.: 1965. "Theory of Atomic Hyperfine Structure," in Löwdin, P.-O. (editor), *Advances in Quantum Chemistry*, vol. 2, Academic Press, New York, pp. 47-91. Employing the Dirac theory of an electron, Blinder constructs the Hamiltonian for hyperfine interactions and discusses the hyperfine structures of various atoms.

Dirac, P. A. M.: 1947. *The Principles of Quantum Mechanics*, 3rd edn., Oxford University Press, London, pp. 252-274. Here Dirac summarizes his Einstein-relativistic theory of an electron.

Rose, M. E.: 1961. *Relativistic Electron Theory*, Wiley, New York, pp. 1-291. In this detailed monograph, Rose develops the theory of spin one-half particles. The approximate Pauli theory is first reviewed. Then the Dirac theory is constructed and applied to free particles. The influences of electromagnetic fields are introduced and the central field problem is solved. Further approximation methods are developed.

Articles

Auvil, P. R., and Brown, L. M.: 1978. "The Relativistic Hydrogen Atom: A Simple Solution," *Am. J. Phys.* **46**, 679-681.

de Groot, E. H.: 1982. "The Virial Theorem and the Dirac H Atom," *Am. J. Phys.* **50**, 1141-1143.

de Lange, O. L.: 1989. "A Simple Solution for the Dirac Hydrogenlike Atom," *Am. J. Phys.* **57**, 883-886.

Epstein, S. T.: 1976. "A Differential Equation for the Energy Eigenvalues of Relativistic Hydrogenic Atoms, and Its Solution," *Am. J. Phys.* **44**, 251-252.

Fanchi, J. R.: 1981. "Critique of Conventional Relativistic Quantum Mechanics," *Am. J. Phys.* **49**, 850-853.

Fanchi, J. R., and Wilson, W. J.: 1983. "Relativistic Many-Body Systems: Evolution-Parameter Formalism," *Found. Phys.* **13**, 571-605.

Glass, S. J., and Mendlowitz, H.: 1989. "Spin-Flips in Dirac Scattering by a One-Dimensional Barrier," *Am. J. Phys.* **57**, 466-467.

Han, D., and Kim, Y. S.: 1981. "Dirac's Form of Relativistic Quantum Mechanics," *Am. J. Phys.* **49**, 1157-1161.

Kapuscik, E., and Uzes, C. A.: 1982. "Dirac Bracket Quantization and Central Force Systems," *Am. J. Phys.* **50**, 1094-1097.

Kim, Y. S., and Noz, M. E.: 1978. "Relativistic Harmonic Oscillators and Hadronic Structures in the Quantum-Mechanics Curriculum," *Am. J. Phys.* **46**, 484-488.

Levy, A. A.: 1985. "Systematic Comparison of the Quantization Rules of Hydrogenoid Atoms in the Old Quantum, Schrödinger, Klein-Gordon, and Dirac Theories, by Means of a Common Set of Three Parameters," *Am. J. Phys.* **53**, 454-459.

Lock, J. A.: 1979. "The Zitterbewegung of a Free Localized Dirac Particle," *Am. J. Phys.* **47**, 797-802.

McKelvey, D. R.: 1983. "Relativistic Effects on Chemical Properties," *J. Chem. Educ.* **60**, 112-116.

Nieto, M. M., and Taylor, P. L.: 1985. "A Solution (Dirac Electron in Crossed, Constant Electric and Magnetic Fields) That Has Found a Problem (Relativistic Quantized Hall Effect)," *Am. J. Phys.* **53**, 234-237.

Ohanian, H. C.: 1986. "What is Spin?," *Am. J. Phys.* **54**, 500-505.

Olsson, M. G.: 1983. "Comment on 'Quark Confinement and the Dirac Equation'," *Am. J. Phys.* **51**, 1042-1043.

Ram, B.: 1982. "Quark Confining Potential in Relativistic Equations," *Am. J. Phys.* **50**, 549-551.

Waldenström, S.: 1979. "On the Dirac Equation for the Hydrogen Atom," *Am. J. Phys.* **47**, 1098-1100.

Waldenström, S.: 1980. "Addendum: 'On the Dirac Equation for the Hydrogen Atom'," *Am. J. Phys.* **48**, 684.

ANSWERS TO PROBLEMS

Chapter 1

1.2. $\alpha = \beta$ = real number. **1.3.** -0.0231. **1.6.** -3.89 eV. **1.7.** 0.600.
1.8. $(\sin \alpha)e^{i\beta} | k\rangle + (\cos \alpha)e^{i\gamma} | r\rangle$. **1.9.** $0.800 | j\rangle - 0.600 | k\rangle$.
1.10. $\sin \alpha | f_1\rangle - \cos \alpha \cos \beta | f_2\rangle - \cos \alpha \sin \beta | f_3\rangle$; $\sin \beta | f_2\rangle - \cos \beta | f_3\rangle$.

1.11.
$$N \begin{vmatrix} \Psi_I(1) & \Psi_{II}(1) & \Psi_{III}(1) \\ \Psi_I(2) & \Psi_{II}(2) & \Psi_{III}(2) \\ \Psi_I(3) & \Psi_{II}(3) & \Psi_{III}(3) \end{vmatrix}$$

1.12. In $\Psi_{II} = f\Psi_I$, $f = e^{i\alpha}$ with α a real constant.
1.13. $|A| = 1$, $|B| = \sqrt{15}/4$. **1.14.** 0.3950. **1.15.** 0.4000.
1.16. No superselection rule; $\langle k | j\rangle$. **1.17.** 0.548. **1.18.** $1.1610 | j\rangle - 0.8746 | k\rangle$.
1.19. $0.231 | k\rangle + 0.973 | r\rangle$. **1.20.** $0.7316 | j\rangle$. **1.21.** $0.99928 | I\rangle$.

1.22
$$\left(1 - \frac{1}{2 N^* N} \right)^{\frac{1}{2}}; 0, 0.7071.$$

Chapter 2

2.1. $2n + 1$. **2.2.** $U = \exp[i \frac{\mu}{\hbar} (\frac{1}{2}v^2 t + vx)]$. **2.3.** A imaginary. **2.6.** 0.

2.8. $-\frac{\hbar}{i} \frac{\partial}{\partial p}$. **2.9.** To $\Psi^*\Psi$. **2.10.** $U = e^{-i\omega b}$. **2.11.** $(aA)^\dagger = a^*A$.

2.14. 0. **2.16.** $Ae^{i(p_x/\hbar)x}T(t)$.

Chapter 3

3.13. $i\mathbf{J}$. **3.14.** 0. **3.15.** 0. **3.16.** $\hbar e^{\pm i\varphi} \left(\pm \frac{\partial}{\partial \vartheta} + i \cot \vartheta \frac{\partial}{\partial \varphi} \right)$.

Chapter 4

4.1. $\dfrac{h^2}{2\mu} \left(\dfrac{n_x{}^2}{a^2} + \dfrac{n_y{}^2}{b^2} + \dfrac{n_z{}^2}{c^2} \right)$.   **4.2.** $\dfrac{1}{r} \dfrac{\partial}{\partial r} \left(r \dfrac{\partial \psi}{\partial r} \right) + \dfrac{1}{r^2} \dfrac{\partial^2 \psi}{\partial \varphi^2} + \dfrac{\partial^2 \psi}{\partial z^2}$.

4.3. $x^{(1)} = z$.   **4.4.** $\sqrt{\dfrac{3}{4\pi}}$.   **4.5.** $x^{(1)} = \dfrac{1}{\sqrt{2}} (x + y)$, $x^{(2)} = \dfrac{1}{\sqrt{2}} (x - y)$.

4.6. $\psi = rR$.   **4.7.** $R = Ae^{-x/2}$.   **4.8.** $Ne^{-6.7\rho}$, $N\rho e^{-1.95\rho}$, $N\rho e^{-1.95\rho}$.

4.9. $\dfrac{\mu e^2}{4\pi\epsilon_0 \hbar^2}$, $-\dfrac{\mu e^4}{32\pi^2\epsilon_0{}^2\hbar^2}$.   **4.10.** $x^{(1)} = \dfrac{1}{\sqrt{2}} [| L, + \mathbf{k}\rangle - | R, -\mathbf{k}\rangle]$.

4.12. $\dfrac{4}{\xi + \eta} \left[\dfrac{\partial}{\partial \xi} \left(\xi \dfrac{\partial \psi}{\partial \xi} \right) + \dfrac{\partial}{\partial \eta} \left(\eta \dfrac{\partial \psi}{\partial \eta} \right) \right] + \dfrac{1}{\xi\eta} \dfrac{\partial^2 \psi}{\partial \varphi^2}$.

4.13. $x^{(1)} = x^{(2)} = z$.   **4.14.** $\sqrt{\dfrac{5}{16\pi}}$.   **4.15.** $x^{(1)} = x$, $x^{(2)} = y$.

4.16. $N(-1)^l r^{l+1} \dfrac{\partial^l}{\partial z^l} \left(\dfrac{1}{r} \right)$.   **4.18.** $Ne^{-5.7\rho}$, $N\rho e^{-1.625\rho}$, $N\rho e^{-1.625\rho}$.

4.19. No.

Chapter 5

5.2. $\hbar e^{\pm i\varphi} \left[\pm \dfrac{\partial}{\partial \theta} + i \cot \theta \dfrac{\partial}{\partial \varphi} \right]$.   **5.3.** $\Theta = A (\sin \theta)^l$.   **5.5.** $\sqrt{2} \left(\dfrac{1}{\pi} \right)^{\frac{1}{4}} x e^{-x^2/2}$.

5.6. $K(k) = 1 - 2E(k)$, $E = v + \frac{1}{2}$.   **5.8.** $y = a^{-3/2} \dfrac{1}{9\sqrt{6}} (4 - x) x^2 e^{-x/2}$.

5.10. 0.0201.   **5.12.** $Y = A\sqrt{2l} \cos \theta (\sin \theta)^{l-1} e^{\pm i(l-1)\varphi}$.

5.14. $-\frac{1}{2}(a^- - a^+)^2 + \frac{1}{2}(a^- + a^+)^2 + b(a^- + a^+)$.

5.15. $\dfrac{1}{\sqrt{2}} \left(\dfrac{1}{\pi} \right)^{\frac{1}{4}} (2x^2 - 1) e^{-x^2/2}$.   **5.16.** $\lim\limits_{\epsilon \to 0} q = 0$, so no.   **5.17.** See Table 4.5.

5.19. $v = \dfrac{1}{2b} - \dfrac{1}{2}$.   **5.20.** 25.   **5.21.** ‘ $\hbar\omega_0$.

Chapter 6

6.1. $E = -\langle T \rangle$.   **6.2.** $\psi = x - (1 + c) x^2 + c x^3$.   **6.3.** $E = 5u$, $21u$.

6.4. 0.5976u.   **6.5.** $-Z^2/2$ hartrees.   **6.6.** (a) 0.0132, 0.0639; (b) 0.1952; (c) 0.

6.7. 0.   **6.8.** $c_1 = c_2 = c_3 = c_4 = \frac{1}{2}$; $c_2 = c_4 = 0$, $c_1 = -c_3 = \dfrac{1}{\sqrt{2}}$;

$c_1 = c_3 = 0$, $c_2 = -c_4 = \dfrac{1}{\sqrt{2}}$.   **6.9.** 0.4721 β.   **6.11.** $\langle T \rangle = 2\langle U \rangle$.

Chapter 6 (continued)

6.12. $\frac{3}{2}$ u. **6.13.** 1.6703 u. **6.14.** 1.8708 u. **6.15.** (a) 0; (b) 0.1135;

(c) 0.2472. **6.16.** $\tan 2\beta = 1.9865\ aeE/E_{1s}$. **6.17.** $\Delta E = -1.4798(4\pi\epsilon_0)a^3E^2$.

6.18. 1.4659 β. **6.19.** (a) β; (b) 0.8284 β.

6.20. (a) $x = -2$: $c_1 = c_2 = c_3 = 1/\sqrt{3}$; $x = 1$: $c_1 = 0$, $c_2 = -c_3 = 1/\sqrt{2}$ or

$c_1 = \sqrt{2} / \sqrt{3}$, $c_2 = c_3 = -1/\sqrt{6}$; (b) $x = -\sqrt{2}$: $c_1 = c_3 = \frac{1}{2}$, $c_2 = 1/\sqrt{2}$; $x = 0$:

$c_1 = -c_3 = 1/\sqrt{2}$, $c_2 = 0$; $x = \sqrt{2}$: $c_1 = c_3 = \frac{1}{2}$, $c_2 = -1/\sqrt{2}$.

Chapter 7

7.2. $u/2$. **7.3.** 0.12 eV. **7.4.** -0.12 eV. **7.5.** (a) 6.5×10^{-9} eV; (b) 0 eV.

7.7. U. **7.8.** -3.6 eV, -0.12 eV. **7.9.** 0.311 β. **7.10.** 0. **7.11.** $E_n^{(0)} - \dfrac{F^2}{2f}$.

7.12. 3.9×10^{-9} eV. **7.13.** $-\frac{1}{2} + 2g$. **7.14.** $l(l+1)\dfrac{\hbar^2}{2I}$.

7.15. (a) $\frac{3}{4}g$; (b) $\frac{15}{4}g$. **7.16.** $\dfrac{\sigma}{\sqrt{8}}[v+3)(v+2)(v+1)]^{\frac{1}{2}}$, $\dfrac{\sigma}{\sqrt{8}}[v(v-1)(v-1)]^{\frac{1}{2}}$,

$\dfrac{3\sigma}{\sqrt{8}}(v+1)^{3/2}, \dfrac{3\sigma}{\sqrt{8}}v^{3/2}$. **7.17.** $(v+\frac{1}{2})\hbar\omega - \dfrac{7}{16}\sigma^2\hbar\omega - \dfrac{15}{4}\sigma^2(v+\frac{1}{2})^2\hbar\omega + \dots$.

7.18 $\pm\frac{1}{2}g$. **7.19.** $\alpha + 2.083\ \beta$, $\alpha + 1.167\ \beta$, $\alpha + 1.000\ \beta$.

Chapter 8

8.1. 0.7023. **8.2.** $\dfrac{1}{\hbar^2}\left|\int_0^t H_{21}\exp\left(\dfrac{1}{\hbar}\int_0^t \gamma_{21}\,dt\right)dt\right|^2$. **8.3.** $\dfrac{(Aq)^2}{2\hbar\omega\mu}e^{-(\omega\tau)^2/4}$.

8.4. $\dfrac{1}{\hbar\omega_{nm}}\int_{t_1}^{t_2}e^{i\omega_{nm}t}\dfrac{d}{dt}H^{(1)}_{nm}\,dt$. **8.6.** $\dfrac{\pi^2\mu A^2}{2\hbar^2E}\dfrac{1}{(2\sin\frac{1}{2}\theta)^2}$.

8.8. 1054, . . . , 5100; yes. **8.9.** 0.783. **8.10.** $\sin^2\dfrac{\pi A}{\hbar B}$.

8.11. $\dfrac{(Aq)^2}{\hbar\omega\mu}e^{-(\omega\tau)^2/2}$. **8.12.** $\dfrac{2^{15}A^2a^2}{3^{10}\hbar^2}e^{-2\omega\tau}$. **8.13.** $\dfrac{\pi^2V_0^2\mu}{4\hbar^2\alpha^4E}\left[1 - \exp\left(-\dfrac{4\mu E}{\hbar^2\alpha^2}\right)\right]$.

8.14. $\dfrac{64\pi\mu^2V_0^2}{3\hbar^4}\dfrac{16k_0^4 + 12k_0^2\alpha^2 + 3\alpha^4}{\alpha^4(\alpha^2 + 4k_0^2)^3}$. **8.15.** $\dfrac{4\mu^2V_0^2}{\hbar^4K^6}(\sin Ka - Ka\cos Ka)^2$.

Chapter 9

9.1. (energy)$^{-1}$. **9.2.** $\nabla\cdot\mathbf{A} = 0$, $q^2A^2 \simeq 0$. **9.3.** $(2/3)\pi\,|\langle n\,|\,\mathbf{r}\,|\,m\rangle\,|^2$.

9.4. $I_0\exp -(k_1C_1 + k_2C_2)z$. **9.5.** 0.578. **9.6.** $r/\sqrt{3}$. **9.7.** $\dfrac{\omega^2(qr)^2}{9c^3\epsilon_0\hbar}$.

Chapter 9 (continued)

9.8. $1.03 \times 10^{-3} s^{-1}$.   **9.9.** $[j_b(j_b + 1) - j_a(j_a + 1)]^2 \hbar^4 \langle b | \mathbf{r} | a \rangle$.

9.10. $\frac{1}{2}\epsilon_0(E^0)^2$.   **9.11.** $(\text{time})^{-1}$.   **9.12.** $4\pi^2 \dfrac{\nu/\Delta\nu}{c} | \langle n | x | m \rangle |^2$.

9.13. $\dfrac{I_2}{I_0} = \left(\dfrac{I_1}{I_0} \right)^r$.   **9.14.** $\left(\dfrac{I_1}{I_0} \right)^r$.   **9.15.** $\dfrac{2^8 a}{3^5 \sqrt{2}}$.   **9.16.** 9.89×10^{-9} s.

9.17. 0.0053 s.

Chapter 10

10.1. $0 \leqslant A \leqslant \frac{1}{2}$, $1 - 2A$ and 1; $\frac{1}{2} < A \leqslant 1$, 0 and 1; $A > 1$, 0 and $2A - 1$.

10.2. More than 2 by 0.598.   **10.3.** $\dfrac{1}{\sqrt{2}} [| 0, 0^a \rangle + | 1, 0^a \rangle]$.

10.4. More than $\frac{1}{4}$ by 0.0665.   **10.5.** $16.558°$.   **10.7.** Coefficients for $+\frac{1}{2}\hbar$ and $-\frac{1}{2}\hbar$ along line: (a) 0.9659, 0.2588; (b) 0.9239, 0.3827; (c) 0.7071, 0.7071.

10.8. More than 2 by 0.319.   **10.9.** More than $\frac{1}{4}$ by 0.127.   **10.10.** $27.37°$.

10.11. More than $\frac{1}{2}$ by 0.1349.

Chapter 11

11.3. No change.   **11.4.** $\mathbf{M..T}$.   **11.6.** $\mathbf{D}$.

11.7. $\begin{pmatrix} \chi_{\cdot} - \phi^{\cdot} \\ \chi_{\cdot} + \phi^{\cdot} \end{pmatrix}$.   **11.8.** Construct $\begin{pmatrix} z & x - iy \\ x + iy & -z \end{pmatrix}$ with $\alpha\delta - \beta\gamma = 1$.

11.10. $(1/k)S^2$.   **11.11.** Rotation by $\mathrm{d}\varphi$ in xy plane.

11.12. (a) $\mathbf{BN..B}^\dagger$, (b) $\mathbf{AM^{\cdot}.B}^\dagger$, (c) $\mathbf{BM^{\cdot}.A}^\dagger$.

Chapter 12

12.1. c^2.   **12.2.** 0, $2i\alpha_z$, $-2i\alpha_y$.   **12.5.** Λ_{nlZA}.   **12.6.** $\epsilon^2 + \dfrac{(Z\alpha)^2}{k^2} = 1$.

12.7. $^2D_{5/2}$, $^2D_{3/2}$, $^2P_{3/2}$, $^2P_{1/2}$, $^2S_{1/2}$.   **12.8.** $\dfrac{\hbar^2}{4m_0^2 c^2} \left(\dfrac{n_x^2}{a^2} + \dfrac{n_y^2}{b^2} + \dfrac{n_z^2}{c^2} \right)$.

12.9. 0, 0.044%, 0.177%, 0.398%.   **12.10.** $\mathbf{A \cdot B} + i\mathbf{\Sigma \cdot A} \times \mathbf{B}$.

12.11. $\begin{pmatrix} 0 & 1 \\ 1 & 0 \end{pmatrix} = \gamma$, $\gamma \mathbf{A \cdot B} + i\alpha \cdot \mathbf{A} \times \mathbf{B}$.   **12.12.** $u_1 = x^{-k}F$, $u_2 = x^k G$.

Index

Ab initio calculation, 200
Abruptness of quantum changes, 331-333
Absorption
  of electromagnetic radiation, 304-306
  Bouguer-Beer law for, 308-310
Action at a distance, quantum mechanical, 350-351
Active operation, 38
Additive combination
  of bras and of kets, 10-11, 14
  of state functions, 4, 10-11
Adjoint $A^\dagger$, 18
  of constant, 47-48
  as operator on bra, 46-47
  of operator product, 47
  of unitary operator, 48-49
Advanced wave, 352-353
Algebra of operators, 23
Alpha matrix α, 385
  radial component of, 412-413
Amplitude factor, for scattering, f, 277
Analytic expression, 2
Angular momentum (Dirac)
  eigenvalues for, 410, 412
  operators for
    orbital, $\mathbf{L}$, 407-408
    spin, $(\hbar/2)\,\Sigma$, 408-409
    total, $\mathbf{J}$, 409-410
Angular momentum (Schrödinger)
  eigenvalues for, 90-92
  matrix elements for, 95-96
  operators for
    due to localizability, $\mathbf{S}$, 94
    due to propagation, $\mathbf{L}$, 74-78
      in spherical coordinates, 92-93
    due to symmetry, $\mathbf{J}$, 81-82

step operators involving, $J_\pm$, 142-144
Angular momentum transition, selection rules for, 320-322
Angular motion
  free
    eigenfunction Y for, 113-116
    separation of, 111-112
  in Dirac theory, 413
Anharmonicity constant, 162
Annihilation operator
  for bosons, b_j^-, 163-165
  for fermions, f_j^-, 165-168
Anticommutator $\{A, B\}$, 143
Anti-Hermiticity, of operator, 137
Antilinear relationship, for bras, 11
Antiparticle, Dirac theory of, 402
Antisymmetry
  in identical particles, 26-27
    and creation-annihilation operators, 167-168
Aperiodic perturber, 259-261
Approximation methods
  perturbational, 211-212
  variational, 177
Asterisk, 4
Atomic parameter $H_{jj} = \alpha$, 190
  empirical determination of, 197-198
  perturbation of, 241-242
Atomic transition, decay constant for, 316
Axioms
  I′—VII′, 2-6
  I—IX, 14-15
  X, 19
  XI—XII, 48-49
  XIII—XVII, 84-87
  XVIII, 98

Azimuthal motion, free, 114-115
  differential equation for, 116
  eigenfunction $\Phi(\varphi)$ for, 116

Band of states
  transitions involving, 270-271, 301-303
  translational
    densities of, 274-276
Basis kets, 14
  for perturbation theory, 224-225
  for variation theory, 181-182
Bell inequality
  for pass-or-fail testing, 345-346
  for two-orientation testing, 339-340
Benzene molecule, 192
  Hückel description of, 193-195
Beta decay, 280-284
Beta matrix β, 385
Binding energy, 222
  of heliumlike structures, 222
Bohr (unit), 120, 203
Boltzmann distribution law, 312
Bond parameter $H_{jk} = \beta$, 190
  empirical determination of, 197-198
Born approximation, 277
Born-Oppenheimer approximation, 117, 237
Bose-Einstein distribution law, 314
Boson, 26, 163-165
Bouguer-Beer law, 308-310
Bra $\langle I |$, $\langle j |$, 9

Center of algebra, 23
Central field (Dirac)
  equations for, 413-414
  operators for, 412-413
  outlying regions of, 414-415
Central field (Schrödinger), equation for,
  112
Characteristic equation, for matrix $\mathbf{A}$, 23
Chemical bonding, 191
  and association quantization, 131
Chemical system, as incoherent assemblage,
  349-350
Classical observable, 23-24
Coefficient c_j
  for basis function, 5-6
  for basis ket, 14, 16
Coherence, alteration of, 348-350

Coherent fermions, correlations between,
  342-343
Coherent photons, correlations between,
  346-348
Coherent state, 6
Colatitudinal motion, free, 114-115
  differential equation for, 116
  eigenfunction $\Theta(\varphi)$ for, 116
Collapse of ket, 335-336
  time asymmetry of, 350
Common eigenkets, and operator commutativity, 21-22
Commutativity
  of H, J_x, J_y, J_z, J^2, 88-89, 408-410
Commutator $[A, B]$, 54
  of H with J_j, P_j, K_j, 55-58, 83
  of H with $\mathbf{r}$, 306
  of J_j with J_k, K_k, P_k, X_k, 82-83
  of K_j with K_k, 83
  of L_j with L_k, K_k, P_k, X_k, 78
  of ϕ with L_k, 75
  of P_j with P_k, K_k, 55-57, 83
  of X_j with P_k, 54, 63-64, 85
Commuting operators, eigenkets of, 22-23
Complementary eigenstates, 3
Completeness, of eigenfunction set, 6
Complex representation, of electric and
  magnetic vectors, 126
Complex variable, differentiation with
  respect to, 188
Complex variable plane, integration around,
  272-274
Composite state, 3-4, 6, 14
Concentration dependence, of absorption,
  308-310
Constraint, effect on Hilbert space of, 187
Continuum approximation
  in decay law, 280
  for temporal Schrödinger equation, 65,
  254-255
Correlation probability $P(\mathbf{a}, \mathbf{b})$, 338-340,
  345-346
Coulomb field (Dirac)
  central structure of, 416-417
  outlying structure of, 414-415
  overall structure of, 417-418
Coulomb field (Schrödinger), structure of,
  120-121

Coupled equations, for rest-mass particle, 380
Coupling by perturbing potential, 225-227
Coupling constant
  for weak interaction field, G, 282-283, 285
  for electromagnetic field, q, 294
Creation operator
  for bosons, b_j^+, 163-165
  for fermions, f_j^+, 165-168
Cross section σ, differential of, 277
Crossing of levels, 237-239

De Broglie equation, for momentum $\mathbf{p}$, 40, 64
Decay law, 280, 316
Degeneracy, 19
Degenerate perturbation theory, 230-233
Delocalization, in chemical bonding, 191
Density of states, $\rho(E)$, 270
  for electromagnetic field, 298
Density operator (matrix) ρ, 15
Derivative, second, numerical approximation of, 184
Detailed balance, 311
Differentiating operator, for space-time, 376-377
Dimensions, in transition-rate formula, 303-304
Dipole moment, transition, 307
Dirac equation, 385, 398
Dirac matrices, 385-387, 398
Dirac state function, 385, 388, 398
Distance dependence, of absorption, 308-310
Distribution law
  of Boltzmann, 312
  of Bose and Einstein, 314
Double vertical line $|\,|$, 12

Eigenfunction Ψ, 3
Eigenket $|\,j\rangle$, 14
Eigenstate, 3, 14
Eigenvalue a_j, 3, 14
Einstein coefficient
  first, A_{mn}, 316
  second, B_{mn}, 306
    and transition moment, 307

Einstein equation, for energy E, 40, 64
Electron wave equation
  in pure-spinor form, 380-383
  in mixed-spinor form, 385
Electron affinity EA, molecular, 198
Electromagnetic field
  energy density in, $u(\nu)\,d\nu$, 298, 312-315
  potentials for, $\mathbf{A}$ and ϕ, 294-295
  sinusoidal forms for, 297
Emission
  of electromagnetic radiation
    induced, 304-306
    spontaneous, 315-316
Energy E
  fluctuations of
    in transitions, 268
  operator for, 64-65
EPR paradox, 337-338
Equilibrium
  detailed balance during, 311-312
  radiation in, 312-315
Evolution operator, 65, 254
  for varying H, 255
Expansion coefficient
  for transitions, c_j, 258
    transformed form b_j of, 260
Expectation value $\langle A \rangle$
  for coherent system, 6, 14
  for incoherent system, 7, 15-16
Exponentiation, 50
Extension of Galilean group, 57-58

Fermion, 26, 165-168
Feynman-Hellmann theorem, 178
Fine structure constant α, 121, 307
Fluorescence, as monitoring signal, 332-333
Four-vector, 372
Frequency (angular) ω
  Galilean transformation of, 41
  as measure of E, 40, 64
Function, as representation of ket, 9-10

Galilean continuum, 1
Galilean group, 51
  extension of, 57, 83
Galilean transformation, 39-40
  of state function, 41, 60
Gaussian orbitals, 125

Geometric operator $U(g)$, 50
  for displacement, 45
  for rotation, 46
Golden rule of Fermi
  for sinusoidal perturber, 303
  for steady perturber, 271
Gradient, physical significance of, 381
Gram-Schmidt orthogonalization, 19-20
Group, 49

Hamiltonian function H, 63
Hamiltonian operator H
  in 1-D space, 63
  in 3-D space
    for 1 particle, 84-85
      including spin, 98-99
    for N particles, 86-87
  uncomplicating part of, 211-212
Hanbury-Brown-Twiss effect, 349
Harmonic oscillator, 118
  eigenfunctions for, 118-119, 148-149
  mean kinetic energy of, 181
  step operators for, 144-148
Hartree (unit), 203, 223
Helicity, 127
Heliumlike structure
  ground state of
    perturbation analysis of, 216-220
      variational adjustment of, 220-223
Hermite polynomial, 118, 149
Hermitian operator, 18-19
  square of,
    diagonal element of, 89
Hilbert space, 9, 14
  sectors of, 24
Hückel description of chemical bonding,
    191
  perturbation of, 240-242
Hybrid spinor, for state function, 385, 388,
    398
Hybridization, caused by electric field,
    236-237
Hydrogenlike atom (Dirac), 416-419
  stability of, 420-421
Hydrogenlike atom (Schrödinger)
  eigenfunctions for, 123
    radial factor for, 120-122, 156-157
    step operator for, 155

perturbation by electric field of
  from first excited state, 233-237
  from ground state, 227-230

Indempotency, 17
Identical particles, 24-28
Incoherent state, 6
Independent-particle model, 124
Indistinguishability, 25-28
Induced transitions, rate of, 304-307
Inertial frame, 38-39, 361
Infinitesimal generator iG, 50
Infinitesimal operator, 50
  for distance, iP, 52, 58
  for phase, iI, 58
  for rotation, $i\mathbf{J}$, 81
  for time, iH, 52, 58
  for velocity, iK, 58
Integral
  of $e^{-\alpha r} \sin Kr$, 278
  of $e^{-\alpha x} x^n$, 157
  of $e^{-2Zr_1} e^{-2Zr_2}/r_{12}$, 219-220
  of $(\sin x)/x$, 272-274
Intensity I, of electromagnetic wave, 308
Interaction
  with electromagnetic field, 293-294
  among particles, 61-63, 84-85
    and pairing of state functions, 337
Interchange operator Π_{12}, 25
Invariances
  of Minkowski, 359-361
  of Newton, 38
  of Schrödinger, 38
Inverse of operator, 44

Jump, quantum, 65, 332-333

Ket $|I\rangle$, $|j\rangle$, 9
Kinetic energy, expectation value for, $\langle T \rangle$,
    179

Laguerre polynomial, associated, $L_j^k(x)$, 121
Laplacian ∇^2
  in orthogonal coordinates, 111
  in spherical coordinates, 111
Legendre function
  associated, $P_l^m(w)$
    differential equation for, 116

Rodrigues formula for, 116
Lie group, 51
Life
  average time τ of, 280
  for half decay, $t_{1/2}$, 284
Light cone, 361-362
  invariance of, 360
  points on, 362-366
Linear independence, 19
Linear relationship, for kets, 11
Linear variation function, 187-189
Locality assumption, 340
Locality in EPR experiment, 338-340
Localizability of particles, 53-55
  during interactions, 61-63, 84-85
  in rotation, 74-75
  in translation, 53-58
Lorentz transformation, spinorial description of, 371-372

Magnitude of state function, $|\Psi|$, 4
Mass μ
  as classical observable, 43
  as universal scalar, 54-56, 84
Matrix element
  for operator A, 12-13
  for perturber $H^{(1)}$, 259
Maxwellian field, 294
Measurement
  classical, 333-334
  quantum mechanical, 334-335
Microscopic reversibility, 311
Minkowski continuum, 359-360
  differential in, 376-377
  vectors in, 362-366, 372-374
Mixed state, 6
Molecular structure as quantization, 131
Momentum operator
  angular, J_j, 75-76, 81-82, 94
  linear, P_j, 55-57, 62-63
    commutator representation of, 306
    off-diagonal matrix element of, 307
Morse potential, 158-159
  Schrödinger equation for, 159
  energy levels of, 161-162
Multielectron atom, 124-125
Multiplicative combination, of state functions, 27-28, 112

Natural units, 56, 200-201
  for angular momentum, 75, 82, 202
  for Dirac electron, 414, 416
  for electronic states, 203
  for harmonic oscillator, 144, 203
  for Morse oscillator, 159
  for radial motion, 120
  for translation, 202
Negative energy solutions, of wave equation, 384, 402
Neutrino wave equation, 378-380
Nilpotency, of infinitesimal generators, 58-59
Non-crossing rule, 237-239
Nondiscrimination, between free and interacting particles, 61, 84
Nonlocality of particle, 2
  justification of, 336
Normalization
  of ket $|g\rangle$, 14
  of state function Ψ, 4
    for Coulomb radial motion, 157-158
    for harmonic oscillation, 148-149
  of step operators, 143, 147, 152-153
Normalization integral
  for basis orbital, S_{jj}, 190
  for complete state, $\langle g \mid g \rangle$, 4, 14, 182
Number operator, 166

Observable, 1-3, 23
  as eigenvalue, 3, 14
Occupancy of disjoint states, 28
Operator acting on
  function, 3
  ket, 14, 137-138
Orbital, 117
Orthogonality, 5, 14
  Gram-Schmidt procedure for, 19-20
Orthonormal set, 14
Oscillation
  between equivalent states, 261-264
  of transition probability, 268-269
Oscillator
  harmonic, 118
    commutation relations for, 144-145
    eigenfunctions for, 119, 148-149
    step operators for, 145-148

Morse, 158-160
  quantization of, 161-162
  radial Coulombic, 120-121, 154
    eigenfunctions for, 121-122, 156-157
    step operators for, 155
Overlap integral S_{jk}, 190

Parameter(s)
  nucler coordinates as, 237
  operator strength as, 50
  perturbation strength as, 211-212
Parity P
  of photon ket, 129
  of state function, 317
Particle, 2
  localizability of, 53
  mass μ of
    conservation of, 42-43
    in infinitesimal generator iK, 57-60
Passive operation, 38
Pauli exclusion principle, 29
Pauli spin matrices, 97, 373-375
Permutation operator Π_{jk}, Π_j, 25-27
Permutation symbol ϵ_{jkl}, 77, 386-387
Perturbation
  stationary-state
    of cluster of states, 230-233
    of single state, 212-213
      to first order, 213-214
      to higher orders, 223-224
      to second order, 225-226
  time dependent
    aperiodic, 259-261, 264-265
    steady, 266-268
      with band of states, 270-271
    sinusoidal
      between single states, 298-300
      with band of states, 301-302
Perturber in operator, 211
  oscillatory, 298-300
Phase of state function, 4
  arbitrariness of, 43
  Galilean transformation of, 41, 59-60
Photon, 126-128
  energy of, 128, 297-298, 304
  ket for, 128-131
  number in radiation field, n_j, 164
Photon pairs, coherence of, 346-348

Pi (π) orbital, 191
Polarization of photon, 128-130
  projection of, 346
Position operator X_j, 53-54, 64
Potential $\mathbf{A}$ ($q\mathbf{A}$), 62-63, 85
  for electromagnetic wave, 295-296
Potential U ($q\phi$), 62-63, 85
  altering spread of, 179-180
  setting depth of, 158-159
Potential energy, expectation value for, $\langle U \rangle$, 180-181
Poynting vector $\mathbf{P}$, 297
Principal quantum number n, 120-121
  effective, n', 124-125
Probabilities
  combining of, 27-28
Probability
  joint, $P(\mathbf{a}, \mathbf{b})$, 338, 345
  of jth state, w_j, 5, 14, 261
  of transition, P_{nm}, 268, 300
    summed over band, 270, 302
Probability density $\Psi^*\Psi$, 4
Processes, 253
  continuum approximation for, 254-255, 280
Projection operator P_j, 16-17
  for boson state, P_S, 27
  for fermion state, P_A, 27
Property, 1-3, 23
  definiteness of, 333-334
Pure state, 6

Quantum jump, in electronic structure, 332-333
Quantum mechanical observable, 23, 334-335
Quantum-relativity principle, 61, 84

Radial momentum operator p_r, 412-413
Radial motion, Coulombic
  of Dirac
    extreme structure of, 414-417
    overall structure of, 417-418
  of Schrödinger
    eigenfunction $R(r)$ for, 112, 121-122
    derivation of, 155-157
Radiation field
  confinement of, 312-314

density of states in, 298
interaction with, quantization of, 128, 163-164
step operators for, 164-165
thermal equilibrium in, 314-315
Range of particle, 2
Rate law, 280
for beta decay, 283-284
Region R, for integration, 4
Relativistic correction, for hydrogenlike atom, 419
Relativity of Einstein, 359-360
Relativity principle, 38-39
Representation of group elements, by matrices, 52
Rescaling
of particle masses, 179
of spread of potential, 180
Resonance, in chemical bonding, 191
Retarded wave, 352-353
Ritz theorem, 186-187
Rotation
by angle φ, 79-80
of orbital, 46
commutation relations for, 82
infinitesimal operators for, 81
spinorial description of, 369-370
Rotational eigenfunction Y, 114-116
derived from $1/r$, 113-114
Rotational transition, selection rules for, 320-322
Rutherford scattering, 279

Scalar product, of bra with ket, 9-10
Scattering, Born approximation for, 275-278
Scattering amplitude f, 277
Scattering potential $V_0 e^{-\alpha r}/\alpha r$, 278-279
Coulomb limit of, 279
Schrödinger equation
fixed-state form of
for N particles, 86-87
for single particle, 63, 85, 87
varying-state form of, 64-65
Screening constant s, 124-125
Second quantization, 170
Secular equation
of matrix $\mathbf{A}$, 23
of perturbation theory, 233

of variation theory, 189
Secular perturber, treatment of, 259-261
Selection rules
for angular momentum changes, 320-322
for vibrational transition, 318-319
Self-consistent field (SCF)
for atom, 124
for molecule, 200
Semiempirical procedure, 200
Separating out
angular factor, 112, 413
azimuthal factor, 115-116
Shelf level, 332-333
Sigma matrix σ, 373
Sigma matrix Σ, 387, 403
Sigma (σ) orbital, 191
Similarity transformation UAU^{-1}, 45
Slater orbital, 124-125
Space
Euclidean nature of, 1, 38
invariance of, 360
measurement of, 360-361
Spanning of Hilbert space, 14
Speed
invariant, c, 361
Spherical coordinates, 74-75
transformation to, 76
of $\mathbf{L}$ and L^2, 92-93
of operator ∇^2, 111
Spin
of particle, 94
projection of, 341-342
of photon, 128
Spin Hamiltonian, 98-99
Spin matrices
of Dirac, 387, 403-406
of Pauli, 96-97, 373-375
Spinor
for general 4-vector, 366-367, 372-373
on light cone, 364-366
Spinorial derivative, physical significance of, 381
Spinorial differentiator, 376-377
Spinorial transformation, 367-369
Spontaneous transitions, rate of, 315-316
State function Ψ, 2-3
State ket $|j\rangle$, 13-14
Stationary expectation value, as eigenvalue, 186-187

Statistical weight w_k, 5, 7, 14
Statistics, as classical observable, 26
Step operators, 138
  commutation relations governing,
    140-141
  differential equation for, 150-151
  normalization condition and, 152-153
Step perturber, 266-268
Structure as quantization, 131
Sudden approximation, 256
Summation convention, 387
Superposition, 4, 14
  change in, 258-259
  constraint on, 24
Superselection rule, 24
Symmetry
  among identical particles, 25-27
  of probability P_{mn}, 268
  in space, 53-55, 82-83
  over time, 55
Symmetry operation, 38
  unitarity of, 48-49

Time t
  homogeneity of, 1, 38, 361
  measurement of, 37-38, 360-361
  projection of, 38, 360
  separability of, 1, 360
Time-dependent perturber, 257-258
  key formulas for, 265
    when sinusoidal, 303
    when steady, 271
Time-dependent Schrödinger equation,
  65, 254
Transaction model, 350-351
Transformation between inertial frames,
  39-40, 361
Transformation operator for ket, 12
  anti-Hermitian part of, 138-139
  Hermitian part of, 138-139
Transition dipole moment, 307
Transition probability
  for fixed perturber, 266-268
    with band of states, 270-271
  for sinusoidal perturber, 298-300
    with band of states, 302-304
  of zero magnitude, 26

Translation, 38
  in angle φ, 74-75, 79-82
  in distance x, 51, 53-54
  in phase ξ, 58-60
  in time t, 51, 55, 74
  in velocity v, 51, 54-55, 74
Translational energy
  of Dirac, 400
  of Schrödinger, 109
Translational eigenfunction
  of Dirac, 399, 403
  of Schrödinger, 109-110
    general 1-D form of, 41
Tritium, disintegration of, electron behavior
  in, 256-257
Tunneling in Stark effect, 229, 235

Unimodular condition, 369
Unitary operator U, 48-49
Units, natural, 56, 200-203, 414-416
  conversion to general ones, 56, 223

Variation theorem, 181-182
Vector, of Minkowski, 372
Vector product $\mathbf{A} \times \mathbf{B}$, 411
  $\mathbf{J} \times \mathbf{J}$, 104, 429
  $\sigma \times \sigma$, 375
Vector representation of function, 7-9
Velocity operator V_j, 56-57, 84
Vertical line $|$, in bra and ket symbols, 9, 12
Vibrational eigenfunction, harmonic,
  118-119, 148-149
Vibrational eigenfunction, harmonic,
  118-119, 148-149
Vibrational transition
  decay constant for, 319-320
  selection rule for, 318-319
Virial theorem, 181

Wave number k, 399-400
Wavevector k, 40, 109
  Galilean transformation of, 41
Weak interaction, 282
Weight, statistical, w_j, 5, 7, 14